NATURA URBAN

Natura Urbana

Ecological Constellations in Urban Space

Matthew Gandy

The MIT Press
Cambridge, Massachusetts
London, England

First MIT Press paperback edition, 2024

© 2022 Massachusetts Institute of Technology

The MIT Press would like to thank the anonymous peer reviewers who provided comments on drafts of this book. The generous work of academic experts is essential for establishing the authority and quality of our publications. We acknowledge with gratitude the contributions of these otherwise uncredited readers.

This book was set in Bembo Book MT Pro by New Best-set Typesetters Ltd. Printed and bound in the United States of America.

Library of Congress Cataloging-in-Publication Data

Names: Gandy, Matthew, author.
Title: Natura urbana : ecological constellations in urban space / Matthew Gandy.
Description: Cambridge, Massachusetts : The MIT Press, [2022] | Includes bibliographical
    references and index.
Identifiers: LCCN 2021008640 | ISBN 9780262046282 (hardcover), 9780262551335 (PB)
Subjects: LCSH: Urban ecology (Sociology) | Human ecology. | Environmentalism.
Classification: LCC HT241 .G36 2022 | DDC 304.2—dc23
LC record available at https://lccn.loc.gov/2021008640

10   9   8   7   6   5   4   3   2

Contents

CONTENTS

Writing a book about urban nature has been an interesting journey, not just to specific places, but also through different fields of knowledge. Reflecting on this topic has involved a kind of excavation or even reinterpretation of my own memories. When I started my graduate studies in the late 1980s—quite a long time ago now—I knew that I was interested in the broad field of urban environmental change and was drawn towards two analytical pathways in particular: the metabolic and the ecological. In November 1989 I had attended a workshop on urban ecology held in the "island city" of West Berlin, just a few days after the fall of the Wall. In the event, however, I moved in a metabolic direction with much of my earlier work, linking in particular with the study of infrastructure, whilst the other strand—the ecological—remained somewhat dormant (to deploy a botanical metaphor) until the early 2000s, after I returned to Berlin as a research fellow at the Humboldt University. Back in Berlin, I became fascinated by the interstitial landscapes that were filled with traces of nature, memory, and cultural artifacts. These *Brachen*, with their distinctive ecological assemblages, became a focal point for a new phase of work on urban nature.

At the time of my doctoral studies, the field of human geography, which remains my primary disciplinary anchor point, seemed to be pulling in several different directions: the spatial science tradition was still very present, newly emboldened by advances in the technical manipulation of data; the confidence of neo-Marxian perspectives that had emerged in the 1970s and 1980s was beginning to ebb; and a host of effervescent new voices was emerging that could be loosely grouped under the umbrellas of cultural, feminist, and postcolonial approaches. From the vantage point of early 2021, however, whilst all of these broad strands of work persist in different forms, the ontological critique of the bounded human subject has become much more fully developed, along with a more incisive (though still only partial) shakeup of the colonial historiographies that underpin the modern discipline of geography.

In writing this book I have been reflecting on various points of intersection between urban ecology and diverse literatures such as ecocriticism, critical legal theory, and forensic science. Part of the challenge in writing this book has been to explore whether a broadly neo-Marxian framework, originating under the metabolic lens of urban political ecology, can withstand a sustained engagement with other intellectual traditions, and whether a new conceptual synthesis for the study of urban ecology and urban nature might be achievable.

My thinking about urban nature has been greatly enriched over the last few years by my involvement in local environmental issues in inner London: working with other enthusiasts to provide data for a "biodiversity action plan" made me reflect on the curious journey of biodiversity discourse from the UN Rio Summit of 1992 to the deprived inner London borough of Hackney; my role as a "citizen scientist" has involved sending in my own records of invertebrates to regional recorders (and experiencing the delight of seeing my dots appear on local distribution maps); my involvement in a campaign to stop a luxury housing development that would have damaged a local nature reserve introduced me to the ecological realpolitik of scientific expertise for sale and the legal labyrinth of a judicial review process; and in giving guided walks and public talks on aspects of urban biodiversity I have gained a sense of what aspects of nature interest people in the heart of the city. In terms of my London-based experiences of urban nature I would especially like to thank the locally based experts who have brought the city to life: Tristan Bantock, Tony Butler, Annie Chipchase, and Russell Miller.

The literature on urban nature is vast, diverse, and rapidly changing, with multiple historiographies, networks, and traditions. I am aware of the frustration that is felt in other language spheres that are routinely overlooked, or even inadvertently replicated, within the increasingly dominant Anglo-American realm of environmental knowledge production. In terms of research on urban nature, urban ecology, and critical landscape studies this is especially significant for debates long under way within the French and German language spheres, and I have tried to ensure some degree of nuance in my account of this wider intellectual field. In terms of the gravitational pull of urban research I am clearly anchored in Anglo-American academic traditions, but over recent years I have developed outlying connections to India, France, Belgium, Portugal, and especially Germany.

My research has relied on many libraries and other resources including the Natural History Museums in Berlin and London, the British Library, Cambridge University libraries, the Graduate School of Design at Harvard University, the Tamil Nadu archives, the Archives Nationales in Paris, and the University College London libraries. The Biodiversity Heritage Library has proved an amazing resource for my research, especially during the closure of many libraries under the Covid-19 lockdowns. I have also drawn widely on semistructured interviews—framed around clearly defined questions—but also on an array of more informal conversations that have enriched many aspects of the work. In writing this book I have realized how much emphasis I place on site visits, serendipitous encounters, and forms of attentive observation as I fill my notebooks with thoughts and ideas. There is an attachment to "pavement ecologies" that permeates my text—perhaps the Americanized "sidewalk ecologies" sounds more poetic—rooted in my fascination with the flora and fauna all around us. As I was writing this book in the summer of 2020, for instance, I noticed a strange-looking wasp trapped against the inside of my kitchen window, so I took a photo and let it go. With the help of Italian entomologists I discovered it was something quite unusual, the "nationally scarce" and "elusive" *Lestiphorus bicinctus* that has a liking for dunes, rough grassland, and "brambly places." I got back to my text with a flush of further enthusiasm.

It is difficult to pinpoint the precise starting point for my book: childhood memories of the wastelands of north London that I revisit in the introduction; a former *Brache* in the Chausseestraße in Berlin, where the Wall had once stood, carpeted with the brilliant blue flowers of viper's bugloss (or *Natternkopf* as it is known in German) that I extensively studied before the site's erasure; the intriguing "nondesigned" landscapes of Gilles Clément, first encountered at an exhibition at the Canadian Centre for Architecture in Montréal; the booming of bitterns on the Paljassaare Peninsula in Tallinn; a pied kingfisher hovering over polluted water in Chennai, its image forever etched in my mind; and my ongoing studies of Abney Park in north London, in collaboration with botanists, entomologists, mycologists, and others, where we have found invertebrates that are rare in Europe, let alone in the middle of a noisy city. My sustained engagement with the Abney site in particular has allowed me to reflect on the threshold of endangerment that must be realized, or is unrealizable, for vulnerable fragments of nature to be protected in an urban context, as well as the kind of heterotopic alliances that can emerge in defense of wild urban nature. If we cannot rely on legal instruments or scientific criteria for the protection of urban nature, where does this leave cultural and political discourses on urban biodiversity?

During the course of my research I was fortunate to present my work at a variety of conferences, workshops, and guest lectures, including at Aarhus University, Berlin University of the Arts, the Brussels Metrolab, the Centre for Research in the Arts, Social Sciences, and Humanities (CRASSH) at the University of Cambridge, ETH Geneva and Zurich, the Goethe University Frankfurt, Harvard Graduate School of Design, the Institute of Advanced Study at the University of London, the Indian Institute of Technology Madras, King's College London, the Ludwig Maximilian University of Munich, Sciences Po Paris, the Synchronizacja conference in Warsaw, the Tallinn Academy of Arts, Taubman College at the University of Michigan, the Technical University Berlin, the Technical University Munich, the University of California Berkeley, the University of Bremen, the University of Darmstadt, the University of Kent, the University of Lausanne, the University of Lisbon, Université Libre, Brussels, the University of Manchester, the University of Oxford, the University of Sheffield, and the Polytechnic University of Turin. I am grateful for all the interesting comments and questions received.

At an early stage in the research I was able to spend some time at the Institute for Ecology at the Technical University Berlin, funded by the Humboldt Foundation, that enabled me to learn more about the practice of urban ecology as a scientific field, led by Ingo Kowarik and his colleagues. A pilot visit to Berlin funded by the British Academy proved hugely helpful. Particular thanks go to the Gerda Henkel Foundation for providing me with the opportunity for a sustained period of writing based at the Universität der Künste in Berlin, between 2013 and 2015, that enabled me to start pulling the different elements of my project together and also share ideas with a fascinating interdisciplinary audience drawn from art history, cultural theory, philosophy, and critical landscape studies, not least through my many conversations with my host Susanne Hauser. Visits to IIT Madras in 2016 and 2019, and interactions with my host Sudhir Chella Rajan, along with Chennai scholars Pushpa Arabindoo, Bhavani Raman, and Niranjana Ramesh, have had a formative influence on my thinking about urban nature in the global South. My invitation to act as an international advisor to the NoVOID project in Portugal, led by Eduardo

Brito-Henriques, proved a wonderful opportunity for me to expand my intellectual horizons. I would like to thank the project team on my European Research Council-funded project Rethinking Urban Nature that ran from 2014 to 2020 for their insights and intellectual enthusiasm: Sandra Jasper, Maroš Krivý, Marcus Nyman, Niranjana Ramesh, Nida Rehman, Mathilda Rosengren, Krithika Srinivasan, along with the wonderful logistical support provided by Danielle Feger, Louise Kay, Rosie Knowles, Louise Moschetta, and Sue Pearce at the University of Cambridge. My thanks also to the Volkswagen Foundation for supporting an international symposium on urban wastelands entitled Horror Vacui held at the Berlin University of the Arts in 2018.

This book has taken quite a long time to complete, and the task would have been impossible without the advice, encouragement, and inspiration of many people: Raúl Acosta, Nikhil Anand, Johan Andersson, Pushpa Arabindoo, Bergit Arends, Stephen Barber, Lucilla Barchetta, Julie Bargmann, Uli Beisel, Solly Benjamin, Christoph Bernhardt, Nick Bertrand, Dorothee Brantz, Bruce Braun, Caroline Bressey, Eduardo Brito-Henriques, Sascha Buchholz, Livia Cahn, Vanesa Castán Broto, Noel Castree, Joëlle Salomon Cavin, Edward Chell, Joanna Coleman, Claire Colomb, Stephen Daniels, Carolyn Deby, T. J. Demos, Seth Denizen, Mustafa Dikeç, Sarah Dockhorn, Rob Doubleday, Paul Draus, Sonja Dümpelmann, Claire Dwyer, Tim Edensor, Henrik Ernstson, Marion Ernwein, James Evans, Somaiyeh Falahat, Teresa Farino, Leonie Fischer, Michael Flitner, Susanne Frank, Timur Hammond, Maren Harnack, Andrew Harris, Susanne Hauser, Michael Heffernan, Wiebke Hofmann, Phil Hubbard, Nadja Imhof, Rivke Jaffe, Tariq Jazeel, Gerry Kearns, Roger Keil, Meaghan Kelly, Alessio Kolioulis, Ingo Kowarik, Christoph Kueffer, Kumiko Kuichi, Joanna Kusiak, Jens Lachmund, Agata Lisiak, Matthew Lockwood, Agata Marzecova, Christof Mauch, Katie Meehan, Leandro Minuchin, Don Mitchell, Jochen Monstadt, Tim Moss, Clare Mouat, Martin Murray, Esther Niemeier, Tim O'Riordan, Chris Otter, Ben Page, Daniel Paiva, Daniella Perrotti, Richard Phillips, Mara Polgovsky

Ezcurra, Sudhir Chella Rajan, Bianca Rinaldi, Martin Ritter, Jenny Robinson, Marit Rosol, Roger Safford, Joachim Schlör, Jonathan Schorr, Birgit Seitz, Avi Sharma, Hyun Bang Shin, James Sidaway, Anindya Sinha, Henriette Steiner, Glen Stoker, Ulf Strohmayer, Herbert Sukopp, Jayaraj Sundaresan, Erik Swyngedouw, Malcolm Tait, Mikkel Thelle, Karen Till, Moritz von der Lippe, Yan Wang Preston, Michael Watts, Jennifer Wolch, Meike Wolf, Suili Xiao, Karin Yabe, Austin Zeiderman, and Benedikte Zitouni. At my (relatively) new home in the Department of Geography at Cambridge University I have benefited from conversations with many people including Bill Adams, Ash Amin, Maan Barua, Michael Bravo, Ulf Büntgen, Karenjit Clare, Mia Gray, Phil Howell, Alex Jeffrey, Esther Kovacs, Charlotte Lemanksi, Emma Mawdsley, David Nally, Clive Oppenheimer, Sue Owens, Olga Petri, Richard Powell, Tom Spencer, and Bhaskar Vira. Thanks also to my current PhD students—Ben Platt, Chloe Rixon, Tiffany Dang, and Jonny Turnbull—and to other graduate students including Adam Searle, Maximillian Hepach, and Chen Qu for their challenging questions and insights. My affiliation with King's College Cambridge has also provided a very welcoming environment: I would like to mention in particular Tess Adkins, John Arnold, Jude Browne, Caroline Goodson, Aline Guillermet, Caroline Humphrey, Surabhi Ranganathan, Chris Prendergast, Felipe Hernandez, Perveez Mody, Nick Bullock, Simon Goldhill, Michael Proctor, Godela Weiss-Sussex, and Nicolette Zeeman.

I would like to thank my acquisitions editor at the MIT Press, Beth Clevenger, for her patience and encouragement, along with Anthony Zannino and my editor Matthew Abbate. A previous version of chapter 2 appeared in the *Annals of the American Association of Geographers* and a section of chapter 3 originates in an earlier essay published by jovis. The cartography for figure 4.2 was completed by Martin Lubikowski. A big thank-you to Kiera Chapman and my three anonymous readers for going through the manuscript so carefully and providing so many helpful suggestions. Thanks also to Mathilda Rosengren and Marcus Nyman for assistance with tracking down some of the sources, and to Mathilda for her help with securing picture rights. I want to thank the Bagel & Culture café in Mehringdamm and Leyla's café in Stoke Newington for keeping me going. And finally, my biggest thank-you to Yasminah Beebeejaun, for supporting me through this long project and for sharing in its many surprises along the way, including the birds of Spreewald and the Lea Valley.

Matthew Gandy

London and Berlin
January 2021

In the summer of 2018, the AirSpace Gallery in Stoke-on-Trent in northern England held an exhibition devoted to postindustrial wastelands entitled "The Brownfield Research Centre." Visitors not only encountered a range of artworks, including video installations, but part of the gallery had been transformed into a study space for urban ecology with books, microscopes, and intricate wall displays depicting fauna and flora found nearby (figure 0.1). The inspiration for the exhibition was an extensive wasteland, or "brownfield" site, located just a few hundred meters away, next to the site of a former greyhound racing track and other abandoned buildings.[1] On my visit to the exhibition, on an overcast June day, I realized that I had already noticed this intriguing space as I approached the gallery: nestling between boarded-up shops and a busy traffic intersection there was a profusion of brightly colored flowers that contrasted with the gray clouds overhead. Stoke-on-Trent once formed part of the fulcrum of the industrial revolution, especially for the pottery industry but also for coal, iron, and steel. In many senses the history of the area is emblematic of the rise of industrial capitalism with its associated energy transitions, social stratifications, and expanding global reach. The city's industrial past holds continuities with the environmental present, including its role in the modification

Figure 0.1 "The Brownfield Research Centre," AirSpace Gallery, Stoke-on-Trent (June 2018). Photo by Matthew Gandy.

of the earth's atmosphere, as part of a shift towards a dangerously uncertain future. Politically, Stoke-on-Trent forms part of the so-called "Red Wall"—a band of formerly industrial and reliably Labour-supporting constituencies that have moved sharply to the right in recent years, electing Conservative members of parliament for the first time in many decades.

Does the recognition of "new natures," exemplified by the cultural valorization of wastelands, offer an alternative to the rise of xenophobic nationalism, and what the writer Anthony Barnett describes as "England without London," rooted in an antimetropolitan, isolationist, and backward-looking political imaginary? Can a different vision of nature, landscape, and society, as both cosmopolitan yet locally distinct, offer a way out of the postindustrial malaise that has blighted so many communities? These developments are not of course a specifically British phenomenon—comparable dynamics are at play in the rustbelt states of the USA, parts of Belgium, France, and Germany, and many other formerly industrial areas.[2] Yet, as this book will show, it is precisely in such dislocated communities that some of the most imaginative interventions in the fields of nature, urban ecology, and alternative environmental futures are taking place.

I grew up in London during the 1970s. There was a hole in my primary school fence next to a partially collapsed wall, and it led into a sun-drenched bombsite covered with weeds. At that time, the topography of inner London was extensively shaped by wartime destruction, in common with many other European cities. I remember this strange paradise where bright red cinnabar moths (*Tyria jacobaeae*) flitted across patches of ragwort (*Jacobaea vulgaris*), their yellow flowers contrasting with purple drifts of rosebay willowherb (*Chamaenerion angustifolium*), a plant that had become a botanical leitmotif for London bombsites. Sprouting among the south-facing ruins were buddleia bushes (*Buddleia davidii*), their honey-scented blossom ablaze with butterflies, hoverflies, and other insects. To enter such a space conjured up images from children's literature, of a magical transition into another world. The postwar ruins created the sheltered ambience of a walled garden in the center of the city, inadvertently connecting with the history of the earliest gardens as a kind of sanctuary or oasis in an otherwise inhospitable environment.[3] Yet in the case of this "accidental garden" there was no planting scheme or trace of human intentionality: the incredible diversity of flowers had sprung up through a chance combination of seeds dispersed by birds, wind, or perhaps even my own shoes.

Sometime around 1975, however, it was announced that the site would be cleared to make way for an office development. I remember writing a letter in self-consciously neat handwriting to a local planning officer, objecting to the destruction of this small patch of urban nature. My first foray into local politics brought the timeless immediacy of my childhood world into contact with a much more remote and abstract realm. It was only years later that I began to understand the connections, processes, and ideas that have shaped the presence and meaning of nature in cities. As for Whitecross Street, where my old school is still located, it seems that everything is very different now: the horses that used to clop past the gates with barrels of beer from the local brewery have long gone, along with the intense winter smogs that occasionally reduced visibility to a few meters. All that remains familiar to me now is the modernist topography of the Golden Lane Estate (a postwar housing complex where I learned to swim) and the futuristic towers of the Barbican, juxtaposed against the earlier, denser, and more deeply shadowed Peabody social housing tenements from the late nineteenth century, which retained a lingering sense of tubercular stigma even in the postantibiotic era of the 1970s.

I IMAGINING URBAN NATURE

What exactly *is* urban nature? One possible starting point is Marx's elaboration on the Hegelian distinction between a "first nature," without human impact, and a "second nature" that has been extensively shaped by human needs.[4] For Marx, however, this Hegelian notion of a first nature disconnected from human intentionality also extends to the second form of nature since modernity remains trapped within the ideological embrace of a "naturalized" set of capitalist social relations.[5] In a similar vein, the relationship between "nature" and "history," and the troubled coupling of "natural history," was a source of reflection for Adorno, who sought to differentiate a historicized understanding of nature from a variety of extant phenomenological, existential, and relativist strands of thinking.[6] Furthermore, the very idea of a "pristine" first nature, which has been so powerful within European romanticism, has been extensively dispelled through archaeological evidence for the human modification of ecosystems that extends far back into deep time and clearly predates the origins of agriculture and early cities. Running through this widely deployed demarcation between first and second nature, however, is an oscillation between various points of mediation between nature as a material entity and a more diffuse set of cultural interactions.[7]

Debates over the meaning of nature have wider resonance for ways in which modernity is understood. As the historian Lorraine Daston shows, the ideological motif of nature has been routinely and arbitrarily used to adjudicate a wider range of disputes over the origins and legitimacy of various social norms. Yet nature in itself clearly provides no kind of template for human societies.[8] As the philosopher Christa Davis Acampora notes, we find in Nietzsche a rejection of the notion that values might be derived directly from nature, since they are "products of human creativity and ingenuity that develop historically and are preserved and transmitted culturally."[9] Rather than a predetermined stage set, nature presents a field of possibilities that are revealed through multiple types of cultural and historical interactions. Yet if we consider the contrast between first and second nature more closely, we find that the richness and variety of nature is often assumed to lie within the realm of first nature, which is pictured lying outside the homogenizing tendencies of modernity. The anthropologist Philippe Descola, for instance, describes a line of demarcation in cultures of nature, implying that occidental modernity is characterized by a hegemonic and univocal interpretation, in contrast to a diversity of premodern cosmologies. His argument has been questioned in turn by the cultural historian Hartmut Böhme, who highlights the multiple and divergent cultural appropriations of nature that have flourished within modernity. In this book I steer closer to Böhme on this point, showing ways in which a looser conception of agency that transcends the individual human subject has enabled diverse cultures of nature to be recognized.[10] Modernity, from this perspective, is not one but many potential pathways.

How do we begin to acquire a measure of urban nature? One starting point is offered by the lens of biodiversity. The term "biodiversity" has, since its appearance in the mid-1980s, become an extremely useful shorthand in discussing the variety of living organisms. The most species-rich cities in the world tend to lie within existing biodiversity hotspots or at the intersection between distinctive ecological zones. Cape Town, for instance, has over 3,000 species of plants, 361 species of birds, and 83 species of mammals.[11] Other contenders for the most biodiverse cities on earth might include Iquitos, Medellín, Mexico City, and São Paulo, on account of their biogeographical setting and diverse topographies.[12] An alternative measure might be those cities that have been the most intense focus of biodiversity research, with dedicated institutes and programs devoted to the study of urban nature: here we might find the historic and ongoing contribution of urban centers such as Berlin, Brussels, Hong Kong, Montréal, New York, Rotterdam, and Warsaw.

Any tendency to picture urban nature as the poor relation of rural flora and fauna is dispelled by a range of studies indicating that levels of urban biodiversity are often higher than those of their monocultural hinterlands dominated by industrialized agriculture, plantation forestry, or other intensely controlled types of landscapes. In Berlin, for instance, over 2,000 species of wild plants have been recorded, many deriving from the original flora of the region, others from centuries of human influence.[13] Furthermore, in Berlin and many other cities, the total number of plant species appears to be steadily increasing.[14] Part of the explanation for high levels of urban biodiversity is the sheer range of potential habitats that cities can offer, including their own distinctive types of socioecological assemblages found almost nowhere else. Urban wastelands, for example, and many other marginal sites are now widely recognized as ecological refugia or islands of biodiversity. It has long been recognized that so-called brown-field sites and other abandoned spaces can host a huge variety of interesting plants, birds, reptiles, and especially invertebrates. In the 1980s, for example, a guide to the natural history of vacant lots in California describes this patchwork of anomalous sites as a series of distinctive ecosystems with complex food chains marked by top predators such as lizards, snakes, cats, and occasionally coyotes.[15] And more recently a string of former industrial sites along the Thames estuary have been referred to as "England's rainforest" on account of the presence of many rare insects.[16]

Both interest in and knowledge of the natural history of cities have been advancing steadily since the nineteenth century. Early contributions include W. H. Hudson's *Birds in London*, published in 1898, a book that marks a wider shift in sensibility away from hunting and taxidermy and towards observation and conservation. Hudson was writing during an era of rapid metropolitan expansion, and worried that many of London's "hidden rustic spots" faced imminent obliteration: "For these little green rustic refuges are situated on the lower slopes of a volcano, which is always in a state of eruption, and year by year they are being burnt up and obliterated by ashes and lava."[17]

Yet Hudson also recognized that London's bird fauna was in a state of flux: whilst some species that were intolerant of urban conditions were being lost, other better-adapted species were steadily arriving. Indeed, Hudson noted how these "changes are not rapid enough to show a marked difference in a space of two or three years; but when we take a period of fifteen or twenty years, they strike us as really very great."[18] In the postwar era we find a number of exuberant celebrations of urban nature such as Richard Fitter's pathbreaking *London's natural history* (1945), only the third title to appear in the recently created New Naturalist book series, and John Kieran's *Natural history of New York City* (1959).[19] Whereas Fitter's survey was inspired in part by the "new natures" emerging out of the rubble of wartime destruction, Kieran's account was based on his intricate knowl-edge of just one site, Van Cortlandt Park in the Bronx, where remnants of the original biodiversity of Manhattan island could still be found.[20]

An interesting counterpoint to these curiosity-driven encounters with urban nature is offered by guides to "useful" plants, with historical prec-edents in the creation of botanical guides to culinary and medicinal plants growing in and around urban settlements. In times of crisis we find a par-ticular emphasis on edible weeds and the need to improve the identification skills of the wider population. In 1961, for example, the Chinese Commu-nist Party published a book illustrating the edible wild plants of Shanghai to help people survive the widespread famine that occurred in the wake of the Great Leap Forward (1958–1960).[21] And more recently there has been an upsurge of interest in foraging, urban gleaning, and other kinds of ver-nacular interaction with wild urban nature as a source of subsistence and solidarity.[22]

Postindustrial transitions and changing cultures of nature have pro-duced an unprecedented degree of fascination with urban biodiversity. There has been a surge of interest in the cohabitants of urban space, with encounters ranging from noticing unfamiliar roadside weeds to the intri-cate soundscapes of the urban dawn chorus. In some cities urban bota-nists have been marking the names of wild plants growing out of walls or sprouting from pavements or other microniches with chalk. Starting with the urban excursions of the botanist Boris Presseq, based at the Museum of Toulouse in southern France, a kind of guerrilla botany has spread through the streets and waste spaces of many other European cities.[23] In Bengaluru (Bangalore), the spider specialist Vena Kapoor has highlighted the secret world of arachnids as a starting point for public appreciation of hidden urban ecologies.[24] Another interesting recent example is the musical col-laboration between the philosopher David Rothenberg and the nightin-gales of Berlin to create a new kind of sonic landscape. Rothenberg's study links with longstanding investigations into the acoustic ecology of cities including the soundscapes of birds, bats, and frogs.[25] Arguments for "wild nature" in cities increasingly combine insights from social psychology, landscape design, and other fields.[26]

An extended characterization of urban nature would embrace those organisms at the edge of the human sensorium. There are many forms of life that lie beyond the threshold of detectability for the human ear, for example, and these hidden soundscapes have been revealed by the work of artists and scientists. The sound ecologist Michael Prime has been incorporating the sound of urban plants and fungi into musical compositions, whilst the artist Lee Patterson, using a hydrophone, has made recordings of aquatic insects that have colonized municipal water features such as ornamental ponds or fountains.[27] Urban life forms include algae that colonize the surfaces of the city as well as the microorganisms present in soil, water, and the multispecies terrain of the human body itself. Until recently, interest in the living surfaces of the city has been focused mainly on the role of these microorganisms in the weathering of buildings. There are ecological assemblages of algae and cyanobacteria that can flourish in urban environments in spite of immense temperature gradients, intense solar radiation, lack of water, and high levels of pollution. These often overlooked microecologies can produce a variety of aesthetic effects on urban surfaces, such as the familiar "ink stripes" associated with leaking pipes and other visible traces of microscopic life.[28] Urban soils form another underresearched dimension to urban nature, revealing the presence of antibiotic-resistant bacteria and other hidden sources of anthropogenic influence.[29] Indeed, the study of urban soils suggests that the differences in biodiversity between cities and classic hotspots such as rainforests are not as great as one might expect.[30]

Urban nature has inspired an upsurge of creative writing. Esther Woolfson's 2013 nature diary, *Field notes from a hidden city*, develops an intricate portrayal of Aberdeen, Scotland, beginning with a chance encounter with an injured bird that is "both momentous and quotidian." For Woolfson, the fledgling pigeon's plight underlines how elements of ordinary nature can pose profound questions about our ethical and sensorial relations to the nonhuman.[31] In Blake Morrison's 2007 novel *South of the river*, set in late 1990s London, it is the figure of the fox that moves effortlessly between the different social and political threads of the story: we encounter garden foxes roaming through the nocturnal urban landscape, a mutant escapee from a laboratory that is accused of attacking a child, and a terrified fox chased by Suffolk hounds.[32] Fictitious animals also stalk the pages of Jonathan Leshem's novel *Chronic city*, published in 2009, including an imaginary tiger that prowls the streets of lower Manhattan.[33] Literary evocations of New York have often alluded to animals such as alligators, both real and imagined, emerging from the city's sewers.[34] Children's literature is full of anthropomorphized animals, perhaps most famously in the German émigré Judith Kerr's book *The tiger who came to tea*, first published in 1968 and now translated into many languages.[35] Another example is Chiyoko Nakatani's illustrated story *The day Chiro was lost*, dating from 1965, which recounts the adventures of a lost dog in Tokyo (see figure 0.2).[36] Animal avatars stalk our thoughts and dreams as well as living and breathing among us as pets, strays, and in the largely unseen worlds of synanthropic ecologies.

Plants and insects are also plentiful in urban literature. The tree of heaven (*Ailanthus altissima*), now a familiar feature of ruderal urban ecologies, appears in Betty Smith's novel *A tree grows in Brooklyn*, first published in 1943, as a portent of neighborhood decline.[37] The ruins and rubble-strewn landscapes of postwar London form part of the setting for Rose Macauley's novel *The world my wilderness*, published in 1950, where she describes the "brambled wilderness lying under the August sun, a-hum with insects and astir with secret, darting, burrowing life."[38] In W. G. Sebald's final book *Austerlitz*, published in 2001, moths serve as a portal into the hidden recesses of modernity. In a poignant passage, the eponymous protagonist reflects on an overgrown Ashkenazi Jewish cemetery near his home in Stepney, East London, which he eventually realizes must also be the source of the moths that regularly enter his house at night. For Sebald, this sacred place serves as a repository of hidden memories within the heart of London: in the postwar era many Jewish cemeteries fell into

Figure 0.2   Chiyoko Nakatani, *The day Chiro was lost* [*Maigo no chiro*] (1965). Reproduced with permission of the publishers Fukuinkan-Shoten, Tokyo. Illustration by Chiyoko Nakatani © Eiko Nakatani.

a state of abandonment and disorder because the relatives who might have tended graves or contributed towards their upkeep had been murdered in the Holocaust.[39] These neglected spaces were transformed into tranquil islands of nature within the city, in an uncanny contrast with those distant places of barbarity that had contributed to their formation. More recently, urban cemeteries have become a focus of intense ecological interest, with many sites serving as havens for plants, birds, and other animals that have been gradually lost elsewhere, especially under the impetus to revalorize ostensibly empty spaces such as wastelands and other marginal sites.[40]

The language of urban nature ranges from the elegiac to the enumerative. Various forms of ecological rhetoric have become ubiquitous in fields such as architecture, planning, and landscape design. The growing prevalence of quasi-technical monikers such as "green infrastructure" or "ecosystem services" betokens a confluence of concerns with air quality, flood control, urban resilience, and other environmental challenges. The term "resilience," for example, has been described by the geographer Bruce Braun as a specific kind of cultural and political *dispositif* (deploying Michel Foucault's apposite term) that signals an ensemble of ideas and practices associated with variants of Anthropocene discourse that hold an underlying faith in the technocratic adjustment of modernity to meet any form of environmental risk.[41] Similarly, the political theorist Alyssa Battistoni draws on a synthesis of feminist theory with historical materialism to consider how utilitarian terms such as "natural capital" and "ecosystem services" disguise the full extent of nonhuman labor. She suggests the concept of "hybrid labor" in order to more fully acknowledge all kinds of work as a collective activity.[42] In a comparable fashion, the geographer Maan Barua highlights the affective and spectacular role of nature within the material and cultural matrix of capitalist urbanization, ranging from companion species to zoos and other kinds of living displays.[43] The insights of Barua, Battistoni, and others signal a wider acknowledgment of the different modes of work and cohabitation that constitute urban space as a series of complex configurations of human and nonhuman labor.

II Delineating the urban biome

In this section I want to reflect on urban nature as an object of study. How might we characterize the medley of cultural and material elements that has been gathered under the aegis of "urban nature"? The field occupies different registers of urban culture, oscillating between the ostensibly mundane realm of urban maintenance and the articulation of alternative ecological imaginaries. But what of the ecological dynamics of urban space itself? There is a domain of nature that coexists with human intentionality, exemplified by roadside weeds, synanthropic organisms, and the increasingly audacious incursions of larger predators on the urban fringe. It is this "other nature," flourishing at one remove from the controlled contours of metropolitan nature with its more familiar cultures of function and display, that forms the focus of this book.

The modern city can be conceptualized as an elaborate synthesis of material and cultural elements that functions as an ideologically charged antithesis of an imaginary elsewhere. As we move from the material to the cultural sphere, the distinction between "urban nature" and other manifestations of an external nature acquires greater interpretative resonance. The idea of "nature" itself, in all its etymological and cultural complexity, is an enduring element in urban political discourse, encapsulating the ambiguities behind the ideological dimensions to the historicity of capitalist urbanization and the articulation of alternative forms of metropolitan nature.

Urban nature comprises several elements: it denotes the socioecological dynamics of urban space at a variety of different scales; it incorporates the material and metabolic dimensions to urbanization, ranging from the domestication of animals to the construction of parks and infrastructure systems; and it lies in cultural tension with "the rural" and other nonmetropolitan spaces as part of an ideological constellation that enables the perceived naturalization of capitalist modernity. Typologies used for the delineation of urban nature inevitably draw these different cultural and scientific strands into dialogue with each other. The urban ecologist Ingo

Kowarik, for instance, introduces a fourfold typology of urban nature in which "first nature" denotes remnants of existing ecosystems, "second nature" includes the managed landscapes of agriculture and forestry, "third nature" incorporates designed features of metropolitan nature such as parks, gardens, and tree-lined streets, and "fourth nature" refers to non-designed elements such as *Stadtwildnis* (urban wilderness) associated with abandoned or marginal sites.[44] Kowarik's typology, first devised in the early 1990s, marks the elevation of spontaneous forms of urban nature to a greater degree of cultural and scientific salience: the phrase "fourth nature" serves as both an ecological marker and also part of the wider historiography of the field. For landscape theorist Vera Vicenzotti, however, the question is not so much one of ecological typologies or spatial demarcations, although these can be useful analytical tools, but of contrasting modes of interpretation. The idea of urban wilderness, for example, that corresponds with Kowarik's "fourth nature" is linked by Vicenzotti to a set of specific yet contradictory cultural and historical associations.[45] In this book, I stress that urban nature is a multilayered material and symbolic entity: it includes the ecological immediacy of the here and now, as a focus of intricate modes of cultural and scientific interpretation, but also connects with the "spectral materialism" of more distant sites in space and time.[46]

Over what scales or temporalities can we conceptualize urban nature? The geographer Erle Ellis has divided the surface of the earth into a series of "anthropogenic biomes," in contrast to the "natural biomes" based on dominant vegetation types for desert, grassland, rainforest, tundra, and so on. Under the revised schema devised by Ellis and his colleagues, some 40 percent of the human population reside in zones categorized as "urban" or "dense settlements."[47] In contrast, the historian Chris Otter offers a tighter definition of urban space set within a wider nexus of global sociotechnical developments that he terms the "technosphere."[48] What Ellis and Otter share is a reorientation of environmental analysis towards what could be defined as "synthetic ecosystems." Ellis, for instance, focuses on a series of modified landscapes that contain nonhuman elements, rather than emphasizing "natural ecosystems" or "wildlands" as a kind of template for the ecological imagination. For Otter, the emphasis is more on the distinctive systems and structures that have expanded the "human niche" under modernity, including those that provide food, water, power, and other necessities. In terms of delineating urban nature, we encounter a tension between the observable elements, as surveyed by urban ecologists, and more remote or diffuse manifestations, including diverse zones and interfaces that extend beyond the conventional parameters of metropolitan space.

Urban landscapes comprise zones of transition interspersed with artificially sustained discontinuities that underline the porosity of the distinction between the urban and the rural. The urban fringe of an expanding metropolis is of necessity a transitory landscape marked by a series of intense juxtapositions. In the case of Buenos Aires, for example, the writer Ezequeil Martínez Estrada describes the edge of the metropolitan area in the 1930s as "the floating city" marked by "a silent and persistent struggle between plants." In this space that Martínez Estrada characterizes as simultaneously city and countryside, "the cans and wood splinters mix with the trash and greenish ponds; these are the metropolis's wastes and the rubbish piles of dreams and opulence."[49] The "rural," for want of a better term, has not disappeared from metropolitan areas: indeed, "greenbelts" and other planning instruments can hold parcels of land in a state of suspended animation so that idealized fragments of "rurality" form part of the urban landscape. The administrative boundaries of contemporary London, for instance, weave in and out of a metropolitan greenbelt first conceived in the late nineteenth century, passing through a jumble of archaeological sites, allotments, copses, farms, golf courses, housing estates, playing fields, infrastructure installations, and other features.[50] In the fast-growing cities of East Asia we encounter a different set of dynamics: in the case of Beijing, for example, the term "greenbelt" forms more of an imaginary or aspirational dimension to urban space.[51] Conversely, in parts of South Asia we can find obverse forms of symbolism used to denote the "city yet to come": in the state of Tamil Nadu, in southern India, for example, I noticed hand-painted signs reading "smart land" springing up next to open fields, as land speculation began to imprint alternative futures onto the landscape.[52]

In other cases, we encounter landscape design projects that involve grafting fragments of nature from one location to another. In the Chinese city of Chongqing, for instance, there are vast urban afforestation projects that involve transplanting thousands of trees from elsewhere, often from destroyed villages or denuded zones. This is depicted in the artist Yan Wang Preston's project *Forest*, comprising a series of photographs completed between 2010 and 2017, in which she chronicles the construction of artificial environments referred to as "ecological recovery landscapes" (figure 0.3).[53] Uncategorizable landscapes also feature in the writer Jeff VanderMeer's rendition of "area X," an imaginary ecological zone that I will return to in chapter 5, where intense forms of hybridity and genetic mutation produce synthetic environments that are neither rural nor urban in any conventional sense.

Global indices of "urban sustainability" attempt to rank and rate urban areas, placing cities such as Phoenix and Zurich at opposite ends of a constructed environmental spectrum. Yet these assessments can only provide a limited measure of difference within the wider dynamics of urbanization. In the case of Phoenix, the sociologist and cultural critic Andrew Ross notes how "the desert metropolis necessarily depends on vast amounts of 'ghost acreage' elsewhere to sustain the daily needs of its population."[54] Unlike the related and more neutral-sounding term "ecological footprint," the concept of "ghost acreage" connects more directly with historical patterns of exploitation in distant locations, including hidden labor and violent forms of resource extraction. The low-density expansion of Phoenix during the postwar era exemplifies a particular kind of racially exclusionary, federally subsidized capitalist urbanization. It was enabled, as the historian Andrew Needham shows, through a specific combination of cheap energy and evolving modalities of political power to access "peripheral nature" in the service of metropolitan growth.[55] Yet the tranquil tree-lined streets of

Figure 0.3 Yan Wang Preston, Longan woodland, University City, Chongqing, China (2011). *Forest* series © Yan Wang Preston.

Zurich (and many other European cities that regularly score highly on ecological sustainability indices) belie their relationship to the maelstrom of global capitalism, both now and in the past. If sustainable urban pathways exist, as much of the environmental literature suggests, then it is important to consider how these local or regional manifestations of urban form intersect with environmental change at a global scale.[56] Recent attempts to develop international indices for the comparison of urban biodiversity illuminate conceptual tensions and data anomalies at the metropolitan scale, let alone at a global level of application.[57] Under what spatial parameters should interventions be evaluated? How do we compare the value or complexity of different types of ecosystems? The question of urban biodiversity clearly forms part of a wider set of ongoing experimental investigations between science, technology, and policy interventions under the accelerating momentum of the Anthropocene.[58] There is now growing interest in the potential power of cities working in combination to advance specific environmental agendas through the creation of international networks, the sharing of expertise, and the amplification of political influence. The urban arena is now increasingly marked by a scaling-up of the experimental field, including the complex terrain of urban biodiversity, so that the boundaries of the urban are undoing a series of epistemological reconfigurations.

III MODELS, SCHOOLS, AND PARADIGMS

In the final part of the introduction I want to introduce my own roughly hewn distinctions between four main perspectives within urban ecological thought: first, the evolution of a series of systems-based approaches, with their roots in the analysis of urbanization as a dynamic interaction between different elements, including energy, materials, and human mobility; second, an intense focus on urban nature as an observational field, with links to early modern botany, marking a shift from medicinal to classificatory idioms, and framed by the evolving scope of a distinctly urban natural history; third, the emerging influence of "political ecology" within urban theory, marked by a synthesis between neo-Marxian and metabolic insights into the production of urban space; and fourth, a more recent emphasis on what I term the "ecological pluriverse" derived from a combination of posthuman and postcolonial conceptions of urban space.

The starting point for conventional histories of urban ecology as a systems-based science is the Chicago school of urban sociology in the 1920s. Insights drawn from vegetation dynamics, and particularly the concept of "succession" coined by the American ecologist Frederic Clements, were used to develop a neo-Darwinian spatial model driven by the competitive outcomes of individual decision making.[59] The sociologist Ernest W. Burgess, for instance, drew on recent advances in "plant ecology" to discuss how urban expansion is driven by a series of concentric zones that displace one another over time through processes of "invasion," "disorganization," and "reorganization."[60] Interestingly, the mixing together of biological metaphors with questions of race, mobility, and neighborhood change, elaborated on by Burgess and his colleagues, resonates with parallel developments in interwar "plant sociology" under way in European botany that I explore in chapter 3. If we look back at this influential body of writings, however, we find that there was actually very little emphasis on urban nature in its own right. Instead, the interpretation of ecology by the Chicago school rested on a dualistic distinction between society and nature within which "nature" was placed outside the urban process as part of a naturalistic framework of analysis.[61]

By the 1960s and 1970s, the emergence of the term "ecological studies," though related to the Chicago school, marked a more elaborate engagement with quantifiable variables that could be correlated across urban space, in an attempt to resuscitate the earlier ecological paradigm now divested of its neo-Darwinian analogies. The 1977 publication of *Contemporary Urban Ecology* by geographer Brian Berry and sociologist John Kasarda represents a notable attempt to reinvigorate the Chicago school approach through the alignment of urban ecology with spatial science. Berry and Kasarda's conception of "human ecology" emphasizes the spatial and organizational dimensions to human-environment relations through the empirical analysis of large datasets. Yet though the book is steeped in the cybernetic promise of computer-driven spatial analysis, it has almost no relation to the cultural and ecological dynamics of urban nature.[62] Indeed, the idea of ecology as a relational science is even more occluded than in the earlier writings of the Chicago school, becoming essentially reduced to the "ecological organization" of spatial phenomena.

By the 1970s, however, we can discern a clear divergence between conceptions of urban ecology based on the study of human adaptation to environmental change and a parallel sphere of field-based science devoted to the socioecological characteristics of urban space. This alternative scientific lineage to the Chicago school is to be found in the Brussels school of urban ecology associated with the botanist Paul Duvigneaud, who began his career with phytosociological analysis of lichens growing in the tropical rainforests of the Belgian Congo during the 1940s and 1950s. In the early 1960s Duvigneaud began to shift his research focus to the rural landscapes of the Belgian Ardennes and then to the city of Brussels, deploying an expanded conception of systems-based ecology based on the measurement of biomass and energy flows. His diagrammatic representation of the urban ecosystem as a "forest flow model," influenced by the holistic theory of the American ecologist Eugene Odum, marked an extension of his analysis of "natural ecosystems" to the modern city, conceptualized as a bounded and measurable socioecological system.[63] Rather than treating the city as an ecological metaphor, as elaborated by the Chicago school, Duvigneaud presented the modern city as an ecosystem in itself, which he termed the *écosystème urbs*, sustained by a series of flows to maintain biomass and energy balance (figure 0.4).[64] His conceptualization of the modern city as a kind of "urban forest" rests on a topographical analogy whereby the scale of the "urban canopy" is derived from the pre-elevator height of nineteenth-century European cities. The transposition of a methodological framework derived from West African rainforests and the Belgian countryside produced a new kind of ecological model for urban space. Yet Duvigneaud's

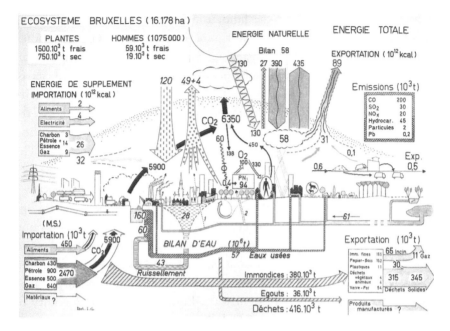

Figure 0.4 The metabolism of the Brussels Ecosystem, from Paul Duvigneaud and Simone Denaeyer-De Smet, *L'écosystème urbs: l'écosystème urbain bruxellois* (1977). The urban trophic imaginary marks an attempt to conceptualize the city as an ecosystem in itself. Source: Paul Duvigneaud Centre, CIVA archive, Brussels.

"trophic imaginary" elides the human and other-than-human elements of urban ecology into an organicist schema that emphasizes functional dynamics over historical contingency. His attempt to elevate the status of ecological science within spatial planning (*aménagement*) for the city of Brussels navigated the limits to a technical conceptualization of a socially and politically fragmented modern metropolis, riven by deep forms of ethnolinguistic difference, made even more intractable through the escalation of regional cultural and political tensions within the Belgian state.[65] The representation of Brussels as a "typical" urban ecosystem could not be reconciled with the particularities facing a divided postcolonial metropolis.

In certain respects, we could view Duvigneaud as an intermediate figure working at the intersection between systems-based urban ecology and the ground-level observational paradigm that I turn to later in this section. Although his analytical emphasis was on the flow-based conceptualization of "urban biomass" à la Odum rather than a direct focus on urban biodiversity, Duvigneaud did recognize the high biological productivity of marginal spaces such as wastelands that could also serve as experimental ruderal sites for the study of what he termed "pure phytosociology" with only limited human intervention. Duvigneaud noted with amazement, for instance, that the "productivity of these *terrains vagues* is 'close to the productivity of an equatorial forest.'"[66]

The field of urban ecology has been marked by repeated attempts to fuse the social and biological sciences. The attempted integration of what the ecologist Paul Sears referred to as "seemingly unrelated fields" has long underpinned the emergence of "human ecology" as a putative synthesis of disparate bodies of work.[67] In 1944, for example, the sociologist Amos H. Hawley lamented that human ecology "remains a somewhat crude and ambiguous concept" and suggested that the prevalent academic "chaos" in the field stemmed from three main sources: the divergence of human ecology from wider developments in ecological science itself; an overemphasis on the effects of competition, derived from a reductionist neo-Darwinian biological standpoint; and a gradual narrowing of the field to the study of "spatial relations."[68] For Hawley, the field of human ecology seemed stranded between two extremes: on the one hand, the attempt to incorporate everything within one conceptual schema; and on the other hand, an insistence that the field represented no more than an applied methodological tool. The post-Chicago school contributions of Hawley and others stressed that interspecies interactions rest on collective modes of behavior rather than the mere aggregation of individual actions.[69] In this sense, there is a movement in this body of work away from atomistic variants of spatial science, yet the parameters of the field remain porous and ill-defined. There are tensions between metaphorical representations of the "city as ecosystem," often diagrammatically represented as a set of flows, cycles, or metabolic interactions, and the socioecological dynamics of historical processes operating across different scales.

An emphasis on quantification and modeling underpinned the rise of "industrial metabolism" within urban ecology, a movement associated with the application of thermodynamics to urban space by the economist and physicist Robert Ayres.[70] By the mid-1980s, these attempts to quantify urban systems influenced the emergence of the so-called Vienna school of social ecology, which established a distinctive research program led by the sociologist and social ecologist Marina Fischer-Kowalski and her colleagues.[71] Working in a similar vein to existing strands of human ecology, this avowedly interdisciplinary body of work aimed to "address social and natural structures and processes on an equal epistemological footing."[72] The emphasis of this field has been on the development of "materials flow analysis" to quantify flows of energy and materials and thereby connect the "socioeconomic system" with the "ecosystem" at different scales of analysis, from the local to the global.[73] Modern societies are conceptualized as an "autopoietic system" in a clear intellectual lineage with the theoretical corpus of systems-based analysis developed in the early 1970s by the Chilean biologists Humberto Maturana and Francisco Varela.[74] The concept of autopoiesis, as a self-replicating system, presents us with another example of a scientific metaphor that has subsequently become embedded in aspects of social theory. In particular, the German sociologist Niklas Luhmann extended the idea of autopoiesis to include the self-organizational and reproductive dynamics of social systems, whereby "cells" might be substituted with "action systems" and other kinds of internally consistent communicative social phenomena.[75] Interestingly, however, Luhmann was wary of extending his conception of the self-maintaining system to that of the "ecosystem," preferring instead to refer to an "eco-complex," since he regarded the complexity of socioecological relations as a separate field to that of social systems.[76]

In fact, relatively few of the studies within this industrial metabolism paradigm specifically analyze cities: exceptions include Peter Baccini's study of urban systems in Switzerland and Ken Newcombe's research on Hong Kong, where urbanization is conceived as an intersecting set of flows amenable to a series of technical interventions in fields such as energy transitions, materials recycling, and the fostering of greater efficiency in urban form.[77] In some diagrammatic formulations the systems-based approach is conceptualized as a series of intersecting spheres: the Berlin-based geographer Wilfried Endlicher, for instance, presents a "sustainability triangle" (*Nachhaltigkeitsdreieck*) comprising ecology, economy, and society in the form of a Venn diagram. The urban arena is framed as a set of interacting systems in which policy emerges from a synthesis of countervailing discourses.[78] The 1980s marks an emerging coalescence between a series of integrative policy-oriented fields such as "human ecology," "industrial metabolism," and "ecological modernization."[79] In parallel with these developments, however, particularly in the wake of Ulrich Beck's "reflexive modernization" thesis articulated in the mid-1980s, an alternative body of work has emerged under the aegis of an ecological critique of modernity that owes a significant intellectual debt to the earlier insights of the Frankfurt School.[80] I will return to this parallel strand of ecological critique later in this section where I outline the emergence of "urban political ecology."

The search for epistemological unity is also reflected in the call by Marina Alberti, John Marzluff, and Steward Pickett to integrate human societies into urban ecology, so that cities are conceived as "complex adaptive systems" driven by a mix of "biophysical and human agents."[81] "Our central paradigm for urban ecology," write Alberti and her colleagues, "is that cities are emergent phenomena of local-scale, dynamic interactions among socioeconomic and biophysical forces."[82] They focus on a series of interactions between human populations and "their biophysical environment" that "generate emergent collective behaviors (of humans, other species, and the systems themselves) in urbanizing landscapes."[83] More recently, Alberti and her research team have stressed the human role in accelerated eco-evolutionary pathways in cities (a theme I will consider in the next chapter), thereby marking a critical point of intersection between urban ecology and evolutionary biology that is largely absent from the Vienna school.[84] Running through these analytical reformulations, however, is an uncertainty about the analytical scope of contemporary ecology in relation to the specific cultural, historical, and material dimensions of urbanization. How, for example, does an emphasis on "emergent collective behaviors" relate to historical understandings of human agency in the urban arena?

In a North American context, the cities of Baltimore and Phoenix stand out, since they were the first metropolitan areas added to the National Science Foundation's long-term study of ecological change across more than twenty sites. Their 1997 inclusion in the project, which began in 1980, marked the formal elevation of urban ecology from "subdisciplinary backwater," particularly in American universities, to a mainstream focus of scientific concern.[85] The Baltimore Ecosystem Study, which draws on the ecological concept of "patch dynamics," has advanced the idea that heterogeneous urban landscapes are comprised of dynamic interactions between different systems, in an echo of wider developments in landscape ecology associated with Richard Forman, Michael Hough, Ian McHarg, Anne Whiston Spirn, and others.[86] The Baltimore-based study, led by J. Morgan Grove, Steward Pickett, and their colleagues, seeks to present the "Baltimore school" as a clear successor to the Chicago school, although they reject the antiurban bias that pervades this earlier body of work.[87] In particular, they develop a connection with the work of botanist Arthur Tansley, who developed the concept of "ecosystem" in the 1930s to include an explicit human dimension.[88] Tansley rejected the prevalent emphasis on the "biotic community," and other organicist or vitalist metaphors, in favor of a series of ecosystems that "overlap, interlock and interact with one another" at a variety of different scales.[89] Yet the systems-based approach of the Baltimore school leaves the historical and political dimensions of the modern city in a state of obscurity. Specific phenomena that structure urban space such as the redlining of housing markets are simply described as "legacies" within a broader context of spatial and temporal complexity.[90] We find that the advent of "nonequilibrium" interpretations of ecological change from the 1970s onwards serves as a conceptual portal during the 1990s into the inclusion of human disturbance within the study of ecosystems. The core problem, however, is that an emerging emphasis on multiple scales and temporalities within the ecological sciences does not easily translate into a historical framework for the analysis of urban change.[91]

The systems-based approach to urban ecology is marked by an emphasis on the measurable characteristics of urban space, and this reliance on various forms of quantification within a broadly technomanagerial analytical framework allows a segue into contemporary concerns with ecological resilience. What most of these approaches share is an integrated analytical framework that is based on "additive" forms of interdisciplinarity derived from the use of ever larger datasets. Even appeals to work more holistically, such as the entomologist Edward O. Wilson's calls for a higher form of epistemological unity, remain rooted in a fundamentally positivist scientific agenda that struggles to make sense of historical change.[92] What remains unclear in terms of the ongoing impetus towards systems-based urban ecology is whether the framing of environmental problems stems from an attachment to an innate belief in an underlying scientific unity in the field itself, or rather emerges from the shifting institutional and policy nexus within which specific questions have been identified and investigated.[93] In contrast to the systems-based paradigm, our next body of work owes more to the historical interpretation of landscape change, and to a particular fascination with shifting assemblages of urban fauna and flora.[94]

This second strand of urban ecological thought has its roots in nineteenth-century encounters with urban nature. From the middle decades of the nineteenth century onwards urban natural history societies began to publish detailed inventories of their local areas that showcased the systematic application of taxonomic knowledge. Early botanical interest in unusual substrates such as walls and ruins, for example, gradually evolved to include ballast hills, construction sites, and other kinds of "ruderal ecologies" associated with human disturbance. By the early decades of the twentieth century we can detect a divergence between the work of "plant sociology," devoted to the study of "natural" patterns of vegetation in places with minimal degrees of human modification, and the creation of elaborate typologies for urban and industrial floras with multiple categories of plants such as local arrivals, long-distance arrivals, recent garden escapes, historic garden escapes, and so on.[95] In the 1920s, for example, the botanists Hans Höppner and Hans Preuss carried out a detailed survey of the flora of the industrial Ruhr district that they presented as "ein getreues Spiegelbild" (a true reflection) of what they had actually found on the ground rather than the search for idealized kinds of "natural" plant communities that had developed without human influence. Their fascination with *Adventivflora* (adventive flora)—note how this term is divested of the ideological connotations associated with other words such as "alien" or "exotic"—revealed a sophisticated grasp of the ecological specificities and complexities of urban and industrial environments. Höppner and Preuss conceived of their study as the latest in a series of regional urban floras dating back to the nineteenth century, such as those completed for Düsseldorf (1846), Aachen (1878), and Bochum (1887), and hoped that their research might tilt the biological sciences away from the general towards the particular, thereby opening up pedagogic possibilities for the enjoyment and identification of urban plants.[96]

Any species that flourished in a human environment was now a focus of potential interest (a theme I return to in chapters 2 and 3). From the 1930s onwards, for example, the Paris-based botanist Paul Jovet extended existing studies of urban botany to emphasize the emergence of novel plant assemblages that reflected multiple global interconnections as well as a variety of local human impacts such as degrees of disturbance.[97] Similarly, from the early 1950s onwards, studies by the botanists Herbert Sukopp, Hildemar Scholz, and their colleagues in Berlin produced some of the most detailed surveys of urban flora that have ever been undertaken.[98] We can refer to a distinctive Berlin school, marked by an emphasis on novel ecological assemblages, that moved beyond earlier inventories to consistently challenge the existing categories deployed in vegetation science (even if the methods did not significantly diverge until later work in the 1980s and 1990s). This close emphasis on hidden dimensions of urban ecology revealed through the painstaking botanical investigation of specific sites was emulated internationally through studies in Bengaluru (Bangalore), Chonju, Kuala Lumpur, Shanghai, Tokyo, and other cities with links to Berlin's Technical University.[99] During the 1980s, for example, the Sheffield-based botanist and lichenologist Oliver Gilbert, who was in close contact with Sukopp, extended many of the insights from Berlin to the urban and industrial landscapes of South Yorkshire. In one study, Gilbert investigated the large number of wild fig trees (*Ficus carica*), a Mediterranean species established in the Don valley south of Sheffield. The trees had clearly originated from seeds distributed in sewage (during heavy rain the city simply discharges raw sewage directly into the river). But why the concentration of trees in just one part of the Don valley? And why were most of the trees over fifty years old? For Gilbert this puzzling situation could be explained by the fact that the fig trees were flourishing in an area that had been the center of the Sheffield steel industry and that the need for industrial cooling water had ensured that this stretch of the river stayed at a temperature of about 20° C all year round.[100] In these and other examples we find that the realm of urban ecology is one of constant surprises—a series of unexpected juxtapositions and accidental encounters.

Although the Berlin school focused principally on plants, a variety of other studies (especially from the 1970s onwards) began to examine additional dimensions to urban ecology such as birds, mammals, and insects, as the field became more global in scope.[101] Examples include decades of research into the ecology and behavior of urban birds undertaken by Maciej Luniak and his colleagues in Warsaw that has contributed to the conceptualization of synanthropic urbanism (a theme I return to in the next chapter).[102] In Los Angeles, by contrast, it is the increasingly bold presence of the coyote during the twentieth century that has become a focal point for the cultural and scientific imagination, especially from the 1960s onwards. By the 1990s, the figure of the coyote had slipped across multiple disciplines, becoming a focus for critical animal studies and attempts to bring nonhuman denizens of cities into urban theory (a development which forms the starting point of my next chapter). The observational tradition now extends to aspects of behavioral ecology, ethology, and more sophisticated readings of human-animal relations.

A third strand of thought has become known as "urban political ecology" and derives from a synthesis between elements of political ecology emerging in the global South and alternative traditions in urban history. During the 1980s a number of critical counterpoints to existing ecological discourse began to coalesce around what has been termed "political ecology," rooted in neo-Marxian insights into patterns of environmental degradation. The pivotal contribution here is Michael Watts's investigation of food and famine in northern Nigeria, which is grounded in a radical synthesis of ecology and history.[103] Although Watts only engages with ecological science tangentially, a key focus of his analysis is "the denouement of the precolonial natural economy through the insertion of capital under the aegis of the colonial state," which he connects with "the historically specific syncretism of capitalist and noncapitalistic production processes."[104]

For Watts, the environmental field is a zone of contestation between different conceptualizations of nature-society relations, including the pivotal interface between an externally imposed European modernity and a diversity of premodern world views. Watts differentiates his study from various cybernetic, organicist, and systems-based conceptions of human-ecological relations that emphasize the adaptive capacities of human societies over the historical and geographically specific dimensions to the production of nature.

In the 1990s we find an extension of political ecology to the study of cities, mirroring aspects of urban environmental history, with systematic studies of pollution, public health, and environmental justice that characterize the urban arena as a field of ecological contestation rather than a locus for homeostatic adjustment.[105] This further elaboration of the field of political ecology encompasses several distinctive elements. First, these works share an interest in hybrid socioecological natures produced under specific historical and geographical circumstances, a perspective which is both influenced by the earlier work of Donna Haraway, Bruno Latour, and other figures from science and technology studies, and in tension with more depoliticized or "horizontal" variants of actor-network theory (including the later work of Latour himself). Second, they analyze the mutually constitutive circulatory dynamics of capital, infrastructure systems, and the production of metropolitan nature, as part of an expanded conceptualization of urban metabolism. Third, they share an empirical focus on a variety of environmental conflicts unfolding within the urban arena, with links to urban history, environmental justice, and other fields. Finally, they highlight the ideological permutations of different cultures of nature and their specific role in the "naturalization" of capitalist urbanization.[106]

Part of the objective of my book is to address some of the lacunae in the "first wave" of urban political ecology. The dynamics of urban nature itself, for example, were relatively underexplored within these earlier formulations.[107] In particular, ecological insights into the distinctiveness of urban space and the changing characteristics of biodiversity over time have not been widely integrated into studies of capitalist urbanization: there have been few studies that seek to link urban botany with neo-Marxian urban theory, for example, or which seek to integrate different kinds of human and nonhuman temporalities. The reasons for this disjuncture are simultaneously epistemological and ideological, in terms of the way in which different facets of material "nature," and the evolving scientific practice of urban ecology, remain poorly represented in comparison with critiques of nature-based metaphors in urban discourse. New dimensions to urban political ecology, either under way or yet to be realized, include critical engagements with multiple types of scale, decentered readings of the human subject that extend to other types of sentience and agency, emerging zoonotic dimensions to the urban arena, the challenge of new materialist epistemologies and their associated critiques, and greater attention to developments in the global South.[108]

This leads me to a fourth and final strand of urban ecology that I will introduce under the term "ecological pluriverse," connecting both to early twentieth-century theories of complexity (William James in particular) and recent developments in anthropology, cultural geography, and other fields that resonate with critiques of Eurocentric science and the bounded human subject.[109] Building on the insights of the anthropologist Arturo Escobar, I want to use the idea of the "pluriverse" to emphasize the shifting ontological and epistemological ground of urban ecology.[110] Important precursors are to be found in cultural geographies of "presence" and "absence" within nature conservation discourse, where urban nature is marked by a degree of provisional complexity and unknowability that has been extensively occluded in the existing literature.[111] The ontological turn in social theory has many ramifications and repercussions for the conceptualization of urban ecology that hinge on questions of living with difference, both material and epistemic, along with the recasting of ethical relations with the nonhuman.

How should we make sense of the many and often disparate cultures of nature that comprise urban space? The anthropologist Philippe Descola emphasizes the significance of seeing different things differently rather than the more familiar emphasis on seeing the same things differently. The recognition that coexisting worlds can be ontologically distinct is not just a matter of acknowledging the diverse perceptual realms of nonhuman nature but also applies to the diversity of human cultures.[112] At the same time, however, there is a tendency to project from cultural specificities towards new forms of universalism as one set of norms risks being replaced by another. Descola is not making an appeal to cultural relativism, but rather a call to recognize an ontological diversity that serves to problematize many of the conceptual categories that are routinely deployed in Eurocentric modes of analysis. But is there a radical incommensurability underlying Descola's position that renders ecological discourse unnecessarily opaque? Can we adopt a less ethnocentric perspective on nature that nevertheless enables connecting threads or regularities to play an analytical role? At the heart of Descola's project is the search for new modes of cohabitation that can accommodate "the whole immense multitude of actual and potential existing things."[113] In a similar fashion, the recent methodological impetus towards "multispecies ethnographies" extends the imaginative scope of environmental research, as part of a wider conceptual shift towards the posthuman defamiliarization of what we think we might already know about entanglements between human and nonhuman others.[114] Implicit, then, within an ecological pluriverse is a recognition of what literary scholars have recently termed a "manifold commons," suffused by the nineteenth-century nature poetry of John Clare, and marked by an intricate exploration of the human and nonhuman assemblages of life, work, and longing that constitute the fragility of the ecological realm.[115] By moving from a singular to a plural conception of nature, we become immediately aware of multiple ontological domains and the interweaving of disparate strands of cultural and material change, including the historical antecedents of colonialism, enclosure, and diverse types of extractive frontiers.

Throughout my text I have tended to use the expression "other-than-human" in preference to "more-than-human" because, although the latter formulation emphasizes a variety of multispecies assemblages, it also leaves the realm of human agency in a state of ambiguity. In drawing on the theme of agency, therefore, I am interested in developing a multispecies account that nevertheless retains a more nuanced sense of the scope of human subjectivities. The idea of the urban sensorium in particular, as a domain that spans human and nonhuman life, has emerged as a point of critical intersection between urban political ecology and a variety of other postpositivist theoretical developments, including critical animal studies, forensic ecologies, and postphenomenological interest in affective atmospheres. Part of my aim with this book is to develop a critical synthesis between different strands of urban ecology, and to consider in particular whether "urban political ecology," broadly defined, might be imaginatively extended to take fuller account of both the historiography of the ecological sciences and a range of more recent developments spurred by feminist, posthuman, and postcolonial interventions in the field of urban nature.

My book is not about "designed nature," though it does engage with examples of ecological simulacra in the place of "wild nature," as part of recent developments in landscape design and attempts to emulate pockets of biodiversity in cities. I have also sought to highlight interesting tensions between human intentionality and the independent agency of nature. In this sense, my conceptualization of "urban nature" moves away from forms of metropolitan nature that are of particular interest to engineers, planners, and other areas of technical expertise. In articulating what might be considered an "ecological sensibility" towards urban space, I am not seeking to advance any kind of analytical chimera for decoding the complexity of the urban process. By implication, then, I am articulating arguments for the study of urban nature for its own sake rather than in terms of what nature can do for "us" as framed by technomanagerial fields of concern with urban resilience and ecosystem services.[116] Yet arguments for the protection of spontaneous forms of urban nature are never reducible to some kind of intrinsic or autonomous ethic, since any fragment of the biosphere under discussion is already enmeshed to a greater or lesser degree with aspects of human consciousness and experience. What I am especially interested in here is the articulation of alternative, and in some cases counterhegemonic, sources of knowledge about urban nature produced by artists, writers, scientists, and in other primarily curiosity-driven encounters, including by children and other voices seldom heard in environmental discourse.

The opening chapter revisits the term "zoöpolis," which was introduced by the geographer Jennifer Wolch at a critical juncture in the development of animal studies. It serves as the starting point for a wider reflection on animal subjectivities, the independent agency of nature, and our ethical relations with the nonhuman. I consider the cultural and scientific significance of animals in the city, ranging from the return of wild nature to varying states of liminality and domestication. The realm of nonhuman urban nature emerges as a dynamic field in its own right, including a variety of evolutionary developments, epidemiological ramifications, and synanthropic interactions.

In chapter 2, I consider spontaneous spaces of urban nature as an interdisciplinary terrain that extends from renewed interest in urban biodiversity to alternative conceptions of vernacular ecologies and landscape design. I begin with early examples of botanical interest in walls, ruins, and archaeological sites that gathered momentum in relation to the novel forms of nature observed in bombsites, postindustrial wastelands, and other anomalous spaces. I explore a range of emerging cultural and ecological insights that seek to valorize specific forms of biodiversity and the serendipitous aesthetic experiences of "nondesign."

Chapter 3 begins with a study of spontaneous vegetation in Berlin undertaken by the French artist Paul-Armand Gette that traces the historical, material, and scientific context for cultural encounters with urban nature. The chapter describes how the island city of West Berlin emerged as an epicenter for new approaches to cultural and scientific experimentation in the field of urban ecology. I consider how the recognition of cosmopolitan ecologies has served as a critique of narrowly defined types of cultural landscapes with wider connotations for questions of race, sexuality, and social difference in the urban arena. I show how the experimental use of botanical methods can contribute towards an expanded conceptual field at the interface between the arts and the sciences.

Chapter 4 develops the theme of "forensic ecologies," drawing on developments in architectural theory, critical legal studies, feminist epistemology, and other fields. New forms of cultural and scientific practice have highlighted how specific organisms can serve as bio-indicators or "sentinels" for the understanding of environmental change and the delineation of specific ecologies of urban endangerment. I suggest that novel strands of grassroots ecological practice are enabling the emergence of more collaborative and counterhegemonic forms of knowledge, including new forms of cartographic representation.

In the final chapter I explore the question of time in relation to urban nature. I consider how the relationship between cities and "deep time" can be conceptualized as an entry point for the interpretation of intersecting temporalities in the urban arena. I show how the Anthropocene debate is marked by uncertainty over the precise role of urbanization in processes of biodiversity decline. The emerging interest in "urban rewilding" and "urban refugia" highlights the role of cities as laboratories for the exploration of future natures. Insights from ecocriticism and other fields reveal that the urban ecological imaginary is in a state of flux between dystopian conceptions of destroyed worlds and the recognition of new kinds of socioecological assemblages. I suggest that the distinction *between* cities and broader processes of urbanization remains significant for a more critically engaged reading of the politics of the biosphere. Indeed, an overemphasis on methodological globalism risks obscuring the differences that matter in the articulation of alternative modernities.

1

ZOÖPOLIS REDUX

When the blackbird flew out of sight,
It marked the edge
Of one of many circles.
Wallace Stevens[1]

One spring morning I heard a rapid splashing sound followed by a brief pause, and then the same noise again. I looked down from the back window of my North London home and saw a male blackbird busy washing itself in a shallow bowl that had filled with rainwater. A few days later the bird was back again, this time with a female companion, taking turns to act as a lookout for cats while the other washed. The Eurasian blackbird (*Turdus merula*) was historically a rather shy forest dweller, but since the early nineteenth century it has been steadily expanding its range into urban areas across Europe and parts of Asia: first recorded moving into southern German cities in the 1820s, the bird did not reach London until a century later.[2] The bird's distinctive song is now one of the most familiar elements of the urban dawn chorus.

How do the diverse agencies of nature intersect with urban space? The question of agency encompasses a spectrum of possibilities ranging from a more restricted sense of human creativity and self-reflection, in the classic sense of bringing imaginary rather than instinctive worlds into being, through to more calculative, responsive, and other-than-human interactions, potentially including the molecular fields of morphogenesis, geological formations, or other types of earth systems, as well as more recent kinds of self-replicating algorithms and emerging applications of artificial intelligence. Many urban situations, including the most fleeting or trivial encounters, comprise interactions between several different types of agency, spanning both human and nonhuman forms of sentience: imagine a crow perched on a rooftop aerial, peering down at passing pedestrians, a few of whom might cast a glance back. Or consider walls and other surfaces that have been colonized by lichens, moss, and other forms of vegetal life; there are complex patterns of agency at work throughout urban space.

Over recent decades animals have been steadily returning to cities. New types of ecological formations have emerged as former industrial sites become havens for wildlife, cleaner rivers regain flourishing aquatic life, and many buildings serve as nesting sites for birds. Longstanding associations with human settlements, including those of rats, monkeys, and crows, have been supplemented by new and evolving sets of relations involving coyotes, foxes, racoons, and diverse birds of prey. The picture becomes even more complicated with the presence of zoonoses, disease vectors, and other elements of epidemiological landscapes. The increasing presence of wild animals in urban areas, including keystone predators, has fostered new cultures of nature including interest in "urban rewilding" and alternative ecological imaginaries.

In this chapter I explore the ambiguous relations between animals and the city, and especially the connections between different conceptions of agency and the articulation of ethical relations towards nonhuman others. On the one hand, animals seem to be regarded as a clearly defined species or population, as categories of life to be controlled or ignored, yet there are a multitude of intermediate states of affective interaction that can operate on an individualized level, including the liminal zones of urban ecology that exist within or beyond the epistemic realm of late modernity. Further-more, by adopting a wide-angle biopolitical lens, which spans both public health concerns and the global remit of conservation biology, we confront a series of ambiguities based on cultural context, classificatory schema, and the wider "political body" within which these encounters are located.

1.1 Decentering the urban subject

My conceptual starting point for this chapter is the geographer Jennifer Wolch's striking neologism *zoöpolis*, complete with its subtle use of a diaer-esis. The 1996 essay in which she coined the term invites an imaginative encounter between the cultural history of modernity and insights from the biological sciences, marking a key intervention in the emerging field of "animal geographies" and the development of "transspecies urban the-ory."[3] Wolch highlights the independent agency of urban nature in order to open up a series of neglected ethical and political questions. "The recovery of animal subjectivity," she writes, "implies an ethical and political obliga-tion to redefine the urban problematic and to consider strategies for urban praxis from the standpoints of animals."[4]

But how can we characterize the "standpoint" of animals? We are famil-iar with standpoint theory in relation to human agency—including that of scientists themselves—but the inclusion of animals marks a significant extension. The notion of a standpoint moves beyond questions of behavior and the established parameters of animal ethology, to emphasize specific forms of nonhuman sentience. The perceptual realm of animals has been a longstanding focus of interest within philosophy and biology even if it has only recently been linked with critical reflections on the human subject. In earlier formulations à la Jakob von Uexküll, for example, the question of the nonhuman perceptual standpoint was not linked with a critique of the human subject or any wider consideration of ethical relations between human and other-than-human forms of life. Indeed, von Uexküll's inter-vention marks part of an organicist strand of twentieth-century European thought that has been consistently ambivalent, or even hostile, towards cities, modernity, and novel ecological formations.[5]

Early contributions to standpoint theory were in tension with homogeneous characterizations of the human subject that effectively replaced one form of universalism with another, as highlighted by postcolonial critiques of "Western" feminist theory.[6] An emphasis on the complexities of epistemic practice is a critical marker for the contribution of feminist epistemology and critical race studies to the study of human subjectivities.[7] A more nuanced reading of objectivity necessarily encompasses the context for knowledge production, the standpoint of the human actors involved, and the status of any putative independent realm of ontological reality to which scientific discourse is oriented.[8] By reflecting on these different dimensions to objectivity we can dispel the apparent tension between the particularism of standpoint theory and the implied universality within dominant scientific frameworks. The feminist philosopher Rebecca Kukla, for instance, building on the insights of Sandra Harding, asks whether we can challenge aperspectival epistemologies—the so-called "view from nowhere"—through an explicit acknowledgment of the additional insights that can be enabled through diverse subject positionalities. A more nuanced form of objectivity would, therefore, question the pervasive dominance of "aperspectival objectivity."[9] Kukla provides an elaboration of the Aristotelian emphasis on human perception as a form of "second nature" derived from a recognition that human faculties of judgment are not innate but acquired through the specificities of context and experience.[10]

Can a multiperspectival variant of standpoint theory include nonhuman others? If we shift our emphasis towards other-than-human nature we enter a contentious ontological terrain where the boundaries between "human" and "nonhuman" become less clearly defined. The rigorous inclusivity of the human subject under feminist and critical race studies, for example, risks becoming more diffuse under the Deleuzian move towards various forms of "subjectlessness."[11] There is a potential reinscription of existing occlusions under the guise of dispersed forms of agency and subjectivity marked by underlying continuities in Eurocentric epistemologies. A key question is the relationship between boundaries, classificatory schemas, and the construction of racial difference under modernity.[12] There remain significant points of tension between the "recursive" dimensions of Eurocentric thought and the articulation of Indigenous subjectivities.[13] How, in other words, can a greater attention to the nonhuman inform a decentering of the human subject that does not reinscribe existing discriminatory framings of what constitutes the human?

There is a degree of conceptual elision between standpoint theory and what Wolch describes as "transspecies urban theory." Transspecies theory seeks to make sense of interspecies understanding and the differences between human and other-than-human forms of sentience.[14] Wolch seeks to transcend objections to this ontological conundrum by emphasizing possibilities for greater kinship between people and animals. She distinguishes this phenomenon of kinship from quasi-spiritual claims that are premised on a human ability to connect with the perceptual realm of animals: we do not need to "know" what other animals think to have ethical concern for them, including the development of sensibilities that extend beyond the domesticated sphere of pets and more familiar mammalian companions.[15] As new media scholar Joanna Zylinska notes, we cannot "unthink ourselves out of our human standpoint, no matter how much kinship or entanglement with 'others' we identify."[16]

How, then, should we incorporate nonhuman others into a more ontologically calibrated analytical framework? The existence of empathy between species rests on a diversity of ethical relations and kinship bonds that develop among humans and nonhuman others. Pivotal to the emergence of critical animal geographies has been an enlarged reading of animal subjectivities that provides new lines of connection between the human and the nonhuman, among a diversity of other-than-human forms of life, and at the same time transcends the limitations of urban ecology as a set of relations among bounded objects, both sentient and nonsentient.[17] For example, the work of the primatologist Anindya Sinha on the behavior of urban macaques suggests a deep complexity to the relationship between monkeys and people that extends to a variety of specific types of relations. In place of the existing ethological and biopolitical emphasis on primate populations, Sinha describes a spectrum of microdifferentiations, behavioral traits, and individualized nonhuman standpoints. For Sinha, the question of human-animal relations must move beyond the elucidation of generalized forms of behavior to explore interaction between individuals. Sinha highlights the presence of intermediate forms of social organization among monkeys such as "commensal macaque troupes" that display variable levels of urban phenotypic adaptation as well as complex territorial demarcations of urban space. The adoption of distinctive gestures, and other forms of cultural evolution, enable macaques to build a much more dynamic set of relationships with people.[18]

The term "zoöpolis," as used by Wolch, encompasses a new phase of urban environmental thought that emphasizes a more active presence for wild nature in cities. Drawing in particular from her research in Los Angeles, where the coyote (*Canis latrans*) now holds a symbolic presence in the ecological imagination, she emphasizes the growing significance of wild animals within urban environmental discourse.[19] The coyote has become part of an "unruly" other-than-human landscape that is interspersed with more familiar elements of metropolitan space. Consider LA's Griffith Park at nightfall, located on the edge of the Santa Monica Mountains, where the "endless city" of artificial light to the south meets the looming darkness of the mountains to the north. The sound of coyotes is now a common backdrop to evening events in the park: a review of an outdoor performance of *Hamlet*, for example, describes how the final act was accompanied by eerie howling.[20] The historian and urbanist Mike Davis notes that Los Angeles has "the longest wild edge, abruptly juxtaposing tract houses and wildlife habitats, of any major nontropical city."[21] A recent field guide to the wild nature of Los Angeles notes how "coyotes cruise the tree-lined streets of Los Feliz and Silver Lake" and "red-tailed hawks perch on light poles along the 10 freeway."[22] Efforts to control coyotes in Los Angeles are stymied by their intelligence: some research suggests that less wily individuals tend to be the ones caught in traps, thereby "skewing the gene pool" in favor of smarter animals.[23] Recent proposals to trap and release animals elsewhere to control urban numbers simply ignore the ecological and territorial dynamics of the urban coyote population, since other individuals will simply move into the area.[24]

For Wolch, the impetus towards "renaturalization" is a process of "reenchantment" that enriches the lived experience of the city for human and nonhuman nature alike. "To allow for the emergence of an ethic, practice and politics of caring for animals and nature," writes Wolch, "we need to renaturalize cities and invite the animals back in—and in the process reenchant the city."[25] The meaning of "renaturalization" ranges from specific measures to encourage plants and animals to the inadvertent creation of new ecologies through the effects of abandonment or neglect (see chapter 2). The prefix "re-" implies a sense of a return or recovery to a former state of nature, exemplified by variants of restoration ecology, but such a view of urban nature also confronts us with multiple ecological imaginaries that include trajectories into unknown environmental futures.

The term "reenchantment" is closely associated with the rise of new materialist and neovitalist interpretations of urban space. In this literature, there is an emphasis on expanding the imaginative scope of encounters with other-than-human nature, including a variety of objects and material artifacts that are routinely ignored or overlooked. The political theorist Jane Bennett, for instance, takes inspiration from the detritus that accumulates in the grate over a storm drain, including a dead rat, as a starting point for her hylozoic claim that "all matter has life."[26] For the geographer Neil Smith, in contrast, writing from a neo-Marxian perspective, interest in the reenchantment of urban space is an ineluctably ideological project suffused by an idealist and romanticist view of nature.[27] Smith is not wrong, of course, to emphasize how nature is produced under capitalist urbanization, along with its symbolic ramifications, but is that all there is? Smith's own delight in urban nature, especially ornithology, is not reflected in his theoretical vantage point: there is a disjuncture between the affective and analytical dimensions to the experience of other-than-human life in the city. The dynamics of urban reenchantment, and the destabilization of the bounded human subject, are clearly something different from established romanticist idioms, even if neovitalist ontologies tend to occlude the ideological contours of nature and aesthetics under capitalist urbanization.

How we conceive nature affects the ways in which we picture our ethical obligations to nonhuman life. Wolch notes that the idea of the individual animal as a subject in its own right "is rarely reflected in eco-socialist, feminist, and anti-racist practices" where nature is regarded as "an active but somehow unitary subject."[28] Indeed, since the time in which Wolch first explored these themes in the 1990s, the shift of emphasis within critical animal studies from nonhuman collectivities such as "populations" or "species" towards that of individual animals has been a marker of critical animal studies.[29] In the next section I want to take the question of animal subjectivities and ethical relations with the nonhuman further by reflecting in particular on the production of meat and its concomitant landscapes of violence and control.

1.2 EDGELANDS

The relationship between animals and people in the modern city involves not only direct forms of empathy, companionship, or fear but also the diffuse and often violent sets of relations that underpin the production of food.[30] Animals form part of what the historian Chris Otter refers to as the "mechanized zootechnosphere" driven by the dietary expansion of the human ecological niche.[31] Global numbers of livestock now exceed the human population, with estimates of around 1 billion pigs, 1.5 billion cattle, nearly 2 billion sheep, and over 20 billion chickens. Selective breeding has changed the physiognomy of animals to increase their productivity and endurance of extreme conditions. Intensified forms of domestication have become part of the larger biopolitical reorganization of both human and nonhuman life under modernity. The large-scale production of protein, including the mass transportation of live animals, incurs high levels of suffering and mortality.[32] Although welfare standards have improved in many countries, and laws against animal cruelty have been widely enacted, there remains substantial evidence for the mistreatment of animals.[33]

Although domesticated livestock have had a longstanding place in human settlements, the extensive presence of feral nature such as stray dogs, pigs, or other animals became a characteristic feature of the fast-growing nineteenth-century metropolis. "Not only were there more people around to own animals," notes the historian Catherine McNeuer in relation to early nineteenth-century Manhattan, "but those people were also creating more garbage that, left uncollected on the streets, fed a burgeoning, free-roaming animal population."[34] Impoverished city dwellers kept goats, hogs, and other animals as a means to supplement their diet in the face of high rents and exorbitant food costs. By the 1820s it is estimated that there was one pig for every five people living in Manhattan, and these large animals regularly barged into pedestrians or obstructed thoroughfares. Legions of stray dogs were routinely joined by pets who were kept outside because of difficulties in controlling fleas. Bourgeois public opinion often viewed the animals roaming city streets as symbolic avatars for their working-class owners. Periodically, city authorities tried to control animals with brutal interventions such as the Manhattan "dog war" of 1836 in which over 8,000 animals were killed.[35] The clearing of pigs, dogs, and other animals from city streets formed part of the regularization of urban space in the nineteenth century: informal piggeries, for instance, were among the existing land uses swept aside during the construction of Manhattan's Central Park.[36] In the case of North America a series of ordinances, culminating in zoning restrictions, sought to control the presence of animals as part of a wider dismantling of existing approaches to food production and refuse disposal.[37] Similarly, in colonial Hong Kong, large numbers of animals—principally pigs and chickens—were kept by recently arrived rural migrants to help supplement their diet and income.

In one official report from 1877 a tenement is described where up to 70 pigs were being raised throughout the building (with some animals found hiding under the beds).[38]

By the middle decades of the nineteenth century the presence of unruly pigs roaming the city streets of North America was displaced by the sad spectacle of serried ranks of animals being marched to their death. In the case of Chicago, for example, William Cronon describes how "drovers sometimes stitched shut the eyelids of particularly obstreperous animals. Once blinded in this way, they could still keep to the road by following their companions, but were less inclined to make havoc."[39] In nearby Cincinnati, known as "Porkopolis" by the 1830s, the meatpacking industry developed new techniques of disassembly for animal bodies that presaged the wider spread of industrialized mass production.[40] "As time went on," notes Cronon, "fewer of those who ate meat could say that they had ever seen the living creatures whose flesh they were chewing; fewer still could say they had actually killed the animal themselves."[41] The increasing separation of urban populations from the production of food mirrors a wider degree of cultural detachment from the sociotechnical systems that provide energy, water, and other basic needs.

It would be a mistake, however, to assume a broadly synchronous and teleological sequence of developments. In the case of Moscow, for example, the new public abattoir completed in the late 1880s was conceived as a large-scale research facility at the leading edge of European modernity and epidemiological science rather than an element of hidden infrastructure.[42] The shift towards the "bacteriological city" was instituted in an exaggerated urban form, in part as a response to Moscow's rapid growth and declining living conditions. In twentieth-century Argentina we find examples of new meat-processing facilities that sought to incorporate visible elements of technological modernism: the seventeen abattoirs designed by the Italian-born architect Francisco Salamone for the province of Buenos Aires during the late 1930s mixed traces of fascist symbolism with a mix of brutalist, futurist, and art deco elements derived from the nascent

International Style. Salamone's distinctive slaughterhouses marked a small-scale precursor to the much larger industrialized facilities that would form the fulcrum of the Pampas meat-processing economy.[43]

It was ultimately biopolitical questions of nutrition and public health that instituted a new phase in the relationship between animals and urban modernity. The rapid growth of nineteenth-century cities necessitated different approaches to the production of food, and in particular the need to ensure the availability of cheap and safe protein for urban populations. Central to this biopolitical reorganization of urban space was a new generation of more efficient and technologically sophisticated abattoirs now operating under the regulatory eye of public health authorities. The architectural challenge for the creation of a more rationalized and hygienic type of slaughterhouse is reflected in the completion of La Villette in Haussmann-era Paris, but the full force of mechanized animal butchery is represented in its most extreme form by the opening of the Union Stockyards in Chicago during the 1860s, where some 200,000 pigs were slaughtered per day.[44] By the late nineteenth century earlier concerns with noise, stench, and congestion had been supplemented by public health revelations, such as the dangers of trichinosis and other hidden threats from unregulated meat production.[45]

The rise of the modern slaughterhouse also brought into focus the inhumane treatment of animals and the exploitation of the largely immigrant workforce in modern food production. The frightening working conditions of the Chicago meatpacking industry were meticulously recorded by the Progressive Era campaigner and writer Upton Sinclair in *The Jungle*, first published in 1906, which recounts how many tasks were so dangerous or noxious that many workers were rendered incapable of work within a few years, their hands reduced to infected stumps or bodies crippled by rheumatism. As for the conversion of cattle into meat:

> Along one side of the room ran a narrow gallery, a few feet from the floor, into which gallery the cattle were driven

47

by men with goads which gave them electric shocks. Once crowded in here, the creatures were prisoned, each in a separate pen, by gates that shut, leaving them no room to turn around; and while they stood there bellowing and plunging, over the top of the pen there leaned one of the "knockers," armed with a sledge hammer, and watching for a chance to deal a blow. The room echoed with the thuds in quick succession, and the stamping and kicking of the steers. The instant the animal had fallen, the "knocker" passed onto another; while a second man raised a lever, and the side of the pen was raised, and the animal, still kicking and struggling, slid out to the "killing bed." Here a man put shackles about one leg, and pressed another lever, and the body was jerked up into the air.[46]

Worse still was the fate of pigs, who from the 1860s onwards were suspended by hooks through their shins while still alive in order to save time in the slaughtering process.[47] The slaughterhouse resounded to a cacophony of squealing so loud "that one feared there was too much sound for the room to hold—that the walls must give way or the ceiling crack."[48] The mechanization of animal slaughter solved one set of public health issues but in so doing intensified the social stratifications underpinning modern food production, to create new spaces of separation and violence. "What makes the history of slaughterhouses particularly interesting," notes the historian Dorothee Brantz, "is that they shed light on the continuous interdependence of human and animal bodies even as livestock was increasingly removed from the streets of cities."[49] Animals not only provided protein but also performed much of the labor needed for the modern metropolis. The horse in particular was widely regarded as a kind of "living machine" that was integral to the functional dynamics of modernity. In 1879, for instance, the *New York Times* referred to New York as "stable city," whilst a survey in 1900 found that there were 7.4 people for every horse in Kansas City.[50] The working horse has left a distinctive imprint on the topography of the modern city: contemporary London, for example, has hundreds of mews and stables where horses were once kept and even tunnels that were designed with limited glimpses of daylight to prevent horses from bolting.[51] The question of animal labor has now become a key element in other-than-human histories of capitalist urbanization, extending metabolic accounts of urban space in new directions.[52]

The periurban landscapes of the global South hold parallels with nineteenth-century cities in Europe and North America in terms of the extensive presence of live animals such as cows, goats, pigs, and chickens. In many cases these animals easily outnumber the human occupants: the city of Dakar in Senegal, for instance, is estimated to have some five million chickens in relation to a human population of around 2 million.[53] In the case of poultry there are often intersections between small- and large-scale producers, so that the distinction between local operations and more globalized systems becomes blurred.[54] In parallel with early nineteenth-century European and North American cities, the many animals roaming the streets are able to feed on uncollected refuse, so their feeding habits effectively constitute part of the waste disposal system. Roadside piles of garbage in many cities of the global South attract a spectrum of domesticated, feral, and wild animals including cows, goats, dogs, rats, and crows. These street-level ecologies are, however, a focus of social and political contestation in relation to noise, dirt, congestion, animal attacks (particularly from dogs), and public health concerns (especially rabies).[55]

The slaughtering of animals remains a focal point for different constructions of marginality, as the political science scholar Zarin Ahmad shows in her study of the sociospatial intersections between meat, ethnicity, caste, and religion in Indian cities.[56] In working-class districts, animals are often slaughtered and butchered in small-scale premises at street level, with most aspects of meat production fully in public view. The production of meat is an ideological terrain that highlights cultural differences over human diet and contributes to intensifying patterns of residential segregation (even the smell of meat is reputed to endanger caste status for Brahmin Hindus).[57]

The geographer Krithika Srinivasan describes how concern with animal welfare in Indian cities is closely associated with upper-caste forms of Hindu nationalism, where death or injury to urban cows, for example, has led to violent forms of retribution.[58] The concentration of abattoirs in predominantly Muslim districts of Indian cities has also engendered specific ecological associations with the black kite (*Milvis migrans*) which thrives on offal and scraps of meat, which are often thrown into the air for good luck or for deliverance from worldly cares.[59]

The slaughtering of animals is not ordinarily considered a dimension of urban ecology, yet the social stratifications that accompany the production of food are related to a series of material and symbolic geographies. Food-processing industries constitute part of the hidden infrastructure of the city, employing predominantly poor or migrant workers, and form an inherent element in patterns of social inequality and environmental degradation.[60] In the film *Killer of sheep* (1978), directed by Charles Burnett, we see a different side to the urban landscapes of Los Angeles that makes explicit the connections between race, poverty, and the industrialization of nature. In Burnett's depiction of the Watts district, the central protagonist Stan (played by Henry G. Sanders) works in an abattoir where the long hours and brutal monotony contrast with the playful adventures of his children in the neighborhood (figure 1.1).[61] Urban nature is present not just in the form of sheep being corralled before their slaughter but also in the gentle swaying of trees in empty lots, intermittent birdsong, and the ubiquitous presence of dogs.

But what of ecological edgelands that fold back onto the epidemiological landscapes of modernity? The prevalence of industrialized slaughter and other forms of food production continues to pose a series of public health risks for modern societies (to say nothing of ethical relations with animals). A series of distinctive epidemiological landscapes are now the focus of research into new and previously unrecognized strains of zoonoses such as swine flu, avian flu, and the coronavirus group (including SARS, MERS, and Covid-19).[62] In the case of the recent H5N1 avian flu mutation

Figure 1.1   Children play in an alleyway in the Watts district of Los Angeles in Charles Burnett's film *Killer of sheep* (1978). Courtesy of Charles Burnett.

there are interconnections between the technoagricultural complex, food-processing industries, new patterns of urbanization, wetland destruction, crumbling public health infrastructures, and pervasive regulatory voids.[63] Similarly, in the case of Covid-19, which was first recorded in a cave-dwelling species of bat in 2013, there is speculation that an intermediate host such as the pangolin, which is traded in "wet markets" where different species of live animals are slaughtered in close proximity, may be involved in passing the virus on to humans.[64] Yet the focus on food markets in China or elsewhere as the main source of Covid-19 or other zoonoses risks infusing epidemiological discourse with xenophobic assertions.[65] Equally, the singling out of individual animals as the source of zoonoses can also provoke forms of violence against nonhuman inhabitants of urban space: following the scientific debate over the presence of SARS in a species of civet cat (*Paguma larvata*) sold in wet markets in East Asia there was widespread persecution of stray cats in Singapore and other cities.[66] Similarly, in the wake of Covid-19, there have been reports of attacks on stray dogs and cats in Wuhan, Guangdong, and other Chinese cities (although it is animals that are at risk of being infected by humans). In contrast to the so-called "spillover model," however, that places the blame on animal intermediaries, there is evidence for a more geographically dispersed set of circulatory dynamics for preexisting zoonoses in human societies. Under this alternative explanation, it is capitalist modernity itself that provides the conditions for a series of "amplifier effects," which increase the chances of genetic mutations facilitating the spread of new and more dangerous pathogens, as has already been observed with the evolution of influenza, chikungunya, dengue, and Zika.[67]

The zoonotic dynamics of capitalist urbanization form part of a wider set of epidemiological developments associated with late modernity. The recent spread of the virus causing African swine fever, sometimes referred to as "pig Ebola," related to international trade in livestock and meat products, has already killed hundreds of millions of animals in China, Cambodia, Vietnam, and elsewhere and may yet escalate into the largest ever recorded disease outbreak to affect global agricultural production.[68] Only wealthier countries with more robust animal health infrastructure systems can afford the biosecurity measures needed to bring the disease under control.[69] The threat of zoonoses is especially associated with zones of transition where urban space blurs into technological landscapes of food production adjacent to complex remnants of existing ecosystems, such as estuaries that support large populations of wild birds. Epidemiological landscapes throw into doubt the cross-sectional "radial geometry" of urban space, as developed in many diagrammatic representations of urban ecosystems, since the "urban fringe" can occur anywhere where pathogens or their insect vectors can flourish, such as blocked gutters, discarded tires, or neglected water tanks.[70]

1.3 ANIMAL TOPOGRAPHIES

Wild animals, including keystone predators, have become an increasingly significant presence in urban areas, exemplified by bears and coyotes in North America, wolves in Europe, and leopards in South Asia. Notions of "wildness" have increasingly permeated metropolitan cultures of nature, enabling new conceptions of human coexistence with animals to emerge.[71] In recent years a range of species including carnivores have become well adapted to urban areas, taking advantage of food sources and opportunities for shelter, and in the case of some animals such as badgers, foxes, and racoons, achieving higher population densities than in most of the rural parts of their range.[72] The increasing presence of coyotes in North American cities in particular over recent decades has provoked intense reflection on the changing cultural and ecological configurations of urban nature. As the ecologist Stanley Gehrt notes, "Because of its mystique among the general public, dramatic range expansion, opportunistic behavior, and role as a top predator in most North American metropolitan areas, the coyote is arguably one of the most controversial carnivores in urban landscapes."[73]

The expansion of the coyote's range across North America has occurred in stages: an initial westward spread from the late nineteenth century onwards was facilitated by the human persecution of wolves. This was followed by further range expansions during the twentieth century towards the north, east, and southeast, including many larger cities. By the late 1990s, coyotes had reached New York's Central Park—perhaps their first presence on Manhattan island for centuries. Yet significant aspects of coyote behavior in urban areas remain unknown. Despite popular mythology there is little evidence that they prey principally on pets, but they can be involved in territorial disputes with other animals such as foxes. Coyotes are known to catch other animals such as rats, squirrels, and birds but are also scavengers of road kill and garbage. They make intricate use of urban space including infrastructure systems as corridors between different habitats: San Francisco's Golden Gate Bridge, for instance, has become a regular conduit between different populations.[74]

In much of Europe it is the reappearance of the wolf (*Canis lupus*) that has captured public and scientific attention.[75] Wolves returned to Germany during the 1990s, probably arriving via western Poland, and have recently been seen for the first time in over 150 years in Belgium, Denmark, and the Netherlands. The increasing frequency of wolves in and around urban areas in Germany and elsewhere has been confirmed by forensic analysis of animal carcasses and the use of infrared cameras to observe animals at night.[76] The latest incursions of wolves appear to have been accompanied by behavioral changes, including reduced wariness of humans. In Germany, wolves have become especially associated with the postindustrial landscapes of "shrinking cities," where demographic decline has been marked by a proliferation of ruins and abandoned infrastructure networks.

The wolf has become a symbolic presence in the urban ecological imaginary: in Nicolette Krebitz's film *Wild* (2016), for instance, set in the former industrial city of Halle in eastern Germany, a young woman named Ania (played by Lilith Stangenberg) notices an urban wolf which lurks on waste ground near her housing complex. Ania manages to capture the wolf by using the traditional Russian method of winter hunting with flags (small pieces of colored cloth) to create a line to restrict the animal's movement. After subduing the wolf with a tranquilizer dart she attempts to domesticate the animal in her apartment. Over time, however, it is Ania who becomes more wolflike in her behavior. She becomes increasingly "feral" in her interactions with the outside world and dreams about allowing the wolf to lick her menstrual blood. The idea of feral in this context denotes a process of "de-domestication" in the adoption of increasingly animal-like traits where the distinction between human and "not human" has become blurred. The English word *feral*, which is of early seventeenth-century origin, is derived from the Latin *ferus* meaning "wild," yet it is this sense of a reverse ecology that is of interest. Krebitz's film serves as a parable for the "human animal" that lies hidden within the strictures of modernity: this is a tale of de-domestication, companion species, and altered subjectivities. The film concludes with Ania and the wolf wandering through a desolate postindustrial landscape, drinking side by side from a water-filled pool left by open-cast coal mining (figure 1.2).

Another animal that has been moving steadily into urban areas is the red fox (*Vulpes vulpes*). In the 1930s the fox began to expand its range into British towns and cities, benefiting from the expansion of low-density suburbs with relatively quiet roads and numerous gardens. In a second wave of expansion, from the 1960s onwards, foxes began to colonize inner urban areas with more challenging types of habitats, and also spread rapidly across urban areas in other parts of Europe, North America, and Australasia.[77] An article about the spread of London foxes, published in the late 1960s, notes that although some people welcomed these animals, others engaged in violent forms of persecution including the burning of live animals with paraffin. Indeed, the presence of foxes had sometimes caused "argument and strife between neighbours who cannot agreed [sic] on a policy of human-vulpine co-existence."[78] More recently, opinion over urban foxes has become more sharply divided in the wake of a small number of

Figure 1.2   Ania (played by Lilith Stangenberg) and the wolf in Nicolette Krebitz's film *Wild* (2016). © Heimatfilm.

reported attacks on people (despite the fact that these are far less frequent than dog bites).[79]

The role of low-density housing estates as an ideal kind of ecological niche for foxes is captured in Lorcan Finnegan's short film *Foxes* (2011), set in one of Ireland's "ghost estates" left in the wake of the protracted 2008 financial crisis, with overgrown gardens and unmown roadside verges.[80] The narrative is framed around the life of a photographer Ellen (played by Marie Ruane) who lives with her partner in the only inhabited house on an abandoned estate. Ellen becomes increasingly drawn to the foxes who gather in their garden at night. After being transfixed by their eerie nocturnal shrieking her affinity with the foxes becomes overwhelming and she eventually disappears into their world: we last see her leaping over a garden fence into the distance. The film's final image of a fox is suggestive of her vulpine avatar lurking outside their home.

The flourishing of wild animals in urban areas can have wider political resonance. The figure of the urban fox, for instance, has emboldened demands to ban the hunting of wild animals. The fox has become a symbolic marker for metropolitan versus nonmetropolitan cultures of nature (even if the political ecology of hunting is far more complex than this distinction allows). In California, an earlier emphasis on the eradication of large predators has been replaced by a form of uneasy coexistence in which these animals can "serve as proxies for broader cultural conflicts."[81] Similarly, in Germany, interest in the "rewilding" of postindustrial landscapes must contend with fear and hostility towards the return of wolves and other animals, and wolves have even been a focal point for the mobilization of far-right antienvironmentalist politics.[82]

A feral urban ecology is suggestive of shifting states of being (including the fluidity of the human subject in the case of the films *Wild* and *Foxes*). Although the word "feral" is derived from a more general sense of wildness, it is the zone of transition out of the domesticated sphere that has been the focus of ecological interest. Feral can be characterized as a liminal state between wild and domestic that unsettles dualistic conceptions of nature and culture.[83] The spatiality of the feral lies outside the home yet resides within modified landscapes that are shared by human and nonhuman life. It resonates with a Nietzschean reading of "cultural zoology" that offers a much more complex middle ground between complete forms of wildness and domestication.[84] The limits of domestication can be framed within a "Nietzschean bestiary" that places the boundary of human, "not human," and "becoming animal" within the realm of the biopolitical, while resonating with contemporary concerns over the limits to the human subject.[85] The term "feral," in an urban context, signifies a space of disorder or even danger that lies beyond the parameters of biopolitical governmentality. For the philosopher Eduardo Mendieta, the blurring of boundaries under Nietzsche's reading of the liminal zone between wildness and domestication points towards a post-Kantian form of "interspecies cosmopolitanism." Mendieta, following Margot Norris, places Nietzsche in a "biophilic tradition" that can be differentiated from romanticist variants of anthropocentrism and neo-Darwinian reductive conceptions of nature.[86] The destabilization of aesthetic and cultural categories under Nietzsche allows for what Norris terms "a phenomenological immediacy of experience" beyond the strictures of domestication. Indeed, Norris aligns the feral or "beastly" imagination with the development of biocentric, as opposed to anthropocentric, theory, rooted in forms of playful experimentation.[87]

This emphasis on feral ecologies as a movement between different sets of cultural and ecological relations can be illustrated by species such as the rose-ringed parakeet (*Psittacula krameria*) that is now naturalized in many European cities since its escape from captivity in the 1960s. In the London area, there are over thirty species of escaped parrots at large, of which four have bred outside of captivity and one—the rose-ringed parakeet—is regarded as a self-sustaining addition to the city's avian fauna (figure 1.3).[88] The flourishing population of London parakeets is popularly traced to the release of a pair of birds in Carnaby Street by Jimi Hendrix in 1968, but there are certainly instances of earlier releases.[89] Similarly, in the Los

Figure 1.3 Rose-ringed parakeets flying over Hither Green Cemetery, London (2014). Photo by Sam Hobson.

Angeles basin, over thirty species of escaped parrots have been observed, with more than twelve now breeding across the metropolitan area.[90]

Feral elements of urban nature can contribute towards novel or "recombinant" ecologies, defined as new mixes of species produced by direct or indirect forms of human activity.[91] The term originated in the work of the biologist Michael Soulé, but in the context of a wider field of concern with "invasive ecologies" as a threat to global biodiversity, in contrast with an approach that treats ecological novelty as a matter of cultural or scientific curiosity.[92] The interest in recombinant ecologies has since evolved into an emphasis on "novel ecosystems," as elaborated by Emma Marris and other proponents of "new natures."[93] Marris, for instance, contrasts conservation biology's preoccupation with intact or pristine ecosystems with "the more nuanced notion of a global half-wild, rambunctious garden, tended by us."[94]

Unruly urban natures have always held an ambivalent relationship with modernity. Feral nature, and especially the presence of stray dogs, has been widely interpreted as an indicator of urban decline or even traumatic historical events. The association between stray dogs and poverty-stricken urban districts is a recurring literary trope. In William Attaway's *Blood on the forge*, for example, first published in 1941, we find ubiquitous stray dogs in the side streets of Pennsylvanian steel mill towns.[95] The narrative is set in 1919 and centers on a trio of brothers who undertake the African American Great Migration to the industrial cities of the North, combining a vivid depiction of the destroyed landscapes with simmering racial tensions. In one instance a stray dog is unexpectedly encountered by the roadside, near to a man lying on the ground who has been brutally beaten:

> A lean hound with staring ribs came out of a hole in a garbage pile. He was full of rotten food but he trotted up to the fallen man and smelled around. He gave short whines, and the concave of his stomach jerked as though it were full of rubber bands. He opened his mouth. The sounds of his coughing broke as he continually reate his gorge.

60

Melody, one of the three brothers, throws a couple of stones:

> With a short yelp the dog fled in a tight circle. It was not the
> stone—he was trying to escape the sudden pain of madness,
> running away from the thing in him that whirled him like a
> top. At last he set a straight line for his flight. Over the prone
> body, across piles of garbage, until he was gone in the hills. His
> faint, crazy yelps hung in the air.[96]

The term "stray" can be ambiguous, however, since it covers a spectrum of feral modes of existence (some authors use alternative terms such
as "street dog" to denote a diversity of animals). Alan Beck's classic study of
stray dogs in Baltimore, carried out in the early 1970s, found that the animals were concentrated in poorer parts of the city.[97] Although Beck regards
his reflections on the reasons for this pattern as "speculative" he nonetheless draws together an interesting set of observations: "Higher reported
crime rates tend to encourage dog ownership; available garbage attracts
and provides food for strays; more open yards mean freer movement for
pet dogs; and the many hiding places created by the urban renewal of poor
areas permit the breeding and survival of strays."[98] A similar connection
between human insecurity and canine ecology is traced in the anthropologist Rivke Jaffe's study of Kingston, Jamaica, where dogs form part of a
heightened sensory apparatus that binds human to nonhuman in a symbiotic kind of relationship. Jaffe extends Beck's earlier insights to explore
the intersections between animal agency and the reproduction (or protection) of socioeconomic inequalities, so that existing insights in fields such
as urban political ecology are extended to include multiple configurations
of human and nonhuman agency.[99]

As the philosopher Clare Palmer notes, however, there is rarely any sense
of ethical responsibility for animals that form part of feral urban ecologies
produced through human actions such as abandonment.[100] Stray dogs pro-
liferate in the wake of social and political upheaval: under the final years
of the Ceaușescu regime in Romania, for instance, the forced rehousing

of people in Bucharest resulted in the mass abandonment of pets, leading to some 200,000 stray dogs wandering the city streets by the late 1980s. Despite repeated attempts at control, the dogs became part of the socio-ecological choreography of the postsocialist city, as the anthropologist Ger Duijzings relates: "As they are territorially based and linked to a particular neighbourhood, they are also called *câini comunitari* or community dogs. One gets to know them, and they get to know you . . . humans and stray dogs share the same space and develop a relationship based on everyday proximity."[101]

These kinds of intricate dog-human relations are also explored by Krithika Srinivasan in Chennai, where we find tensions between the affective relations that surround more familiarized everyday interactions between people and stray dogs, especially in poorer parts of the city, and periodic biopolitical interventions to reduce the number of dogs. The eradication ethos holds parallels in the Indian context with the intermittent removal of informal human settlements: there is a logic of disappearance used against human and nonhuman forms of "encroachment" under the aegis of "urban beautification," which may partly explain the greater kindness exhibited towards stray dogs and other animals in poorer communities. Srinivasan notes important differences between the conceptualization of the "stray dog" as the animal that is not identifiable as human property, and an understanding of the "street dog" (in Chennai and elsewhere) that recognizes the innate right of an animal to live in urban areas, irrespective of human claims or interests. For Srinivasan, echoing elements of Sinha's study of commensal macaques, the figure of the street dog living outside a biopolitical framework of intervention and control provides an illustration of the possibilities for cohabitation between human and nonhuman life.[102]

Stray dogs create distinctive urban ecologies: rats, for example, can benefit from the presence of dogs since they often chase away potential predators such as cats but rarely attack rats.[103] From an epidemiological perspective, the presence of stray dogs in Africa and South Asia is the main source of human infection with rabies (in Latin America transmission is principally via bats). In India, thousands of people die every year from rabies after being bitten by infected dogs, whilst the number of stray dogs in many cities has risen significantly since the decline of culling in the early 2000s, and also because of decreased food competition with vultures. Vulture numbers have declined following the use of the anti-inflammatory drug diclofenac on cattle which is transmitted through the food chain to carrion-feeding birds.[104] Interestingly, the presence of keystone predators in Indian cities such as the leopard (*Panthera pardus*) reduces the number of stray dogs and indirectly helps to reduce the threat of rabies. In Mumbai, for example, the leopard is the main predator of stray dogs: a population of around 40 leopards in and around the Sanjay Gandhi National Park is estimated to have saved many human lives from rabies infections. Yet these leopard-dog ecologies may be a transitional phenomenon because of development pressures in and around the park (figure 1.4).[105]

Another example of canine ecologies of abandonment is the Chernobyl exclusion zone in Ukraine, marked by novel ecological formations comprising dogs, wolves, jackals, and other animals. The large numbers of stray dogs, many descended from abandoned pets, have become the focus of contrasting ethical and biopolitical regimes of care and surveillance, ranging from an initial emphasis on eradication to more recent interest in adoption schemes with input from animal welfare charities. The last pack of feral dogs is thought to have roamed the area in 2011; since then, a gradual process of partial "redomestication" has taken place, with animals coming into contact with increasing numbers of people, both scientists and visitors, who have returned to the site.[106]

Wild animals also feature in postapocalyptic ecological imaginaries. In Terry Gilliam's film *Twelve Monkeys* (1996), for instance, we encounter a world where a virus has wiped out most human life, leaving animals in control of the surface of the planet whilst the human survivors live like "worms" in a labyrinth of underground bunkers. In an early portent of the "animal city," before the virus has taken hold, a group of activists release all the animals from the Philadelphia zoo: we see giraffes on the freeway,

Figure 1.4 Leopard sauntering along a wall next to a residential housing complex, Mumbai (2018). Photo by Steve Winter.

a tiger stalking the steps of a municipal building, and a cloud of flamingos taking to the sky. Something not unlike Gilliam's cinematic depiction occurred in 2015 following severe flooding in Tblisi, Georgia: following the inundation of the city zoo many of the surviving animals managed to escape, including a hippopotamus, big cats, wolves, bears, and hyenas, all of whom wandered through the streets. And in the wake of coronavirus lockdowns numerous animals have ventured into towns and cities: there have been many reports of goats, deer, and other creatures nonchalantly grazing roadside verges and front gardens.

1.4 SYNANTHROPIC ECOLOGIES

Synanthropic organisms can be defined as species of animals and plants that thrive in relation to human environments: a category that ordinarily excludes domesticated organisms, or those that have been chosen as companions or elements of planting schemes, but includes the many species that have adapted themselves to living with, among, or even on human beings.[107] Archaeological evidence points to the increasing incidence of synanthropic species associated with early efforts to store grain and other foodstuffs. These early links with human settlements were later joined by new waves of organisms spread by expanding trade routes or military conquests.[108] The synanthropic realm includes some species that are regarded as pests, pathogens, or vectors for disease but many more that are welcomed, little-known, or simply overlooked. A recent guide to the fauna of Zurich, for instance, includes not just a familiar range of birds, mammals, and insects associated with European cities but also the ubiquitous dust mites (*Dermatophagoides* species) and human hair follicle mites (*Demodex folliculorum* and *D. brevis*).[109]

The conceptualization of urban nature as a series of distinctive zones or biotopes has also been applied to the synanthropic realm of human dwellings: the entomologist Bernhard Klausnitzer divides the fauna of modern buildings into a series of microecologies extending to cellars, floors,

wardrobes, balconies, and roof spaces.[110] In the entomologist George Ordish's remarkable book *The living house*, published in 1960, we are treated
to what could best be described as a natural history of one house, detailing the shifting assemblages of nonhuman life that had colonized the "new
environment" after its completion in 1555. "The pattern of living in the
house may be said to be a series of circles," notes Ordish, "slowly moving, expanding, changing, declining and merging with that of human
beings . . . so that the history of the house is a figure resulting from the
inter-movement of the various circles of the life forms in the house."[111] One
of the few detailed studies of the "indoor ecology" of houses, carried out in
Raleigh, North Carolina, found a high level of arthropod diversity, with
many homes supporting hundreds of species (principally flies, beetles, spiders, and ants). The vast majority of these species are harmless and often
include significant predators of pest species such as minute parasitoid wasps
that are almost invisible to the naked eye and feed on the eggs of cockroaches and other insects. Many of the species encountered such as book
lice (Liposcelididae) and clothes moths (Tineidae) have made an ecological
transition from naturally occurring niches such as birds' nests to thrive on
other sources of organic matter within the home.[112]

The ecology of the human home ranges beyond more familiar companion species to regular encounters with birds, bats, and other animals
living nearby. Even a single outside light can develop its own microecology of lizards, spiders, bats, and night-flying insects. In warmer climates a
water-filled saucer under a flower pot on a window sill can support mosquitoes such as *Aedes aegypti* and *A. albopictus*, which are the main vectors
for chikungunya, dengue, yellow fever, the Zika virus, and many other
diseases.[113] Among the most significant contributory elements to urban
microecologies favoring mosquito-borne disease are discarded tires that
enable the development of *Culex* and *Aedes* species.[114] Dengue fever is one of
a number of insect-borne diseases that are increasingly prevalent in urban
areas and is now spreading under the influence of climate change—in some
areas the insect vectors are already present, though dengue itself has not

yet been recorded. The rapid postwar expansion of dengue has been driven in particular by a combination of new patterns of urbanization, increased international mobility, and misdirected mosquito control efforts towards politically high-profile insecticide spraying of whole neighborhoods, rather than painstaking searches for insect vectors that live inside homes. In most of the larger cities of the global South dengue is now considered to be "hyperendemic with multiple virus serotypes co-circulating." As the epidemiologist Duane J. Gubler writes,

> dengue is an insidious disease that is always present in large tropical urban centers, lurking in the shadows out of sight and out of mind of physicians and public health officials during interepidemic periods, waiting for the opportunity to strike, either by mutating to a genetic subtype that has greater epidemic potential and virulence, or by catching a ride to another city where the population is more susceptible to that particular serotype.[115]

Dengue and other infectious diseases form part of what the geographers Vinay Gidwani and Rajyashree Reddy refer to as a "strange geography of encounters and contact zones" emerging from the combination of parasitic, speculative, and technoecological modes of urbanization. They single out dengue in particular as indicative of "the emerging hazardscapes of eviscerating urbanism" that accentuate unequal landscapes of epidemiological risk in the cities of the global South.[116] The corporeal dimensions to Gidwani and Reddy's analysis point towards the affective aspects of negative encounters with nonhuman inhabitants of urban space. Beyond the immediate epidemiological threat of dengue, malaria, Zika, and other diseases lies the realm of unwelcome cohabitation.

Among the most detested and ubiquitous synanthropic organisms is the bedbug (principally *Cimex lectularius* and *C. hempiterus*). The presence of bedbugs was noted in the early natural histories of Aristotle and Pliny but undoubtedly troubled earlier sedentary human societies, and most likely their cave-dwelling ancestors: the insect is thought to have made an evolutionary switch from bats to people at some point in the distant past. Though there are medieval records of infestations, the organism became much more widespread from the early modern period onwards. In the early seventeenth century, for example, the Italian entomologist Ulisse Aldrovandi had noted that infestations were worse in poorer habitations where linen and bedding materials were changed less frequently. A range of historical records suggests, however, that bedbugs were just as numerous in royal palaces as in the simplest of dwellings: almost any human habitation could serve as a potential breeding ground.[117] By the early decades of the twentieth century the scale of the problem had grown steadily due to factors such as changes in building design and heating systems, with very high levels of infestation documented across Europe and North America. The "bug crisis" led to a desperate array of control measures such as burning, heating, freezing, poisoning, trapping, and various modifications to the design of beds: among the advice for tormented individuals was the placing of beds in the middle of rooms, with the base of bedposts placed in pots of water or oil to hinder the advance of the bugs (they often arrived anyway by dropping off the ceiling).

The bedbug is also a recurring presence in modern literature, stalking the pages of Anton Chekhov, Joseph Heller, and Ralph Ellison, to name but a few. Consider, for example, George Orwell's description of the Hôtel des Trois Moineaux in *Down and out in Paris and London*, first published in 1933:

> The walls were as thin as matchwood, and to hide the cracks they had been covered with layer after layer of pink paper, which had come loose and housed innumerable bugs. Near the ceiling long lines of bugs marched all day like columns of soldiers, and at night came down ravenously hungry, so that one had to get up every few hours and kill them in hecatombs.[118]

Bedbug infestations in slum housing had become so prevalent by the 1930s that specious claims of greater "resistance" to bugs among the urban poor were offered as a form of corporeal essentialism to explain away failures in housing and public health policy.[119] The postwar use of DDT (dichloro-diphenyl trichloroethane), replacing the earlier use of the highly toxic hydrogen cyanide, led to a steady decline in the incidence of bedbugs yet also heralded a concomitant decline in communal control efforts. First tested in wartime to disinfect barracks and other military facilities, DDT was regarded by the US Bureau of Entomology as "the perfect answer to the bed bug problem" and became such an affordable control measure for the general population that it was even incorporated into products such as paints and wallpapers.[120] Interestingly, the decline in bedbugs appears to have begun before the widespread use of DDT, which was curtailed in the wake of environmental concerns from the early 1970s onwards. Yet the reliance on such individualized technical solutions laid the basis for a resurgence of the problem.[121] By the 1980s, the bedbug had declined so significantly that it began to fade from cultural and political consciousness. From the 1990s onwards, however, there has been a sharp increase in bedbug infestations almost everywhere, driven by factors such as increased resistance to insecticides, more international travel, higher levels of poverty and overcrowding, and an associated need for more homeless shelters and other types of temporary accommodation.

Other synanthropic pests include cockroaches, although only about five out of the 30 or so species associated with human settlements cause significant infestations. As the tropical ecologist William Beebe notes in his book *Unseen life of New York*, published in 1953, the cockroach made a switch at some point in human history from the detritus of the forest floor to the synanthropic realm of human habitations.[122] Cockroach infestations are especially prevalent in poor housing, an effect exacerbated by the impact of racism and other forms of discrimination. In the poorly maintained public housing projects of North America, for example, the majority of children suffering from asthma have been found to be reacting to cockroach-related allergens.[123] Perhaps even more than the bedbug, the figure of the cockroach has been used to enact forms of social and political denigration, notably towards the Jewish population in Nazi-era propaganda, and more recently against the Tutsi minority in the Rwandan genocide of 1994. In some cases, however, the symbolism is reversed to signify forms of urban insurrection, as in the writer and political activist Oscar Zeta Acosta's account of the Chicano political fightback in postwar Los Angeles.[124] Similarly, the Puerto Rican poet Pedro Pietri draws on the figure of the cockroach to describe low-income housing conditions in New York, except that in one of his poems he presents a suicidal insect's point of view, driven to distraction by loud television playing "at full blast."[125] In *Cockroach* (2020), the Chinese artist Ai Weiwei's recent documentary about the Hong Kong protests, the title marks an ironic reversal of the authorities' derogatory characterization of political opponents and also a recognition of the role of dispersed and dynamic forms of antiauthoritarian action and mass mobilization.

The synanthropic realm thrives in the largely unseen ecologies of urban technological networks. The "tubular ecologies" of infrastructure systems support a distinctive assemblage of organisms. Water pipes, for instance, can develop "biofilms" that harbor numerous species of bacteria, fungi, crustaceans, and nematodes, especially where there are corroded surfaces or higher ambient water temperatures (there are some pathogens that specifically thrive in hot water systems such as *Naegleria* and *Legionella*). Most of the organisms associated with pipes and water distribution systems are harmless, but some do present a threat to human health such as *Mycobacterium avium* (causing similar symptoms to tuberculosis), *Acanthamoeoba* species (a source of serious eye infections for contact lens wearers), *Naegleria fowleri* (a cause of amoebic meningoencephalitis), and several *Legionella* species.[126] In the case of *Naegleria fowleri*, known as the "brain-eating amoeba," which was first recorded in Australia in the 1960s, a number of serious infestations of public drinking water supplies have been reported from the USA, most recently in Florida, Texas, and Louisiana, and also in the

Karachi metropolitan region where there is speculation that the organism has adapted to more saline conditions, as well as benefiting from climate change (figure 1.5).[127]

Another example of infrastructure-related disease is the occasionally fatal respiratory condition known as Legionnaire's disease, caused by the *Legionella* genus of bacteria and first identified from an outbreak at a convention held for the American Legion in Philadelphia in 1976. The *Legionella* group of freshwater bacteria normally parasitize protozoan organisms but can also infect human phagocytic defense cells, sometimes facilitated by disruption to water disinfection practices, as occurred during the water crisis facing Flint, Michigan, from 2014 to 2015.[128] Synanthropic ecologies highlight not only the interdependencies of urban socioecological assemblages but also the adaptive capacities of nonhuman nature. The processes of adjustment to human environments can even involve changes in the composition of nature itself: it is here that we encounter the phenomenon of "new natures" emerging at the genetic or epigenetic level of individual species.

1.5 EVOLUTIONARY PATHWAYS

Urban environments are marked by a distinctive kind of dynamic complexity that can foster accelerated forms of evolution. For the evolutionary biologist Menno Schilthuizen, the urban arena is enabling faster rates of evolution than that originally envisaged by Darwin.[129] A striking example of urban evolution is offered by the genetic divergence of subterranean mosquitoes living in different parts of public transport networks. In London, a subspecies of the *Culex pipiens* complex, named *Culex pipiens molestus*, has evolved after just a few hundred generations to feed on human blood in the underground network, whilst the aboveground *Culex pipiens pipiens* feeds mainly on birds. The *molestus* form of the mosquito first came to public prominence in 1940 with the use of subterranean tunnels to shelter people from aerial bombing raids during World War II and has continued

71

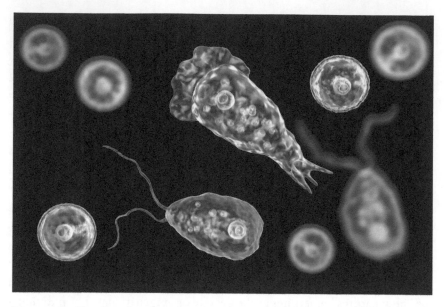

Figure 1.5 Flagellate forms, trophozites, and cysts of the "brain-eating amoeba" *Naegleria fowleri*. Photo: Kateryna Kon / Adobe Stock.

to trouble maintenance workers ever since. Similar genetic changes in subterranean mosquitoes have now been observed in underground transport systems in Amsterdam, New York, Tokyo, and other cities.[130] The *molestus* form is now quite widespread in a variety of subterranean environments, such as waterlogged building foundations and sewer networks across Europe and North Africa.[131]

Some of the first empirical studies of urban evolution in action include the gradual preponderance of darker-colored moths that are better disguised against soot-stained walls and other surfaces.[132] The phenomenon of industrial melanism was initially observed in nineteenth-century industrial cities, where the dark form of the peppered moth (*Biston betularia*) quickly began to outnumber the existing paler form. In the 1890s, the Manchester-based entomologist J. W. Tutt proposed that this was an example of Darwinian evolution in action, a hypothesis recently reaffirmed by the biologist Michael Majerus in the face of methodological critiques of earlier studies.[133]

Moving from melanin to melatonin, a contemporary field of investigation that mirrors the earlier interest in adaptations to industrial environments examines the effect of artificial light at night on the phenology, physiology, life history, behavior, and population dynamics of many different organisms. Recent studies of nocturnal insects, for instance, suggest that behavioral adaptations are emerging to avoid increased levels of light pollution, such as more limited movement with wider implications for plant pollination.[134] Global levels of light pollution have increased dramatically in recent decades, so that many celestial features of the night sky have become imperceptible in metropolitan areas. Intense levels of light pollution place urban ecological systems under a constant glare, affecting not only insects but also birds, fish, and plants, as well as human circadian rhythms.[135]

Birds have also featured widely in studies of urban evolution. In the case of the blackbird (*Turdus merula*), for example, a variety of behavioral adaptations to urban environments, including alterations in the pitch and timing of its song, have been detected alongside specific genetic changes. It is even suggested that urban blackbirds are in the process of evolving into a new species that will no longer interact with rural birds to any significant extent.[136] Studies of urban birds have also revealed an island effect within relatively isolated populations: a recent study of the great tit (*Parus major*) within the parks of Barcelona found patterns of genetic variation emerging across the metropolitan area fostered by habitat fragmentation.[137] Yet the implications of elevated levels of genetic drift among small or isolated populations, and the precise dynamics of "gene flow" within urban environments, remain only partially understood.[138] Plants also exhibit evolutionary responses to urban environments. Examples include the advantages conferred by reduced seed dispersal so that fewer offspring are lost to concrete surfaces and other inhospitable substrates.[139] Even urban soils show signs of evolutionary change, with increasing concentrations of drug-resistant bacteria and other organisms producing an "antibiotic resistome" that is now an integral part of urban ecologies.[140]

The dynamic response of animals, plants, and other organisms to urban environments has also prompted interest in epigenetic dimensions to urban ecology. Epigenetics can be considered an intermediate zone between short-term behavioral adaptations and inheritable change, so that phenotypic variations are expressed without alterations in the underlying nucleotide sequence of the genome.[141] These epigenetic changes in gene expression can be activated in response to specific conditions such as noise, light, temperature, pollution, and other factors.[142] A study of the leaf-cutter ant (*Atta sexdens*) in São Paulo, Brazil, for example, found that inner-city colonies displayed a greater tolerance of higher temperatures caused by the urban heat island effect. The research team were careful to stress that observed differences could not yet be ascribed to genetic change, although there is clear evidence that urban conditions can affect phenotypic variation.[143] Similarly, a recent study of the lizard *Anolis cristatellus*, found throughout Puerto Rico, showed phenotypic variations in urban areas towards longer limbs and more "grippy" feet to enable more acrobatic forms of ecological parkour in the built environment.[144] Recent studies suggest that genes may have some kind of "memory" of environmental pressures that can be passed onto future generations: although many epigenetic characteristics are transient in their expression, others are clearly not.[145]

The urban environment forms part of the history of evolutionary science itself, with influential studies discussing the effects of industrial melanism, light pollution, habitat fragmentation, and many other factors. Accumulated evidence shows that urban environments can generate their own forms of biodiversity including the early stages of speciation. For advocates of the "adaptive Anthropocene" these accelerated patterns of urban evolution are an integral element of the "sixth genesis," a process of evolutionary divergence and speciation in response to human influence that is leading to the emergence of life forms that are better adapted to new types of environments (a theme I return to in chapter 5). An emphasis on urban evolutionary change can clearly form a biological corollary to contemporary resilience discourse. To what extent, though, can intimations of the future city, and its associated ecological imaginaries, divert attention from structural dimensions to environmental change?

1.6 IMMUNOLOGICAL PARADIGMS

Cities illustrate the limits to human control over nature. Furthermore, efforts to control nature illuminate the ethical ramifications of relations with nonhuman others. In the case of urban mammals, for instance, we find a tension between the cherished individual, exemplified by pets, and the more abstract concept of a population that may be subsumed within a biopolitical framework of enumeration and control.[146] These dilemmas are revealed through interventions such as the culling of squirrels, stray dogs, and other animals. Even rats have become a focus of ethical concern in Paris and other cities, where there are demands to simply leave them alone (there are some parallels with the decline of mole catching and other historical pest control activities).[147] The organization Paris Animaux Zoopolis

(PAZ), formed in 2015, campaigns for the protection of all animals "without distinctions between species" as part of an ethical commitment to a multispecies urban society. The organization derives inspiration from the Canadian philosophers Sue Donaldson and Will Kymlicka, with their expanded conception of citizenship extending to "liminal organisms" that live in proximity to human settlements.[148] The PAZ campaigns include the removal of animals from circuses, a prohibition on fishing, and the protection of rabbits, rats, and other organisms from extermination. A series of PAZ posters on the Paris metro, calling for a halt to "the massacre of rats," highlights the intelligence, sociability, and altruism of rats as cohabitants of urban space (figure 1.6). For PAZ, the right of nonhuman organisms to be regarded as "denizens" of the urban arena is rooted in a utilitarian emphasis on the capacity to feel pain or suffering, thereby emphasizing degrees of sentience or the sophistication of a central nervous system as the basis for animal ethics.

Critical insights from urban epidemiology have revealed tensions between technomanagerial immunological paradigms and structural dimensions to public health, such as poverty, poor housing, and crumbling infrastructure systems. The limits to control can be illustrated by emerging threats such as mutating zoonoses, drug-resistant strains of tuberculosis, and newly recognized pathogens. In the case of the Zika virus, first isolated in 1948, many epidemiological models assume a homogeneous interaction between insect vectors and human populations. Yet the major outbreak in the Americas and parts of Southeast Asia during 2015 and 2016 included high concentrations of cases in cities.[149] As previously mentioned, the principal insect vectors are the mosquitoes *Aedes aegypti* and *A. albopictus*, which are well adapted to urban environments and flourish in areas with intermittent water supply, for example, where households are forced to store large quantities of water. Indeed, *Aedes aegypti* is among the most synanthropic of all insect vectors for disease, biting by day, and closely associated with domestic interiors and the complex and often inaccessible interstices of urban space. Recent research in São Paulo suggests that the urban heat

Figure 1.6 PAZ (Paris Animaux Zoopolis) campaign poster on the Paris metro against the killing of urban rats. Source: Paris Animaux Zoopolis.

island effect is a further contributory factor in the abundance of potential insect vectors for disease.[150] Furthermore, other mosquito-borne diseases such as dengue and chikungunya render the human body more vulnerable to infection with Zika, thereby generating compound forms of corporeal vulnerability in affected areas.[151]

Methods of mosquito control have also added to the intractable complexities of urban epidemiology in Brazil and elsewhere. Decision making over disease control can be remote from grassroots public health expertise and is often driven by high-level lobbying from the agrochemical industry through entities such as Fundación Mundo Sano and Chemotecnica, which have powerful connections to the World Health Organization, the Pan American Health Organization, and the Brazilian Ministry of Health.[152] Similarly, the fumigation of urban areas to kill adult mosquitoes often uses Malathion, a probable carcinogenic organophosphate that has been banned in Europe. The Brazilian Association for Collective Health (ABRASCO) has demanded urgent epidemiological research into the effects of mass chemical control of urban mosquitoes. The emphasis on "hegemonic strategies," such as the high-profile helicopter spraying of urban neighborhoods, or aggressive ground-level incursions by hazmat-clad insecticide teams, enacts a kind of performative show of biopolitical strength (figure 1.7).[153] As the Argentinian public health advocacy organization REDUAS points out: "Mass spraying is not the solution to a problem; it's merely generating a business within a problem."[154] The immunological paradigm creates its own frontiers for speculative biopolitics within the urban arena. Not only are such mass spraying programs ineffective (as the *Aedes* vector is so well adapted to domestic interiors and other types of inaccessible spaces) but indiscriminate control methods facilitate vector resistance.[155] The use of synthetic chemicals consequently produces a series of corporeal vulnerabilities that perpetuate existing inequalities by class, gender, and race.[156] The inescapable materialities of urban space unsettle any attempts to restrict epidemiological discourse to the parameters of laboratory-based science alone. The generalized use of insecticides in urban areas

Figure 1.7   A woman tries to protect her child from insecticide spraying against the Zika virus in the neighborhood of Imbiribeira, Recife, Brazil (2015). Photo by Ueslei Marcelino.

contributes towards mass invertebrate decline and the emergence of new
ecological assemblages within which a smaller number of more dangerous species are present.[157] New approaches to mosquito control such as the
mass release of genetically modified insects, predominantly in poor areas,
continue to ignore the underlying relations between poverty and urban
epidemiology.

For the philosopher Roberto Esposito, drawing heavily on Foucault,
the articulation of a nonutilitarian ethics towards nature is grounded in
a critique of the biopolitical denigration of the nonhuman.[158] Esposito
emphasizes a diversification of modes of governmental intervention,
including the presence of an affirmative strand of biopolitics that occupies
conceptual ground between the starkly contrasting positions of Giorgio
Agamben and Antonio Negri. Of interest here is the shifting definition of
a community that might extend to the inclusion of nonhuman occupants
of the urban biosphere. There is an implicit extension of an immunological
logic towards the defense of all or part of this elaborate constellation of life
forms from real or perceived sources of threat and disturbance. Following
Esposito, an immunological logic is concerned with the integrity of a perceived community, but how is the edge or limit to this politics of protection to be defined? What modes of relation within an extended community
of life forms can be articulated? Esposito argues that the only way to prevent biopolitics from becoming a form of thanatopolitics is to value all life,
shorn of its violent demarcations. Yet his ostensibly all-encompassing position has elicited criticism for its overextension of rights discourse to facets
of nature that are inimical to human well-being.[159] In defense of Esposito,
however, we can argue that his concern is with the flourishing of conditions for life in general, rather than an individuated emphasis on any and
all life forms: his formulation might include a biodiversity hotspot even if
it contains some elements of threat such as the presence of the tetanus bac-
terium in soil, or an as yet unknown zoonosis.[160] The evidence from recent
destruction of relatively undisturbed global biomes suggests that human
health and ecosystem health are intertwined at every level.

At issue here is the possibility for a relational ethics that can encompass the complex interactions between species, populations, and the viability of life over different scales and temporalities. In practice, attempts to control or protect aspects of nature under a biopolitical frame span a number of different discourses, ranging from biosecurity and human health to strategically oriented interventions on behalf of endangered species, ecosystems, or genetic resources. The disavowal of any collection of "voucher specimens" for biodiversity surveys, for example, might undermine the possibilities for the production of environmental knowledge with sufficient legal and scientific veracity to play a role in the protection of specific sites (a theme I consider in chapter 4). For critics, however, there is a residual humanism underpinning even the most sophisticated immunological paradigms that leaves the ethical status of the nonhuman in a state of precarious uncertainty. We encounter shifting categories of life, so that some fragments of nature—both human and nonhuman—are deemed worthy of greater degrees of care and protection.[161] Furthermore, the line of demarcation between different species is not necessarily a clear indicator of underlying biopolitical momentum, as evidenced by the socioeconomic dimensions to urban epidemiological discourse as well as the porous boundaries afforded to nonhuman others that encompass companionship, cultural resonance, food production, scientific curiosity, and many other domains of interaction.

1.7 Itineraries

In October 2017 a terrified but otherwise unhurt stag sought refuge from hunters with hounds in the front garden of a house in the village of Lacroix-Saint-Ouen to the northeast of Paris. Despite public protests (that can be clearly heard on eyewitness video recordings) the hunters proceeded to kill the animal and drag its carcass into nearby woods for their dogs to devour. Incidents of this kind illustrate direct confrontations between different cultures of nature: between the killing of animals for sport and the attempt to enact alternative approaches to the sharing of space. The politics of human relations to nonhuman life is marked by spaces of proximity and separation, the propinquity of specific encounters, as well as the willful maintenance of distance or invisibility. Cruelty towards animals as a source of entertainment—badger baiting, dog fighting, and other instances of inhumane treatment—has not disappeared but become more attenuated or occluded in relation to metropolitan cultures of nature. Brutal attacks on "nuisance animals" are not infrequent, especially if couched in terms of public health as we saw in the case of misidentified intermediate hosts for zoonoses, so that nonhuman denizens of urban space are often vulnerable to varied forms of hostility and stigmatization.[162]

Running through this chapter has been the question of ethical relations with the nonhuman. Legal scholars such as Taimie L. Bryant have reflected on what renders certain organisms "killable" and under what circumstances.[163] In the case of animals, the distinction between pets, strays, livestock, vermin, and wild nature has varied across space and time, moving between different spheres of domestication and control. The question of cohabitation and the conceptualization of urban space as an enlarged community of living organisms raises questions about the scope and purpose of protection for nonhuman others.[164] Demands to stop the poisoning of urban rats are one thing, but calls for the ethical protection of mosquitoes and other insect vectors for disease is quite another, let alone pathogens such as *Legionella* lurking in poorly maintained plumbing systems. Some kind of biopolitical distinction has to be made between what forms of life should be protected and what elements of nature might reasonably fall under the immunological paradigm of control. It is in this context that an affirmative biopolitical paradigm must acknowledge the occasional need to mediate relations between human and nonhuman nature to protect or modify specific dimensions to the web of life.

The status of foxes, pigeons, squirrels, and many other cohabitants of urban space has gone through a series of permutations. The eastern gray squirrel (*Sciurus carolinensis*), for example, has played a variety of roles within the American city since its first appearance there in the nineteenth century: a source of everyday curiosity or indifference; a supplementary source of food for the urban poor; a focus of scientific research; a persecuted "trash animal" categorized as vermin; an object of cruelty through "urban hunting" practices; a tame acquaintance capable of building affective individual relations with regular human feeders; an ecological intruder capable of outcompeting other species (especially outside of its American range); and most recently, a source of prey for newly returned hawks and falcons, as part of a wider cultural acceptance of "urban rewilding."[165] The urban pigeon has similarly undergone a process of cultural reframing as a "problem species" and has become much more closely associated with dirt, disorder, and even epidemiological threats to human health.[166] Like the squirrel, the pigeon is now entrained in urban rewilding discourses, with raptors swooping through the sky to catch an easy meal in public view. The earlier discourses of urban pest control, with their attendant technocapitalist apparatus, have been partially displaced by an emphasis on a new kind of urban keystone ecology.

Rewilding is sometimes characterized as a projection of metropolitan cultures of nature onto more distant or economically deprived localities. An emphasis on urban rewilding, by contrast, brings these cultural tensions into close proximity. The arrival of animals in urban areas, including predators such as coyotes and wolves, has provoked interest in the possibilities for urban rewilding. In some cases, however, these new ecologies have provoked fear or hostility in the wake of wolves attacking domestic dogs, elks wandering across highways, or wild boar invading gardens and allotments. The return of the wolf, in particular, remains a focus of anxiety: there is unease over waiting at remote bus stops in Finland, for example, where "wolf taxis" have been introduced (for people not wolves) and there are now regular reports of attacks on sheep in France, Italy, and elsewhere.[167] How can these contrasting cultures of nature be reconciled? Can a synthesis between these alternate perspectives be crafted that also protects the web of life as a whole?

An emphasis on the "return" of animals is potentially misleading. Not in the sense that these predators on the urban fringe were previously undetected but rather because backward-looking ecological imaginaries fail to capture the dynamic of unfolding socioecological assemblages that are distinct from what might have existed before the pervasive human modification of the landscape. Furthermore, the question of how humans might relate to animals, and the array of nonhuman others that cohabit urban space, moves the question of ethical relations beyond forms of "belittling" that accompany softer varieties of kinship or domination over nature.[168]

The agency of nature in the urban arena has been widely feared as a source of disorder—exemplified by the presence of pests or weeds—or simply downplayed since it unsettles existing modes of analysis that are framed around the bounded human subject. Since Wolch's original intervention, a range of studies have made the independent agency of nature a focal point for analysis.[169] Her term *zoöpolis* implies an ironic reversal of the "alienated theme park" as a model of urban nature, exemplified by zoos, the passive absorption of wildlife imagery, or other controlled experiences that rest on the commodification and spectacularization of nature.[170] Rather than the relegation of wild nature to designated zones or more distant types of ecosystems, the presence of spontaneous nature in cities has been extensively reframed. In the next chapter I turn to some of these spaces of discovery in greater detail.

MARGINALIA

It is not the parks but railway sidings that are thick with flowers.

Richard Mabey[1]

Our generation—I was born in 1967—has very clear memories of our childhood, which today's youth don't have, because they no longer have the open spaces and wild nature of the *terrain vague.*

Community activist in Mantes-la-jolie, Île-de-France[2]

The British writer Richard Mabey's paean to the spontaneous exuberance of urban nature, inspired by the marginal landscapes of deindustrializing London in the early 1970s, presents an alternative sensibility towards nature that eschews either a narrow scientism or a romanticist attachment to the idea of pristine wilderness. Mabey's observations connect with a heterogeneous ground of nature writing that draws together aspects of popular science, vernacular landscape culture, and a wider sense of curiosity or enchantment with everyday objects and spaces. At about the same time that Mabey was exploring the landfill sites, towpaths, and other hidden reaches of the London landscape, the French artist Paul-Armand Gette was studying the floristic diversity of wastelands in several European cities, the Japanese architect Arata Isozaki was creating collages for ruined cities, and teams of botanists were busy compiling some of the most comprehensive urban floras yet produced. The marginal spaces of Berlin, London, Mont-réal, and other cities were becoming a significant focus for cultural and scientific attention, reflecting a series of developments such as the emergence of new art practices, increasing levels of ecological awareness, and the changing characteristics of cities themselves.

A focus on spontaneous forms of urban nature transcends the merely speculative or utilitarian potentialities of ostensibly empty spaces. By regarding nature differently, in both cultural and scientific terms, a set of counterdiscourses can be articulated that question the pervasive emphasis on wastelands as sites simply awaiting erasure and redevelopment.[3] An engagement with the independent agency of nature enables intellectual threads to emerge between new understandings of urban ecology and philosophical developments within the epistemology of science. The sense of nature as active, dynamic, and constitutive of the cultural and material characteristics of urban space reveals the metropolis to be both unfixable and to a significant degree unknowable.

The word *wasteland* is defined by the *Oxford English Dictionary* as an "empty or barren area of land" and originates via French from the Latin *vastum* meaning "unoccupied" or "uncultivated."[4] Yet this essentially practical rendition disguises a poetic or rhetorical undercurrent to the term, predating even T. S. Eliot's poem of 1922, that reveals a certain allure towards spaces of emptiness within the European cultural imagination. An interest in the "void" as a powerful scientific metaphor has undergone various architectonic and philosophical permutations since Blaise Pascal's essay of 1647 on "emptiness within emptiness."[5] Here, Pascal elaborates on the classical nostrum of the *horror vacui*, a preoccupation that created ripples through architectural thought as well as becoming an established focus for psychoanalytic ideas by the early decades of the twentieth century. "Pascal's resistance to the open transparency of rationalism," writes the architectural historian Anthony Vidler, "was seen as a way of symbolically and affectively exploiting the ambiguities of shadow and limit, remaining a sign of potential disturbance beyond and within the apparently serene and stable structures of modern urbanism."[6] Yet this interest in "the void" is by no means restricted to European philosophical traditions—its counterpart exists, for example, in the Korean word *gong* (공) that combines the physical and metaphysical connotations of emptiness—and the idea has been repeatedly rearticulated in relation to the proliferation of "empty spaces" that litter urban and industrial landscapes.

The other widely used term, "brownfield," which carries a more technical timbre, is similarly defined as a site which has been previously used and is especially associated with traces of industrial contamination, in contrast with the verdant connotation of "greenfield" sites at or beyond the urban fringe. The brownfield is a hybrid aesthetic-technical category and has widely served as a cultural descriptor for the changing landscapes of the post-Fordist transition, with its epicenter in the North American rustbelt as well as the former industrial zones of Europe and elsewhere.[7] Indeed, this rhetorical brown-green antinomy is intensified by patterns of intervention in urban land markets, so that the precise cultural or political connotations of wastelands cannot be disentangled from the wider dynamics of metropolitan change, the history of planning, or the latest incarnations of urban boosterism. In addition to these widely used Anglo-American terms we can add, for example, the German *Brache*, the French *friche urbaine*, the Japanese *arechi* (荒れ地 or 荒地), the Chinese *fei* (废) and *huang-di* (荒地), and the Korean *hwang-mooji* (황무지) and *gong-teo* (ji 공터). Each of these terms contains elements of ambiguity: the Japanese word *arechi*, for instance, harbors connotations of uncertain ownership and is used as a cartographic category for land use mapping, whilst the German *Brache* acquired heightened significance during the 1990s in relation to the postunification *schrumpfende Städte* (shrinking cities) debate.

The designation of a void space, along with its symbolic and legal connotations, has also served as a recurring tool of colonial governmentality and modes of forcible enclosure.[8] The Tamil word *poramboke*, for example, which has its origins in the colonial classification of marginal lands unsuitable for cultivation has recently been reappropriated as a focus for cultural and political demands to protect "urban commons."[9] These multifarious and sometimes intersecting etymologies belie a common emphasis on the "unproductive" characteristics of these sites in relation to agriculture, industry, or other former land uses.[10] But these are often also places of unease and symbolic signification, less easy to categorize or identify, that connect with memories, inspire Ballardian psychogeographies, or function as spaces of autonomous social and cultural life.

Alternative vocabularies for wastelands that begin to move away from an emphasis on their utilitarian characteristics include "edgelands," "interim spaces," "interstitial landscapes," and especially the term *terrain vague* with its connections to radical architectonic discourse.[11] This diverse alternative lexicon for marginal spaces also intersects with emerging ecological insights that seek to valorize specific forms of biodiversity or the serendipitous aesthetic effects of "nondesign." More recently, for example, conservation biologists have sought to replace the term "brownfield" with "open mosaic habitat" as part of a scientifically driven effort to modify dominant attitudes towards void space in land use planning.[12] This recognition of the richness of urban biotopes owes much to the efforts of ecologists and urban nature conservationists to reveal the full complexity of wild urban nature. Places that appear useless to the momentary glance of passing commuters may nonetheless be spaces of adventure, imagination, and self-discovery for artists, children, filmmakers, and other explorers of the urban realm.

2.1 City of weeds

Interest in spontaneous forms of urban nature stems from early distinctions between "wild plants" growing in and around cities and "cultivated plants" largely confined to parks and gardens. Early studies of wild urban flora include Pitton Tournefort's treatise on the plants of Paris and its environs, published in 1698, which places particular emphasis on species with medicinal properties. Using a pre-Linnaean nomenclature Tournefort meticulously recorded all the plants that he could find, along with references to over 60 other botanical works to provide a vibrant snapshot of the state of scientific knowledge at the time.[13] The seventeenth century also saw some of the first botanical "rambles" through semiwild places near cities, producing records which remain of significant ecological and historical interest.[14] The apothecary Thomas Johnson's guide to the plants of London's Hampstead Heath, for example, published in 1629, represents the first local flora compiled in Britain.[15] Johnson describes how he set out with some excitement, along with a group of friends, on the morning of 1 August 1629 for a botanical excursion "undeterred by the lowering sky" and a rainstorm on the way.[16]

During the nineteenth century we find growing scientific attention devoted to the distinctiveness of urban nature including the botanical characteristics of walls, ruins, and unusual plants growing on "ballast hills" near ports. In 1855, for example, Richard Deakin compiled *The flora of the Colosseum*, recording some 420 species of plants growing on the 2,000-year-old ruin, including "some plants so rare in western Europe that they may have arrived as seeds caught in the fur of gladiatorial animals from North Africa."[17] Deakin laments the loss of many species due to various "alterations and restorations," mourning the loss of the building's "wild and solemn grandeur," and contrasting the barbaric spectacles of the past with the quiet profusion of life amidst the ruins.[18] In the event, however, it would prove to be the consolidation of the modern Italian state after 1861 that led to the "tidying up" of the site and removal of many of the interesting plants.

By the late nineteenth century several guides to wild urban nature had been published reflecting factors such as improving transport connections and the burgeoning membership of scientific societies. In Edmond Bonnet's

Petite flore parisienne, for example, published in 1883, he exhorts Parisians to delight in "the knowledge of spontaneous vegetation."[19] In 1893 the nature writer Richard Jefferies published a collection of essays entitled *Nature near London*, in which he notes how his "preconceived ideas were overthrown by the presence of so much that was beautiful and interesting close to London."[20] By the early twentieth century the unusual plants growing on disturbed ground had become a sustained focus of botanical curiosity: in 1912, for example, an article describing the vibrant flora of London construction sites speculated on the reasons why these ostensibly empty spaces were transformed so rapidly into diverse botanical assemblages. The survey notes various factors such as the arrival of birds, spilled forage for horses, and the presence of garden escapes, but singles out the "clouds of seeds" carried by the wind that "could be seen floating in every direction" as the determining element in many of these ruderal ecologies.[21]

The botanist Paul Jovet, from the Muséum National de l'Histoire Naturelle in Paris, is widely regarded as the first modern scientist to devote his full attention to the spontaneous flora of cities.[22] Jovet recognized that urban vegetation was a distinctive kind of ecological *mélange* (mixture), comprising a bewildering array of plants from all over the world that not only disrupted existing conceptions of plant associations but was itself in a state of constant flux through the impact of human activities such as construction or *piétinement* (trampling). He carried out a series of meticulous studies of the flora of urban wastelands during the 1930s and 1940s, which were formative for a later generation of botanists including Herbert Sukopp, who led the study of West Berlin from the 1960s onwards and remains the most influential figure in the field. Sukopp's initial fascination with urban ecology was spurred by early studies of the unusual flora associated with postwar rubble landscapes. However, as geopolitical division intensified, including the building of the Berlin Wall, Sukopp and his colleagues increasingly turned their attention to the flora of the island city, which provided an elaborate complex of experimental field sites.[23] The studies by Sukopp and his colleagues, based at the newly created Institute of Ecology at the Technical University in Berlin, produced some of the most detailed surveys of urban flora that have ever been produced, pushing modes of cartographic representation to their limits (a theme I will return to in later chapters).[24] Not only did these studies rework existing phytogeographic approaches to botanical research, as elaborated by Josias Braun-Blanquet and others, but they also provided a welter of socioecological insights into the changing structure of urban space over time.

The emergence of urban ecology as a distinctive subfield within the biological sciences has developed significantly since the early 1970s. The botanist Paul Duvigneaud, for example, defined urban ecology from a metabolic perspective, revealing continuities with organicist conceptions of the nineteenth-century metropolis as well as with the systems-based engineering of Abel Wolman and his contemporaries.[25] Yet the uncertain relationship between the science of urban ecology and the nascent environmental movement revealed tensions between the modeling of biophysical processes and the production of space. In contrast to the more confident assertions of Duvigneaud about the coherence of urban ecology as a scientific field, other authors questioned the lack of any clear theoretical basis for the study of urban nature. In a provocative article for the journal *Capitalism, Nature, Socialism*, for example, the Berlin-based ecologist Ludwig Trepl found that although knowledge of biophysical processes in cities had advanced significantly, the nature of ecological relationships with social, cultural, and economic processes remained in a state of flux and confusion.[26] In particular, Trepl, echoing Adorno, suggested that the presence of weeds and other spontaneous forms of urban nature posed a challenge for ecological thinking that was rooted in bourgeois conceptions of unintentional nature as the antithesis of urban space.[27]

A distinctive aspect to the ecological dynamics of cities is the transitory or frequently disturbed character of many sites, which has long formed a focus for both cultural and scientific explorations of urban space. These marginal spaces are typified by an array of so-called "pioneer species," specially adapted for the colonization of new substrates, which can engender rapid and unexpected changes in the appearance of urban land-scapes. Examples include the yellow-flowered crucifer known as London rocket (*Sisymbrium irio*), which spread quickly in the burnt spaces of London after the Great Fire of 1666, and the purple spikes of rose-bay willow herb (*Chamaenerion angustifolium*). The latter had been considered relatively scarce under natural conditions, sporadically appearing in burnt patches on heathlands, for example, until it suddenly filled the bombsites of London and other European cities in the 1940s (figure 2.1).[28] "At the end of the war," writes W. G. Sebald, "some of the bomb sites of Cologne had already been transformed by the dense green vegetation growing over them—the roads made their way through this new landscape like 'peaceful deep-set country lanes.'" In Hamburg, there was a "second flowering" of chestnuts, lilacs, and other trees in the autumn of 1943, just months after the devastating firestorm that had laid much of the city to waste.[29] In Sebald's hands, the term "natural history" invokes powerful connections between materiality, the production of meaning, and the limits to representation: precisely those elements that have yet to emerge within the field of urban ecology itself.

Studies of the natural regeneration of damaged sites in cities such as Berlin, Bremen, and Kiel, and the emergence of new ecological patterns and assemblages, also began to reshape aspects of ecological science and the classification of different vegetation types.[30] As a consequence, the idea of the "cultural landscape" as a recognizable and regionally specific aesthetic unity was irrevocably altered, with both methodological and ideological implications for the study of nature and landscape. The emerging emphasis on the "cosmopolitan ecology" of cities, highlighting the role of adventitious or introduced species, served as an implicit critique of nativist approaches to landscape design and pervasive antiurban sentiments within conservative strands of environmental thought (a theme I return to in the next chapter). Traces of spontaneous nature acquired a double significance as markers for an explicitly urban ecological paradigm as well as symbolic indicators for shifting ideological contours in the urban landscape. The

Figure 2.1 Botanists looking for wildflowers on a bombsite, Gresham Street, London (1943). Source: Getty Images.

geographer Gerhard Hard, for example, used studies of ruderal vegetation in the city of Osnabrück as part of his detailed critique of the aesthetic, epistemological, and ideological limitations of existing approaches to the interpretation of landscape.[31] For Hard, a close engagement with both the ecological and ideological complexity of ordinary landscapes not only carried implications for the understanding of the socioecological dynamics of urban space but also enabled a more broadly framed critique of the reactionary legacies of botany, geography, and other disciplines.

Some increasingly ubiquitous plants are now termed "invasive" on account of their ability to outcompete other species and effectively reduce biodiversity: in some cases these are indigenous species that have dramatically expanded their range in response to changing urban ecologies, such as traveler's joy (*Clematis vitalba*), and in other cases they are accidental introductions or ornamental escapes. In a European context examples include Japanese knotweed (*Reynoutria japonica*) from Asia and giant hogweed (*Heracleum mantegazzianum*) from the Caucasus. A recent study of the Paris region lists ten invasive species which are already regarded as having a deleterious ecological impact, along with a further twelve potentially invasive species.[32] In Chinese cities a number of invasive plant species have been a focus of attention including the sumac (*Rhus typhina*) from North America, which was initially chosen as a fast-growing addition to afforestation and landscape design projects, and also a type of ornamental spurge (*Euphorbia dentata*) that escaped from the garden of Beijing's Institute of Botany and originates from the Americas.[33] In some cases, there are conflicting conceptions of what constitutes an invasive species: the so-called "butterfly bush" or buddleia (*Buddleia davidii*), of nineteenth-century Chinese origin, is a very widespread and characteristic flowering shrub in many European cities, with its distinctive panicles of purple flowers, but its localized dominance can exclude many other species and even cause structural damage to walls and other surfaces. Yet it is the global characteristics of urban ecologies, and their myriad potential socioecological configurations, that underpin their cultural and scientific interest: the question of

Figure 2.2   Chausseestraße, Berlin. A typical *Brache* or urban wasteland (now lost) where the Berlin Wall once stood (2009). Photo by Matthew Gandy.

value is no longer reducible to a nativist aesthetic or preexisting ecological template.

The intersections between ecological science and changing conceptions of urban nature are perhaps most strikingly illustrated by the emphasis on wastelands as the focal point for high levels of biodiversity in cities. Previously neglected or overlooked spaces have been reframed as safe havens for nonhuman forms of urban life. These marginal spaces are now recognized as part of the ecological infrastructure of the city, extending to roles such as flood control, water purification, and the mitigation of the urban heat island effect.[34] An extensive survey of brownfield sites in the Hauts-de-Seine area of the Paris region, for example, found that they contain nearly 60 percent of all species recorded in the region—far higher than parks, gardens, and other typical elements of urban green space.[35] Rather than clearly differentiated vegetation zones, as postulated by the pioneers of ecological science, urban ecologists have emphasized how the city comprises a lattice of microniches varying by substrate, aspect, time, and other factors. Linear spaces such as roadside verges or railway embankments can form green corridors or "eco-ducts," a term first used in a Dutch context, that allow small populations of vulnerable species to be connected while also playing a role in the dispersal of new species, so that plants, for example, may spread their seeds in a radial pattern across the city.[36] Eco-ducts serve as a kind of infrastructure network for the nonhuman and can even be intentionally incorporated into ecologically oriented variants of landscape design.

The concept of biodiversity contains elements of ambiguity between the idea of ecosystem diversity (which is high in urban areas), species diversity, and genetic diversity. However, as David Takacs shows, the very idea of biodiversity is as much a mirror of entanglements between different cultural and scientific discourses as any putative representation of external nature.[37] In an urban context, the concept of biodiversity becomes even more difficult to determine, especially when used in relation to wider conservation objectives, such as the protection of rare species or vulnerable habitats. How, in other words, do we apply the concept of biodiversity to what Bernadette Lizet terms the "ordinary nature" encountered in cities? The production of inventories of rare or threatened species in order to influence policymaking dates from the first so-called "Red List" produced by the International Union for Conservation of Nature in 1963. Since that time the use of Red Lists has extended to encompass different tiers of government as well as an expanded range of life forms. An increasing number of cities now have their own Red Lists for plants, animals, and even invertebrates, as insights from urban ecology become connected with scientifically inflected strands of urban environmental discourse. Ecologists in Berlin and elsewhere have successfully argued for the inclusion of nonnative species in lists of endangered urban species on the basis of their vulnerability rather than their geographical origin.[38] Yet these emerging intersections reveal tensions between the role of technical expertise—in this case urban ecology—and the contested political exigencies of the urban arena.

2.2 FLUX, MIMICRY, AND NONDESIGN

The emerging fascination with the aesthetic and ecological characteristics of spontaneous urban nature does not preclude an element of human design or intentionality. Indeed, the independent dynamics of nature have been an increasingly significant element in alternative approaches to urban design since the 1970s that seek to combine the enhancement of biodiversity with a less regularized or formulaic aesthetic experience. Whereas the eighteenth-century creation of semiwild aesthetic experiences, exemplified by a fascination with the picturesque, rested on a visual simulacrum of an idealized nature, recent interest in the design possibilities engendered by spontaneous forms of nature stems from a synthesis between a metropolitan "wasteland aesthetic" and developments in scientific knowledge. The somewhat elusive, yet rhetorically powerful, concept of biodiversity serves as an organizational focus for new approaches to urban design and the development of ecologically oriented approaches to the utilization of urban space. Leading exponents of "natural design" and nonintervention such as Gilles Clément and Louis Le Roy work within a conceptual framework of "guided" landscape dynamics to produce specific aesthetic or ecological effects.[39] The innovative Irchelpark in Zurich, for example, completed in 1986, uses a minimal mowing regime with no herbicides to foster a high degree of seminatural biodiversity in the heart of the city. Despite its naturalistic appearance, however, the park is an intricately engineered landscape crisscrossed with walkways, water bodies, and other design features. Specific elements such as weathered stone steps produce a floristically rich type of "ruin aesthetic" with an abundance of plants adapted to rock crevices and other microbiomes.[40]

Many distinctive urban habitats mimic natural features: in the eyes of raptors, swifts, or other birds high buildings become inaccessible mountain ledges or cliffs, derelict buildings may play the role of caves or hollow trees for bats and spiders, and green roofs may resemble flower-rich meadows.[41] Both horizontal and vertical urban surfaces can harbor high levels of biodiversity either by accident or design. In a study of "living roofs," the geographer Jamie Lorimer explores a Deleuzian-inflected notion of ecological "striation" as a way to combine the spontaneous agency of nature with elements of conservation-oriented design. Lorimer identifies a "fluid biogeography" that combines an emphasis on a more self-reflective scientific practice with new understandings of materiality within urban ecological discourse.[42] Urban botanists are often interested in enhancing the prospects for early-stage ecological succession since this exhibits the highest levels of biodiversity, including many uncommon species or those that require highly specialized habitat niches.[43] "Temporal suspension" or "stilled time" is deployed for a combination of aesthetic and scientific reasons to modify aspects of ecological change. Rather than repetitive mowing to maintain various types of lawns, there is interest in more infrequent or strategically timed mowing regimes that can encourage specific plants to flourish in an urban context. The use of staggered mowing regimes can also protect elements of botanical archaeology: the gardens of the Charlottenburg Schloss in Berlin, for instance, contain remnants of the original seventeenth-century planting scheme, preserved as part of a carefully maintained botanical palimpsest.[44]

The innate hybridity of urban landscapes challenges the pervasive emphasis on native species or landscape authenticity in design practices such as ecological restoration. Yet what is the historical reference point to which the objective of ecological authenticity might relate? Is it a distinctive cultural landscape in the sense that Hansjörg Küster and others have described, or is it an early Holocene state of nature with minimal human impact as evoked by rewilding enthusiasts?[45] Is it an intervention to retain the species-rich early stages of ecological succession or the creation of some other type of specific habitat that has a more diverse cultural or scientific potential in comparison with traditionally managed green spaces? A recent study of the Chicago waterfront, for instance, reveals four different points of historical reference for ecological contestation, ranging from pre-European nature before the 1830s to more recent modifications that have become naturalized within the public imagination.[46] In some cases the ecological imagination encompasses a "double temporality" that combines a reconstruction of the past, and the origins of specific socioecological formations, with intimations of a future nature marked by the arrival of new species.[47] Specific cultures of nature have emerged over time, along with their distinctive combinations of aesthetic sensibility and human subjectivity, culminating in an emphasis on a seminatural aesthetic linked to more polyvalent conceptions of public culture. The recognition that "cities can create their own flora" has been accompanied in a few instances by the scaling back of control measures for "weeds" and other forms of spontaneous vegetation.[48]

In some cases urban and industrial wastelands have been transformed into spaces of leisure, with multiple forms of nature ranging from semiwild landscapes to closely maintained elements of conventional park design. Yet this focus on the multifunctionality of "optimal landscapes" poses significant differences in emphasis and context.[49] Recent examples such as

Duisburg Nord in the Ruhr, Parc André Citroën in Paris, and the High Line in New York City display marked differences in terms of their continuity with earlier approaches to park design. The High Line, for example, which has been constructed along a disused section of elevated railway in Manhattan, has recreated aesthetic aspects to spontaneous vegetation through the replanting of birch trees to produce a distinctive kind of ecological simulacrum of what occurred on the derelict structure before its extensive landscaping (figure 2.3). In this instance, the wasteland as artifice becomes a cultural motif that serves to underpin real estate speculation, and the boundary between private and public is reworked in the form of a neopastoral urban spectacle.[50] The park, in this context, is a designed fragment of nature that inscribes social and political power into the urban landscape. More rarely, the specific characteristics of an abandoned site have been retained in order to preserve aspects of spontaneous urban nature—examples include Berlin's Südgelände park—but these remain exceptional cases and have usually only occurred after years of intense political and scientific lobbying underpinned by extensive public support. The presence of a wasteland aesthetic shows that spaces which may appear superficially similar, even in biotic terms, may nonetheless owe their existence to markedly different processes.

2.3 THE SEARCH FOR AN ECOLOGICAL AESTHETIC

But what kind of landscape aesthetic is invoked by a focus on spontaneous spaces of urban nature? Wastelands are not readily identifiable cultural landscapes in the conventional sense but something more ill-defined in relation to public culture. Indeed, their aesthetic appeal lies precisely in their cultural and scientific complexity, thereby raising questions concerning the relationship between knowledge and experience in the perception of landscape. The more detailed and engaged our knowledge of such spaces—and their ecological dynamics—the more sophisticated our grasp of their aesthetic characteristics. For the philosopher Allen Carlson we

Figure 2.3   The High Line, New York (2011): an ecological simulacrum of elements of what once existed on the site. Photo by Matthew Gandy.

cannot reduce the aesthetic appreciation of nature to phenomenological or subjective experience alone: there is a synergy between scientific and aesthetic understanding that heightens our awareness of nature whereby the aesthetic appreciation of nature is "informed and enriched" by advances in scientific knowledge.[51] Yet the idea of knowledge about nature, which requires years of patience and dedication to acquire, lies in tension with a culture of immediacy in which the science of biodiversity becomes subsumed within more vapid discourses of resilience, sustainability, or other related fields. In such circumstances, how can cultural or scientific complexity be effectively communicated? What happens when more autonomous criteria for cultural or scientific evaluation conflict with externally imposed agendas for reshaping knowledge?

The sociologist Pierre Bourdieu calls for the defense of the "inherent esotericism of all cutting-edge research," yet he also insists on appropriate strategies for the scientific enrichment of the public realm.[52] In the case of urban biodiversity, there is a disjuncture between specialized scientific understandings of urban space and mediated discourses of nature for consumption or recreation.[53] The political ramifications become apparent where science, in this case urban ecology, serves to enhance and protect alternative social and cultural discourses about urban nature, just as archaeological or art historical insights might prevent the destruction of ostensibly esoteric or insignificant cultural artifacts. This is especially significant for urban biodiversity, where the most noteworthy concentrations of so-called Red List species and other categories of scientific interest may be little studied groups such as wasps, beetles, or other biota that have little conventional aesthetic appeal.[54]

In what sense does urban nature engender a different set of aesthetic sensibilities than classic objects of contemplation, such as cultural landscapes or wilderness? The geographer Natalie Blanc calls for an aesthetics of nature that is derived from a shared sensibility, rather than a phenomenological emphasis on individual experience. For Blanc, the aesthetics of nature is inherently political because it fosters political dialogue and raises public awareness of environmental issues.[55] But where does her normative reading of aesthetic experience leave the "useless," ineffable, or more esoteric realms of cultural engagement with nature?

The recent emphasis on an "ecological aesthetic" is often marked by attempts to reconcile the irreconcilable, so that insights from the biophysical sciences are simply grafted onto other fields such as behavioral psychology.[56] Yet the idea of the aesthetic in these science-led reformulations rests on a cognitive or behavioral model of human interaction with landscape that lacks any clearly articulated engagement with either the historical production of cultural meaning or the symbolic resonance of space within the social imaginary. A different approach is offered by the philosopher Cheryl Foster, who seeks to emphasize how sensuous aspects of nature have been downplayed in favor of a narrative which ensures the dominance of preexisting understandings of science and cultural value: she enlists Gaston Bachelard, John Dewey, and other figures in order to reengage with the "ambient" dimensions to human experience.[57] A greater emphasis on the acoustic, tactile, or olfactory texture of space rather than fleeting visual encounters expands the critical scope of aesthetic theory in relation to marginal landscapes. The aesthetics of nature can be disentangled from associations with an existing view or vista, so that sensory immersion in nature takes precedence over the enframing of nature as a space of spectacle.[58]

The emphasis on pleasure in nature has recently been developed through the concept of "vital beauty," drawing on nineteenth-century observations of plant life by Gustav Fechner and especially John Ruskin, to build a phenomenological approach to the aesthetics of nature. These nineteenth-century attempts to identify an independent aesthetics of nature sought to transcend natural theology or "vaguely pantheistic celebrations of nature's mystical power" through corporeal immersion in the natural world.[59] For Ruskin, the life force represented by the growth of vegetation was "a realm of strange intermediate being."[60] Yet this search for "inherent value" differs from vitalist conceptions of agency that seek to dispel the ontological centrality of the bounded human subject. In this sense, nature is divested of a humanist reductionism so that it can be considered in terms of its own dynamics, albeit through the lens of human intuition or perception. Henri Bergson is arguably the key figure here in his emphasis on the "life force" as a structure of meaning independent of human intentions or values.[61] It is this emphasis on an inherent force that provides the philosophical lineage from Bergson to Deleuze, and the opening up of new possibilities for the interpretation of material reality and nonteleological conceptions of nature.[62] Under Bergson's reading of positivism, distinctions between order and disorder are replaced by an affirmation of material processes of change within which human beings are themselves an integral element. For Bergson, materiality is bound up with human finitude and our perception of time: hence the poetic eloquence engendered by the independent dynamics of nature.[63]

For philosophers such as J. Baird Callicott, the anthropocentric emphasis on aesthetic pleasure in nature distorts priorities in nature conservation away from less spectacular or awe-inspiring places and serves to strengthen aesthetic values over ethical values.[64] Similarly, the historian Ronald Rees notes how the pervasive aesthetic emphasis on "wild" or remote spaces has served to undermine concern with "local environments" and the articulation of an environmental ethics that takes nature itself into consideration.[65] The difficulty, however, with the critique of an anthropocentric aesthetics of nature articulated by Callicott, Rees, and others is that a workable alternative cannot be located within nature as an autonomous realm lying beyond human interests. This is the basic contradiction that underlies ecocentrist or "deep ecological" critiques of modernity. The ecocentrist position ultimately reiterates existing dichotomies between nature and culture through its search for external sources of "truth." This misplaced search for the "essence" of things, as Richard Rorty noted, remains a critical undercurrent for much of what we might term "environmental aesthetics" articulated on behalf of a putatively autonomous nature serving as a repository of "truth."[66] Yet Rorty's philosophical pragmatism, with its extensive deployment of irony, poses its own limitations in terms of the extent of possible dialogue between different cultural and scientific domains.[67] If an innate incommensurability between fields of knowledge is accepted, then the possibilities for bringing an array of disparate ecological arguments closer together are rendered much more difficult.

The possibility of an "ecological aesthetic" remains complicated by the lack of any necessary relation between the "scenic" (or other forms of aesthetic delight) and what is ecologically significant.[68] As the London-based botanist Nick Bertrand points out, "aesthetics has nothing to do with conservation."[69] A rotting carcass swarming with maggots is integral to ecology but not ordinarily considered to have any relation to the culture of nature as a source of pleasure.[70] Over time, however, ostensibly incongruous elements of urban nature such as fallen trees or rotting wood can become part of a scientifically enriched public culture in which the enhancement of biodiversity becomes a more pervasive element in urban design.[71] Equally, there are aspects of environmental change that, by virtue of time, space, or scale, lie beyond the realm of human perception: the experiential dimensions to nature are of necessity limited, partially prefigured, and open to multiple interpretations.

2.4 SPACES OF AMBIVALENCE

Spontaneous spaces of urban nature can be sites of unease as well as places of freedom or creative expression. The multiplicity of cultural responses to wastelands is partly related to the diversity of such sites and their varied origins: whilst some spaces have developed spontaneously within ostensibly empty sites, others have emerged out of the neglect or abandonment of previously maintained spaces such as lawns or parks. These neglected spaces reflect a late modern dislocation between designed landscapes and capitalist urbanization, since municipal parks in particular have been badly affected by the fiscal crisis of the state since the 1970s. Labor-intensive municipal landscapes have their origins in a nineteenth-century reformulation of metropolitan nature that encompassed developments in landscape design, infrastructure improvements, zoning law, and biopolitical interventions in the sphere of public health.[72] The intensified control of "weeds," and the regularization of urban nature, form one element in this distinctive interface between nature, science, and society that reached its apogee in the middle decades of the twentieth century. The partial unraveling of these relationships poses implications for the design, maintenance, and meaning of urban landscapes, yet at the same time suburban lawns or front gardens have become subject to greater degrees of control in parts of North America and elsewhere from so-called "weeds ordinances."[73] Legal disputes over the maintenance of lawns have in a few cases been resolved through the articulation of a "naturalistic aesthetic" in which it can be shown that the presence of weeds is an intentional dimension to alternative garden design.[74] There is an apparent divergence between landscapes of tighter control (and surveillance) and new forms of "loose space" within which greater degrees of aesthetic, biotic, or social heterogeneity are tolerated or even encouraged.[75] These tensions can coexist within the same space: in Berlin's recently created Park am Gleisdreieck, for instance, elements of the existing topography of wild urban nature associated with abandoned railway infrastructure have been retained, evocatively named "track wilderness" (Gleiswildnis), along with the creation of an officially sanctioned "graffiti wall" and a concrete bowl for skateboarders. Shortly after dawn, park staff can be seen removing litter, and in hot weather banks of discreetly hidden sprinklers emerge out of the subterranean infrastructure system to water grassy areas. The cultural valorization of spontaneous nature is set within an intricately controlled type of urban landscape.[76]

There is an unfixed dimension to the aesthetics of spontaneous nature that requires a greater degree of imaginative engagement or reflection than conventional components of metropolitan nature. Such spontaneous spaces are characterized by a multiplicity of "aesthetic worlds" reflecting the heterogeneous characteristics of urban space and the dissolution of an imaginary "landscape unity."[77] The study of modern landscapes has been marked by a tension between various kinds of pattern-oriented analysis

Figure 2.4   Tempelhof airport, Berlin (2011). The dry grasslands between the abandoned runways have become important nesting sites for the skylark (*Alauda arvensis*), a bird that is now much diminished in its natural habitat of open country across much of Europe. Photo by Matthew Gandy.

that downplay cultural and historical aspects of human experience, and the inherited legacy of "landscape indicators" and other ideologically charged delineators of cultural landscapes.[78] The recent reworking of the sublime, most notably through the "technological sublime" in relation to urban and industrial landscapes, is difficult to disentangle from romanticist readings of aesthetic disorientation as the cultural antinomy of beauty. These aesthetic formulations rest on the capacity of contemporary landscapes of marginality or technological excess to unsettle or even overwhelm the human observer. In the twentieth-century the term "neoromanticism" has been used to denote a combination of eighteenth- or nineteenth-century interest in awe-inspiring landscapes with more modern preoccupations with psychological unease or feelings of estrangement.[79] In terms of urban space, however, an emphasis on the sublime effectively displaces ecology with aesthetics: there is an implicit spectacularization of urban landscapes that is far removed from more tactile or direct forms of cultural or scientific practice. To disavow the neoromanticist aestheticization of space is to demystify its production and introduce a different aesthetic register rooted in the scale of corporeal human experience. It also opens up the possibility of a postphenomenological reading of urban space that can incorporate multiple subjectivities and other-than-human sensory realms. Following the insights of the cultural theorist Sianne Ngai we might ask whether existing aesthetic categories such as the "sublime" have now lost much of their analytical utility.[80] In the place of the generic conceptual terrain inherited from European modernity, what kind of aesthetic language might correspond with the dispersed subjectivities and material displacements of contemporary urban space?

Abandoned urban landscapes have frequently been venerated as places of reverie. In the case of London, for example, Patrick Keiller's film *Robinson in ruins* (2010), or Chris Petit's collaboration with the novelist Iain Sinclair for *London Orbital* (2002), present marginal landscapes as estranged or mysterious. For Keiller's *Robinson in ruins*, the narration (spoken by Vanessa Redgrave) follows the eponymous Robinson as he seeks out "marginal and

hidden locations." Robinson, we are told, is suffering from an unspeci-
fied "malady" that he will purge by creating "picturesque views on journeys to sites of scientific and historic interest." A close-up of an aluminum
street sign reveals a cellular surface pattern encrusted with lichens and
other traces of life, so that the familiar is rendered mysterious. The largely
deserted locales recall science fiction scenarios in which modernity has dis-
sipated into a new kind of wilderness. There is a quasi-mystical emphasis
on uncovering layers of meaning or ghostly traces of collective memory.
Similarly, the cities of the North American rustbelt have come under a sustained neoromanticist gaze, most notably in the photographic depictions
of Detroit by Yves Marchand and Romain Meffre, which provide an eerie
echo of earlier representations of urban ruins in the wake of the 1943 race
riots.[81] Buried beneath these representations, however, are significant dislocations in terms of class, gender, and ethnicity between the "deserted"
characteristics of these spaces, as they are encountered by the figure of the
late modern flâneur or male wanderer, and those communities cut adrift
within the marginal spaces of the contemporary city. This is not to argue
that aesthetic interactions with wastelands are necessarily gendered—at
least not in essentialist terms—but to draw attention to the specific contexts in which they are encountered.[82]

Although wastelands often elicit a degree of ambivalence, their cultural
malleability has enabled their appropriation into what the botanist Oliver Gilbert terms the "urban commons."[83] This politically charged ecological formulation extends the "right to nature" beyond municipal park
provision or Lefebvrian conceptions of public space to encompass a more
broadly defined realm of cultural and scientific imagination. Furthermore,
these vernacular spaces of "new wilderness" tend to be concentrated in
precisely those areas that often have the least access to more formal elements of designed nature.[84] In Andrea Arnold's film *Fish tank* (2009), for
example, we encounter a contrast between the concrete landscapes on the
eastern edge of London and an extraordinary interlude focused on a visit
to a fenced-off space of wild nature by a lake: for a brief second the camera

pauses on a blue damselfly (*Enallagma cyathigerum*) resting on a reed—a motif of entomological detail widely deployed by Arnold—which serves to introduce a momentary element of wonder for her cinematic protagonists. In *Fish tank* we are reminded that marginal spaces such as wastelands form a fundamental element in public cultures of nature for the poorest urban communities: an association exemplified by the childhood recollections of the community activist from the French *banlieue* cited at the start of this chapter.

The ambiguous connections between nature, science, and public culture have been the focus for a variety of cultural interventions since the 1970s. The artist Lara Almarcegui's photographs of marginal spaces, for example, make an explicit connection with the concept of *terrain vague* and a defense of space against the "excesses of architecture."[85] A politics of "antienclosure" is combined with a poetic celebration of marginal spaces. Her poignant photograph entitled *To open a wasteland, Brussels* (2000) depicts the blurred figure of a child in the foreground rushing into this newly opened space (figure 2.5). Another interesting example of artistic practice inspired by marginal urban spaces is the Berlin-based Wasteland Twinning Network, founded in 2011, which seeks to create cultural dialogue between wastelands in Amsterdam, Bengaluru, Berlin, Kuala Lumpur, Sydney, Yogyakarta (Indonesia), and many other locations. The twinning ceremonies, often involving artists and writers, are framed as an ironic challenge to neoliberal forms of "urban twinning" and the role of art in gentrification (figure 2.6). A key figure in the wasteland twinning network is the German artist Matthias Einhoff, who has been exploring ways of bringing cultural life into ostensibly empty sites since the mid-2000s, notably as part of the artist collective KUNSTrePUBLIK, which evolved out of intense interactions with one wasteland in central Berlin.[86] Using a combination of digital and material events, the diverse program of actions is described as "networked transdisciplinary practice," with accompanying workshops and other pedagogic events. The exchange of knowledge often extends to the use of botanical surveys (or other means) to evaluate the presence

Figure 2.5 Lara Almarcegui, *To open a wasteland, Brussels* (2000). Courtesy of the artist.

Figure 2.6 Wasteland twinning ceremony between Nottingham, UK, and Yogyakarta, Indonesia, held at Zentrum für Kunst und Urbanistik, Berlin, featuring (left to right) Tonya Sudiono, Ansgar Reul, Alex Head, Will Foster, and Lars Hayer, 23 September 2012. Photo by Matthew Gandy.

of spontaneous nature, along with the collection of found objects that reveal hidden traces of human culture. In these instances, close observation, or the "botanical eye," becomes a specific form of cultural-scientific practice that can reveal new insights into the production of space and the often arbitrary assignment of cultural or economic value. We can enlist the "botanizing" impulse of Walter Benjamin, himself a keen observer of urban nature, to evoke an alternative dimension to urban flânerie that eschews the psychogeographic impasse of neoromanticist detachment or late modern masculinist malaise.[87] Imaginative interventions by artists, writers, and scientists remind us that looking, thinking, and representing the familiar in an unfamiliar way can also be a kind of radical cultural and political praxis.

2.5 RECONFIGURATIONS

Urban wastelands unsettle the familiar terrain of cultural landscapes, designed spaces, and the organizational logic of modernity. Much of the conceptual vocabulary we have available is geared towards idealized landscapes—with or without human influence—or stems from a neoromanticist or phenomenological preoccupation with the aesthetic experience of the bounded human subject. To reframe marginal spaces of nature as a vibrant dimension to urban life introduces a different kind of complexity into the socioecological landscape of cities, where questions of access, design, and land ownership are radically juxtaposed with insurgent forms of cultural and scientific practice. The recognition of *terrain vague* within the public realm introduces possibilities for cultural and scientific autonomy that invert or unsettle bourgeois conceptions of nature. Yet what is "nature" anyway in an urban context? The distinction between the natural and the unnatural, in terms of landscape aesthetics, is historically produced: when Marx mocks Feuerbach's perception of the cherry orchard as a "natural" feature of the German landscape he is also questioning the limitations of philosophical idealism.[88] A materialist reading of urban landscape makes

113

the connections between aesthetics and the historical production of space explicit, yet it also harbors its own lacunae in terms of the tensions between restricted readings of human subjectivity and the communicative ethos behind shared cultures of urban nature.

The term "biodiversity," as a cultural construction of nature, holds similarly ambiguous implications in an urban setting. Although cities have high levels of biodiversity—in some cases greater than their immediate hinterland—the relationship between cities and nature becomes more problematic at wider scales of analysis.[89] Urbanization is itself a major cause of habitat destruction at a global level, so that any emphasis on urban biodiversity needs to be set in a wider context: this emerging paradox is marked by higher levels of regional biodiversity associated with increasingly diverse urban ecological assemblages but declining levels of global biodiversity as endemic, vulnerable, or yet-to-be-described species are lost at an increasing rate. In making this distinction between urban and "non-urban," however, we should be careful not to fetishize the city as a discrete entity, since the process of urbanization is increasingly ubiquitous and encompasses spaces that lie far beyond the administrative confines of metropolitan boundaries.

An emphasis on "ecological cosmopolitanism," in terms of vibrant concentrations of global biodiversity in cities, contains a double-edged aspect: it emphasizes the variety and vitality of urban ecosystems but may downplay the wider ecological impact of urbanization. Indeed, the ecologist Charles Elton's negative use of the term "cosmopolitan" in the late 1950s signals ideological tensions running through the historiography of ecology that persist today.[90] Most contemporary ecologists would acknowledge a spectrum of "invasiveness" ranging from mere scientific curiosity to real threats to the well-being of existing ecosystems.[91] For the biological ethicist Ute Eser, however, the core issue remains the relative lack of critical reflection within ecological discourse, so that the "politics of neophytes" is as much an epistemological question as a practical challenge for nature conservation.[92]

What does the radical elevation of ecological science within urban discourse imply? Are there ways in which urban ecology can enrich public culture on its own terms yet avoid the types of epistemological elision that have marred previous attempts to build socioecological understandings of urban space? The decline of what the geographer Karl Zimmerer terms "ahistorical systems ecology" has shifted the analytical emphasis towards the dynamic and heterogeneous characteristics of biophysical systems.[93] Leading urban ecologists such as Marina Alberti, John Marzluff, and Steward Pickett have called for a "new ecological paradigm" that directly incorporates the "human dimension" into ecological processes.[94] Running through these reformulations, however, is a continuing uncertainty about the analytical scope of contemporary ecology in relation to the specific cultural, historical, and material dimensions of urbanization. Despite the ambitious research agendas of post-Rio Summit "applied ecology" to encompass the full gamut of social and ecological processes, the scientific underpinning for a unified socioecological model for the study of cities does not yet exist.[95] Although a range of research has now been carried out on urban nature, there remain significant gaps in current knowledge for many cities, especially in the global South, where the status of nonhuman nature is especially precarious or little known. And within the field of urban ecology itself there are concerns that the predominance of observational studies risks placing ecological research outside the most significant developments within the biophysical sciences, where emphasis is increasingly at the molecular or even submolecular level of analysis.[96] The molecularization of the life sciences holds analytical and methodological implications for ecology as a field science that may yet presage a new phase of distanciation between scientific practice and public culture.

But why should we be interested in urban wastelands? What is at stake culturally, politically, or scientifically when we argue for their appreciation or protection? We have seen how these spontaneous manifestations of urban nature connect with an array of cultural and scientific discourses, but their political implications remain only partially explored. The promotion

115

of urban biodiversity holds implications for the development of a scientifically enriched public realm that rests on the combination of greater affinity for nonhuman life in cities along with new forms of knowledge production. Furthermore, a closer engagement with the socioecological dynamics of urban space may help to dispel aspects to the ideological opacity of the urban arena itself.

There are points of intersection between urban ecology and other cultural fields that may be used to create an alternative approach towards marginal spaces of spontaneous nature, producing insights very different from functionalist or utilitarian perspectives. Wastelands exist in dynamic tension with human intentionality, whether in terms of their preservation—the slowing of time—or their erasure to make way for the new. As sites of discovery and experimentation, wastelands also challenge unified conceptions of the "cultural landscape" and other ideological motifs that pervade contemporary urban thought. These marginal sites of spontaneous nature allow cultural and scientific explorations of the city itself, in contrast with the outward focus of much environmental discourse towards spaces and places that lie elsewhere. Above all, wastelands are islands, in cultural, material, and political terms, which pose an ideological as well as practical challenge for the utilitarian impetus of capitalist urbanization. The intrinsic worth of the ostensibly useless is as much a political question as an aesthetic or scientific one.

Ailanthus altissima Swingle
Aesculus hippocastanum L.
Robinia pseudoacacia L.
Acer negundo L.
Galinsoga ciliata Blake
Galinsoga parviflora Cav.

These are the scientific names of plants that the French artist Paul-Armand Gette photographed for his exhibition entitled "Exotik als Banalität" (Exoticism as banality), held in West Berlin in the autumn of 1980 (figure 3.1).[1] The photographs featured the spontaneous vegetation that Gette had observed during walks through districts such as Kreuzberg, Tiergarten, and Wedding, along with a few locations in Mitte in the eastern part of the city.[2] Eschewing conventional elements of urban nature such as parks, gardens, or planted street trees, Gette focused only on those plants that had colonized the interstitial spaces of the city, such as pavements or waste ground. All of the plants depicted in Gette's photographs share one thing in common: they are not native to the Berlin area. Taking a meticulously geobotanical approach, Gette organized his photographs by region to show

Ailanthus altissima Swingle   Voltastraße, Berlin Wedding

Paul-Armand Gette

EXOTIK ALS BANALITÄT

daadgalerie
4. Oktober – 9. November 1980

Eine Ausstellung des Berliner Künstlerprogramms des Deutschen Akademischen Austauschdienstes

Figure 3.1 Promotional postcard for Paul-Armand Gette's exhibition "Exotik als Banalität" held at the DAAD gallery in Berlin (1980). Source: DAAD archive, Berlin.

that *Ailanthus altissima* originates from China, *Aesculus hippocastanum* from the Balkans, *Robinia pseudoacacia* and *Acer negundo* from North America, and *Galinsoga ciliata* and *Galinsoga parviflora* from South America (figures 3.2 and 3.3). *Galinsoga parviflora*, for example, is a probable nineteenth-century escape from Berlin's Schöneberg Botanical Garden. Originally from Peru, this small white-flowered plant, which acquired the misleading vernacular name of *Kleinblütiges Französenkraut* (Small-flowered French herb), was once the focus of intense horticultural and scientific interest but had become a largely unnoticed "weed" nestling in the interstices of the urban landscape and commonly encountered throughout the city.[3] This shift in sensibilities from the "exotic" to the "banal" illustrates an oscillating ecological imaginary in which certain species of plants or other erstwhile natural features are emphasized for aesthetic, didactic, or symbolic purposes. *Galinsoga parviflora* is now an increasingly ubiquitous presence in cities worldwide: its successful adaptation to urban biotopes, and its cultural appropriation by Gette as a symbolic indicator for urban nature, illuminates the ambivalence towards many "weeds" within botany, and particularly within the field of *Pflanzensoziologie* (plant sociology), whose taxonomic categories have been historically developed in opposition to both the presence of nonnative species and the putative artificiality of urban space.

Gette's use of scientific names in the exhibition, along with the names of the species' original describers, suggests not only a certain ordering of the natural world, but also a historical link to the origins of modern taxonomy through the Linnaean binomial system.[4] That many of these plants, which were first described from colonial surveys, should form part of the spontaneous vegetation of Berlin is a poignant indicator of the reflexive global reach of modernity. Gette's project represents, therefore, a living historical archive that dispels artificial distinctions between the city and its constitutive elsewhere. The ubiquitous presence of an unruly cosmopolitan ecology serves to unsettle the very boundaries and distinctions that have enabled the uneasy coevolutionary dynamics of cosmopolitan discourse to emerge in relation to the expansion of European science, knowledge,

Figure 3.2 Paul-Armand Gette, photograph of a Berlin street corner with *Galinsoga parviflora*, originally from Peru, featured in the exhibition "Exotik als Banalität" (1980). Courtesy of the artist.

Figure 3.3   Paul-Armand Gette, photograph of a Berlin street with *Ailanthus altissima*, originally from China and North Vietnam, featured in the exhibition "Exotik als Banalität" (1980). Courtesy of the artist.

and imperial conquest.[5] Yet we should be cautious in setting up a simplistic distinction between a singular and "rootless" variant of cosmopolitan ecology and a prevailing nativist idiom of landscape authenticity: of particular interest here is the manner in which an empiricist genre of vegetation science moves between these two biogeographical imaginaries in the urban arena.

The exhibition "Exotik als Banalität" marks a continuation of studies that Gette had made since the early 1970s of spontaneous vegetation growing in the wastelands and assorted *terrains vagues* of London, Paris, Rotterdam, and other European cities, including an earlier visit to Berlin in 1971. For Gette, the uncontrolled presence of nature contrasted with the domesticated difference of botanical gardens, where fragments of nature are categorized and labeled for didactic purposes. Though ostensibly only a cursory survey of traces of nondesigned urban nature, Gette's project poses a range of interesting questions about the cultural politics of landscape and the ideological parameters of ecological art. His botanical flânerie not only connects with classic tropes of walking the city but also incorporates scientific methodologies such as transects, quadrats, and other ways of surveying the biophysical characteristics of nature.

It is arguably Gette's photographs of adventitious street trees that form the most poignant dimension to his project. The presence of these spontaneous urban trees provides a surprise embellishment to poorer Berlin neighborhoods that have historically lacked the extent of greenery of wealthier districts.[6] Several of his photographs depict the tree of heaven (*Ailanthus altissima*), known as *Götterbaum* (tree of the gods) in German, which became one of the most characteristic trees across the city during the postwar period, forming dense stands alongside railway tracks and abandoned buildings.[7] As recently as the early 1960s, *Ailanthus* was in transition between its former role as an ornamental tree and its characteristically spontaneous presence in waste spaces.[8] Its presence reveals an alternative cultural physiognomy of the city to the native lime, *Tilia x europaea*, for example, most memorably evoked by Berlin's tree-lined boulevard Unter den Linden, dating from 1647, and a longstanding iconographic element within conservative conceptions of urban nature exemplified by "organic urbanism."[9] Gette's focus on *Ailanthus* also differs markedly from the German artist Joseph Beuys's choice of native oak trees for his installation *7,000 Eichen—Stadtverwaldung statt Stadtverwaltung* (7,000 Oaks—urban afforestation instead of urban administration) for the Kassel documenta of 1982. Beuys's use of oak for his *Stadtverwaldung* (urban afforestation) project took place against a contemporary backdrop of widespread concern with *Waldsterben* (forest die-back) as a result of acid rain, which proved a significant spur for the formation of the German Green party in 1980 (which Beuys publicly supported). Beuys's neoromanticist attempt to "Germanize" the urban landscape provides a poignant contrast with Gette's interest in adventitious street trees as indicators of a global ecology that lies closer to the existing characteristics of urban space.

3.1 IDEOLOGICAL CONTORTIONS

During the 1970s West Berlin gradually emerged as an experimental enclave, in an apparent reprise of aspects of the cultural ebullience of the Weimar era. The city's cultural life became part of a loosely constituted urban archipelago that extended to other cities such as Cologne and New York, connected through a network of curators and gallery owners such as René Block.[10] The complex relationship between the postwar West German state and the geopolitical island of West Berlin became manifest in a highly subsidized municipal enclave that mirrored the strategic significance of the adjacent GDR capital of East Berlin. It is in this unusual cultural and political context that Gette began his initial discussions with the Berlin-based Artists-in-Residence Programme funded by the Deutsche Akademischer Austauschdienst (DAAD), referring to his project under the working title "Some waste grounds in Berlin."[11]

Unlike many examples of ecological art that are not in fact rooted in any aspect of ecological science, and merely attached to the rhetorical allure of ecology, Gette's work is marked by a sustained engagement with the practice of science. Through his use of existing scientific methods he develops a "parallel science" based on what the art historian Günter Metken terms "a parody of scientific conventions."[12] His mimicry of scientific practice is accentuated by his choice of locales such as parking lots, street corners, or towpaths that have not been a conventional focus for botanical study.[13] The experimental dimension to his work (he exchanged letters with André Breton) connects with aspects of the twentieth-century avant-garde, through the deployment of science against itself, in order to produce ironic inventories or unexpected taxonomies. Gette has expressed his delight, for example, in the spontaneous dynamics of urban nature that exist in spite of bureaucratic efforts to exert "botanical control" over space, yet his quixotic scientific investigations have an ideological dimension, connecting with distinctive postwar urban ecological discourses that seek to challenge nativist conceptions of vegetative order that had hitherto dominated *Pflanzensoziologie* (plant sociology) and its counterparts in landscape design.[14] By conducting what Lucius Burckhardt describes as a form of "science without questions," Gette's work evokes a certain aimlessness to scientific enquiry while simultaneously provoking reflection on the underlying purposes of mapping, taxonomy, and other attempts to produce a knowable world.[15] Gette's use of survey techniques recalls the geopolitical legacy of scientific reconnaissance, yet his work is unconnected to obvious utilitarian objectives beyond his own practice as an artist.

We can trace the modern origins of vegetation mapping to the German geographer Alexander von Humboldt, who sought to record every conceivable facet of the natural world, in a radical elaboration of the scope of natural history and early advances in the earth sciences. Humboldt's interest in banded vegetation patterns on mountains, for instance, is a realm of fascination that links the development of the empirical sciences with the emergence of European romanticism and the expansionary impetus of capitalist modernity. The recasting of previously feared "wild landscapes" as a source of scientific allure, cultural inspiration, and also potentially utilitarian value forms part of an emerging modern, and to a significant degree metropolitan, sensibility towards nature.[16] In the 1920s, the earlier biogeographical insights of Humboldt and his followers were further elaborated by the Swiss botanist Josias Braun-Blanquet to produce a distinctive strand of botany known as the Zurich-Montpellier school of plant sociology, based on an intricate spatial elaboration of the Linnaean binomial system of classification. Braun-Blanquet's approach was eagerly taken up by a new generation of German botanists such as Heinz Ellenberg, Erich Oberdorfer, and Reinhold Tüxen, who adopted an increasingly regionalist, and ultimately nationalist, interpretation of naturally occurring plant communities. By the 1930s, taxonomic distinctions between different vegetation patterns became increasingly regarded as the innate expression of regionally specific human cultures, and the ideological preoccupation with boundaries acquired an increasingly geopolitical edge. The earlier scientific insights of Humboldt and his successors into the distribution of plants became drawn into a nationalist schema within which the geopolitically and biogeographically defined entity of Mitteleuropa would serve as a bulwark against the perceived threat of biological and cultural contamination.[17]

From the early 1930s onwards, the nativist preoccupations of plant sociology found increasing resonance with the practical and ideological needs of the National Socialist project. With the 1939 establishment of a new national vegetation mapping center, the Zentralstelle für Vegetationskartierung des Reiches (ZVR), the science of plant sociology was now directly geared towards the strategic needs of the Nazi state. The Hannover-based ZVR promoted vegetation mapping to assert the innate relationship between nature and culture within distinctively German landscapes, a politically motivated research program that gradually extended to the study of regions with German-speaking minorities in Alsace, Poland, and elsewhere.[18] Tüxen, as the first director of the ZVR, had begun his work in the 1930s with a comprehensive vegetation mapping project for northwest Germany.

It is in the elision, however, between empirically oriented plant sociology and the increasingly significant concept of *Heimat* in German landscape studies that the ideological commonalities between these different fields of study became increasingly apparent. The strategic role for plant sociology would be to warn landscape planners against "intolerable combinations" of different species.[19] Tüxen, for example, not only condemned the presence of exotic plants such as the balsam, *Impatiens parviflora*, in the German countryside, but also likened its spread to the threat of Bolshevism from the east (where the species had originated). Although Tüxen and other nativist botanists sought "to cleanse the German landscape of disharmonious foreign bodies," the greatest biological threat from invasive species was actually to be found in the more distant colonial landscapes that had been disrupted through the expansion of European conquest and control (a theme I return to in chapter 5).[20] These botanical obsessions found echoes in the social and political sphere with the banning of Jewish people from parks, swimming pools, and also forests as part of the meticulous combination of politicized taxonomy and racial hygiene.[21] Alternative voices in the fields of botany and nature conservation were systematically sidelined, or murdered in the case of Benno Wolf, the Berlin-based Jewish lawyer who had led the drafting of Germany's first nature protection legislation, enacted in 1920.[22] By the early 1940s, the work of the ZVR had been extended to "botanical reconnaissance" in strategically significant zones such as the oil-rich Caucasus. The prominent Berlin-based botanist Kurt Hueck sought to identify "German" landscapes within the newly occupied territories of Poland, whilst another leading figure in plant sociology, Heinz Ellenberg, worked on the design of "camouflage plantings" for military bunkers.[23]

Key figures involved in the Nazi-era mapping projects persisted with nativist conceptions of landscape after the war. Both Ellenberg and Tüxen saw their international scientific stature grow significantly from the 1950s onwards whilst their more racist publications quietly slipped from view.[24] Kurt Hueck became director of Berlin's Institute for Agricultural Botany from 1946 to 1948 and presented a nativist "replanting" scheme for Berlin's postwar reconstruction based on a list of "acceptable plants" that divided the city into a series of "natural zones," in an early variant of what is now widely referred to as "restoration ecology."[25] By 1949, however, Hueck had switched his attention from the increasingly complex landscapes of Berlin to the "primordial" *Urlandschaft* of subtropical northern Argentina and Venezuela.[26]

The 1950s marks a critical divergence in the continuing practice of *Pflanzensoziologie* in Berlin: whilst some botanists such as Wolfgang Müller-Stoll continued to focus their attention on plants growing at or beyond the city limits, others became fascinated by the novel ecological assemblages forming in the postwar ruins of the city itself.[27] By the late 1950s the city's extensive *Trümmerlandschaften* (rubble landscapes) had evolved from sites of collective forgetting into an array of vernacular public spaces for disparate subcultures ranging from sexual dissidence to new forms of ecological fieldwork.[28] At the same time, they had begun to display unique ruderal assemblages of plants that held cultural and scientific implications for the nascent field of urban ecology.

The first systematic study of ruderal vegetation in the city was undertaken in the early 1950s by the botanist Hildemar Scholz. Scholz uncovered a "revolution in the plant world" within the ruins of the city that necessitated the addition of new taxonomic categories for plant associations beyond the existing scope of *Pflanzensoziologie*.[29] With the geopolitical separation of West Berlin this scientific divergence between "urban" and "rural" ecology intensified as the island city itself became an urban laboratory under intense scrutiny. The division of Berlin also acquired ecological significance over time, through the proliferation of anomalous spaces and the restriction of development pressures beyond the barbwire and watchtowers of the city limits. Studies by Scholz, his colleague Herbert Sukopp, and other urban botanists identified typical Berlin plants such as *Dysphania botrys*, a fragrant weed from the eastern Mediterranean, which had become so characteristic of the city's wastelands that an entire issue of a botanical journal was devoted to the species in 1971 (figure 3.4).[30]

Figure 3.4 Botanical study site with *Dysphania botrys*, Potsdamer-straße, Berlin-Tiergarten (1968). This plant, originally from the eastern Mediterranean region, spread rapidly through the *Trümmer-landschaften* (rubble landscapes) of postwar Berlin. Source: Institute of Ecology, Technical University, Berlin. Photograph: Ursula Hennig.

From 1973, the newly created Institute of Ecology at Berlin's Technical University embarked on a research program that would culminate in an ambitious mapping exercise that divided the whole of West Berlin into 57 distinctive vegetation zones, or urban biotopes (I return to the cartographic theme in the next chapter). This intense focus on urban ecology finds resonance with earlier attempts to produce a modernist synthesis of nature and culture in the reconstruction of postwar Berlin, exemplified by the architect Hans Scharoun's short-lived tenure as Stadtbaurat (director of urban development) in the late 1940s.[31] Yet the attempt to replicate a plot-by-plot analysis across the whole city had to contend not just with the shortage of resources to carry out the detailed survey work but also the paucity of scientific expertise to identify critical species correctly.[32] Gette's concern with the limits to scientific knowledge mirrors the dilemma facing Sukopp and the other urban botanists at Berlin's Technical University in the 1980s: how to generate enough data to ensure the protection of vulnerable fragments of urban nature in the face of development pressures.

The precision of scientific naming and biotope classification lies in tension with the unknowable complexity of urban space, since even the most sophisticated survey can remain only a partial representation of a specific moment in time. Gette's application of scientific methods is also a demonstration of their limits: by focusing his attention on a few marginal spaces in intense detail he reveals what the landscape theorist Lucius Burckhardt has referred to as the "vast blank spaces" of scientific knowledge.[33] By highlighting only a few plants in his photographs Gette illustrated the impossibility of a complete survey of urban nature. Like Gette's intervention, however, the analytical scope of the Institute of Ecology's research program did not extend significantly beyond plants and the physical characteristics of the environment (with the partial exception of occasional collaborations with entomologists and ornithologists). In this respect the Institute's work simply elaborated on the empirical scope of existing *Pflanzensoziologie* but in an urban setting. The ill-defined presence of human agency reveals a tension between an analytical framework originating within the biophysical sciences, including urban ecology, and more recent attempts to address the political aspects of urban nature.[34] The science-based approach to urban ecology had paradoxical consequences: on the one hand its ostensible impartiality enabled urban ecological discourse to distance itself from the ideological tentacles of Nazi-era *Pflanzensoziologie*, and thereby facilitated an influential role for ecology in the emergence of progressive environmental politics in the 1970s and 1980s; on the other hand, the lack of a historical framework beyond that of human influence over land use and other biophysical parameters proved both a political and epistemological lacuna.

A key insight of Sukopp and his colleagues was their emphasis on the biophysical characteristics of urban landscapes within which the erstwhile "inauthentic" becomes authentic. Their work allowed previously ignored elements of nature to become a legitimate focus for ecological science. Sukopp, for example, succeeded in extending Berlin's Red List of rare or endangered plants to some nonnative species, in the context of recognizing the scientific significance of urban biotopes in a radical challenge to existing conceptions of nature conservation.[35] The criterion of endangerment, rather than species origin, enabled the valorization of certain urban plants on both cultural and scientific grounds. To emphasize the divergence of Berlin's postwar urban ecology from the earlier nativist emphasis of *Pflanzensoziologie*, the Institute even depicted an *Ailanthus* tree on the cover of its first ten-year report in 1983 (figure 3.5).[36] The drawing by Jörg Eling, entitled *S-Bahn Landschaft mit Ailanthus* (S-Bahn landscape with *Ailanthus*), presents a distinctive postwar Berlin landscape to a knowing audience of urban botanists.

Yet the Institute's critical engagement with the biophysical dimensions of urban nature did not extend to any wider reflection on the nature of urbanism: the concept of nature itself remained tightly contained and nondialectical in order to protect their version of urban ecology from unwanted ideological intrusions. The attempt to separate science from politics in the field of urban ecology reflects the *ideologische Verrenkungen*

Figure 3.5 Jörg Eling, *S-Bahn Landschaft mit Ailanthus*, used on the cover of the first ten-year report of the Berlin Technical University's Institute of Ecology (1983). Source: Institute of Ecology, Technical University, Berlin.

(ideological contortions) running through postwar German environmentalism. The attempt to produce a nonideological urban ecological discourse in Berlin did not resolve these questions but rather deflected them epistemologically. An urban botany emerged which was devoted to the protection of distinctive types of *terrain vague* but could not connect with cultural and political analysis of capitalist urbanization, beyond the formation of strategic institutional alliances to protect specific sites. Apart from the work of a few individuals such as Ludwig Trepl, the botanists associated with Berlin's Technical University did not pursue the wider political implications of their work beyond an implicit disavowal of the past or an attempt to protect spaces of spontaneous nature from utilitarian erasure: by anchoring their critique within an essentially positivist conception of scientific practice, the historical dynamics of capitalist urbanization were rendered obscure, and the work remained close to the dominant systems-based approaches that had emerged within postwar ecological science.[37]

3.2 ALIENS AND OTHERS

The political implications of the focus on spontaneous vegetation in Berlin and other cities resonated not just through ecology but also other fields such as geography, planning, and landscape design. In cultural geography, for instance, we find some of the first explicit attempts to explore the ideological dimensions to *Pflanzensoziologie* in an urban context. The Osnabrück-based geographer Gerhard Hard, for example, published a series of articles that explored the political significance of ruderal vegetation. He also set out his argument in the inaugural issue of *Landschaft+Stadt* in 1986, alongside an article by Sukopp, demonstrating the wider significance of spontaneous urban vegetation as a challenge to organic conceptions of landscape. For Hard, wastelands served as the "ordinary terrain" over which wider intellectual struggles would be played out.[38]

In a documentary film made for German television in 2008 we see an excerpt from an earlier film in 1972 showing Sukopp on a botanical excursion to Berlin's Teufelsberg (Devil's Mountain)—an artificial hill comprised of wartime rubble towards the western boundary of the island city.[39] This stony landscape, not unlike the semiarid lower slopes of a volcano, was the focal point for a series of scientific studies by Sukopp and his colleagues because of its unique flora. The 1972 film emphasizes the global origins of plants growing on the Teufelsberg site, and in the commentary for the more recent documentary the theme is further emphasized by referring to the unusual biotope as *Multikulti*, or "multicultural." This explicit linkage between urban ecology and multiculturalism is interesting in a Berlin context because it draws together the material characteristics of urban space with emerging political discourses about the changing characteristics of postwar German society. The term *Multikulti* first emerged in Germany in the context of progressive movements in the 1970s and 1980s; its use in relation to spontaneous urban vegetation holds wider implications for the realignment of public culture towards greater *Weltoffenheit* (world-openness).[40] The acknowledgment of a *Multikulti* society implied a more polycentric conception of cultural identity that conflicts with conceptions of German identity as relatively homogeneous, place-bound, and linguistically circumscribed. The use of the term has altered over time from its earlier association with a more explicitly political agenda to a more nebulous contemporary formulation based on essentialized conceptions of cultural difference. An earlier emphasis on racialized and discriminatory labor practices, poverty, and citizenship rights has been transmuted into a narrower focus on the valorization of cultural difference in the context of festivals, food, and other domains, within which hegemonic attitudes lurking beneath the exoticist gaze remain unchanged and to a significant degree unchallenged.[41]

The putative reaffirmation of West Berlin as an open city in the 1960s and 1970s resonates with the outward-facing public culture of the Weimar era and the role of the city as a magnet for waves of international migration.[42] Writing in 1971, the novelist and long-term Berlin resident Günter Grass juxtaposed jaded stereotypes of the island city with the emergence of a new kind of vibrant cosmopolitan culture, based on migration from Turkey, Croatia, Spain, and elsewhere. Grass recast the Kreuzberg of the 1970s, a key site for Gette's project, as a "multiethnic utopia" rather than a geopolitically sustained urban backwater.[43] Yet this apparent eulogy from Grass displays a certain form of cultural condescension that resonates with the politically pallid notion of "tolerance" and a carnivalesque sensibility towards urban difference. Furthermore, as the geographer Brenda Yeoh has argued in a Singaporean context, the very idea of a cosmopolitan order rests on a celebration of difference that is superficially alluring when perceived from a lofty vantage point, yet nonetheless serves to reinscribe specific strands of globalized identity within the context of existing forms of social and cultural stratification.[44]

Changing attitudes to neighborhood change in postwar Berlin are also found in the contemporary street photography of Max Jacoby, Wolfgang Krolow, Hartmut Schulz, and others in the 1970s and 1980s. The exoticist gaze is clearly present in the work of Jacoby, for example, who refers to one of his Kreuzberg street scenes as "little Istanbul," and in another photograph describes "dark-eyed children" who inhabit "alien streets" (an odd observation considering that most, if not all, the children he depicted were born in Berlin).[45] The contrast with the photography of Krolow, who was also active in Kreuzberg during this period, is especially illuminating. Rather than taking the perspective of an outside observer, Krolow was immersed in the social and political context of his subject matter, with numerous depictions of the squatter movement and street scenes including riots. In one untitled image we encounter two girls adopting a mirrored dance pose in the street (see figure 3.6). The photograph has subsequently acquired symbolic significance as part of an alternative cultural iconography for postwar Berlin, and the image was later used for a school design competition as part of the Kreuzberg-based IBA (Internationale Bauaustellung/International Building Exhibition) in 1987, which itself marks a critical moment in the urban politics of the city. The IBA highlighted the *grüne Mitte* (green center), an interconnected green space comprising many existing ruderal sites,

134

Figure 3.6   Wolfgang Krolow, untitled street scene from Kreuzberg, West Berlin (1981). Courtesy of the Wolfgang Krolow archives.

and *behutsame Stadterneuerung* (careful urban renewal), an approach associated with urban planners such as Hardt-Waltherr Hämer, who emphasized the protection of existing inner-city neighborhoods through consultation and renovation rather than the destruction of existing housing stock.[46]

The political implications of a "rooted" definition of citizenship find their most obvious postwar expression in the figure of the *Gastarbeiter* (guest worker). The federal republic's guest worker program consisted of a series of bilateral labor recruitment campaigns beginning in 1955 with Italy, followed by Greece, Turkey, Morocco, Portugal, Tunisia, and finally the former Yugoslavia in 1968. By the end of the successive labor recruitment programs in 1973 there were five million guest workers and their families living in West Germany who were denied full citizenship rights. In contrast, the *Aussiedler* were ethnic Germans from elsewhere in Europe or even further afield, who could prove distant ancestry. They retained a permanent right to return, whereas the *Gastarbeiter* program instituted an intergenerational form of discrimination against a significant part of German society.[47] The "return rights" of the *Aussiedler* owe their origins to a change in the law dating from 1913, when the earlier "ten-year rule" for retaining German citizenship was overturned in an effort to encourage long-term emigration to newly acquired colonies. The question of modern German citizenship is rooted, therefore, in the strategic calculus of German imperialism and the use of demography as a political tool for geographical expansion.[48] By the late 1970s we find a political divergence between various grassroots initiatives, concentrated in Berlin, Frankfurt, and other larger cities, focused on helping foreign workers integrate into German society, and a shift in federal policy towards a more hard-line populist stance under the Kohl administration from 1983 onwards to encourage *Ausländer* (foreigners) to leave the country.[49] Even if one of Gette's trees of Balkan origin, *Aesculus hippocastanum*, had been known from the Berlin region for many centuries, the growing presence of people from Greece and the former Yugoslavia under the postwar *Gastarbeiter* program provoked very different reactions.[50] The spontaneous trace of "the Other," represented by the plants that Gette photographed, forms an ecological counterpart to the "provincialization" of Berlin and other European cities. These ecological traces exemplify the interweaving or "concatenation" of different histories that comprise postcolonial urban space.[51] The fragments of nature captured in Gette's photographs represent the living materiality of "elsewhere" within the rarefied cultural milieu of the metropolitan center.

The late 1970s marks a crucial juncture in the politics of West Berlin. Various single-issue campaigns or *Bürgerinitiativen* (citizens' initiatives), radical newspapers, and other elements of the extraparliamentary left (including remnants of the Maoist KPD) coalesced around the formation of a new political party in October 1978 called the Alternative Liste für Demokratie und Umweltschutz (Alternative List for Democracy and Environmental Protection).[52] This new political constellation was enmeshed in tense negotiations with Die Grünen (the Greens) in the federal republic, formed in 1980, and also sought to differentiate itself from conservative strands of environmental politics within Berlin such as the neo-Malthusian outlook of the Grüne Aktion Zukunft (GAZ) led by Herbert Gruhl. The Alternative Liste developed a radical set of policies including a much more inclusive stance towards immigrants, greater protection for alternative forms of social organization within the city such as housing collectives, new forms of feminist politics including the protection of women from domestic violence, and an explicit recognition of the cultural and scientific value of nondesigned elements of urban nature.[53]

A key component in the campaign agenda of the Alternative Liste was the protection of vernacular spaces of urban nature or *Wildwuchs* (wild growth), later referred to *Stadtwildnis* (urban wilderness), thereby bringing together several strands: the cultural valorization of interstitial spaces exemplified by the work of Gette and other artists; the political acknowledgment of scientific interest in distinctive aspects of urban ecology; and the incorporation of wider grassroots demands for the *Erhalt* (preservation) of "urban wilderness" as a distinctive element of public space. The focal point for the Alternative Liste's political commitment to urban ecology was the demand to convert an 18-hectare stretch of abandoned railway sidings, called the Südgelände, into an urban nature park. The eventual success of the campaign to protect the Südgelände marks perhaps the most significant material outcome of this distinctive phase in the city's environmental politics. Ultimately, however, the Südgelände represented only an island of nature rather than the extensive green network or *grüne Mitte* envisaged in the Internationale Bauaustellung of 1987.[54] After 1989 many other similar spaces were lost under the gathering momentum of urban redevelopment in the new German capital: the fragile socioecological political constellation that had developed during the 1980s was subsumed within a new postunification phase in the city's politics.

After the fall of the Berlin Wall, *Ailanthus altissima* was chosen as the symbol of the reunified city in recognition of its ubiquitous presence, especially in the anomalous spaces produced by geopolitical separation. This officially sanctioned gesture reveals a significant shift in cultures of nature for Berlin, underpinned by both scientific understanding of urban ecology and wider recognition of the cultural ambiguity of urban landscapes.[55] It is especially ironic that the grasslands of the award-winning Park am Nordbahnhof, opened in 2009 along a former "death strip" that divided the city, should be described as "steppe-like," given that prominent botanists in the 1930s had fretted over ecological (and political) threats from the East.[56] These projects attest to the influential legacy of the Technical University's Institute for Ecology in the field of urban landscape design. But to what extent did Berlin's postwar urban ecological discourse represent a self-conscious repudiation of the past or a coincidental contribution to a new kind of socioecological political constellation in the city?

The emergence of what we might term "ecological art" since the 1970s poses a series of difficulties ranging from the metaphorical elision between incommensurate fields of knowledge to naïve misunderstandings of the institutional context for scientific research. Art, like science, has its patrons and its webs of influence. Art that is self-consciously "ecological" must be considered in relation to its adopted epistemological stance, not least because ecology itself is a contested intellectual terrain: it does not represent a scientific consensus, still less a coherent field of political action.[57] In this sense, much environmental art effectively replicates the anomalies or contradictions within its borrowed frame of reference. Consider, for example, the American artist Kathryn Miller's use of "seed bombs" that she lobbed into a variety of marginal sites in southern California between 1991 and 2002. Miller describes how these compacted balls of soil and carefully selected seeds "match the native ecology of the area where they will be used."[58] Implicitly, however, these forms of art as ecological action serve as a counterpart to discourses of "renaturalization" that seek to reenact specific socioecological assemblages. By taking a more measured stance, both methodologically and metaphorically, Gette succeeded in highlighting precisely the types of tensions that are frequently occluded by self-consciously political forms of ecological art.

The boundary between the *aperçu* (noticed) and the *inaperçu* (unnoticed), to use Gette's distinction, introduces a scientifically charged phenomenological dimension to the perception of space, since the mere recording of urban landscapes in intricate detail is both a cartographic and political intervention (a theme I will explore further in the next chapter).[59] For Gette, then, spaces of urban nature represent a dynamic and transitory socioecological constellation that illuminates the historical, material, and scientific context for cultural practice. By drawing on Adorno's description of the "bourgeois landscape" as one that cannot abide "weeds" or other kinds of spatial disorder, we could argue that critical forms of ecological art can use insights from the biophysical sciences to counter the ideological tentacles of nature as artifice in the urban arena. Gette's work marks more than a metaphorical engagement with ecological science and clearly avoids any kind of lapse into neoromanticist obscurantism. Yet can the cartographic or diagrammatic representations of marginal spaces in his work allow us to differentiate a positivist epistemological stance from a more incisive critique of the landscape idea in an urban context? A range of more recent cultural interventions in relation to the marginal spaces of Berlin and other cities suggests that the art-science interface can transcend positivist notions of ethical neutrality to encompass critically reflective perspectives on the socioecological dynamics of capitalist urbanization.[60]

3.3 QUEERING THE TRANSECT

Though Gette's use of the botanical transect and other scientific methods as a form of radical artistic practice raises many interesting questions about the cultural and ideological resonance of spontaneous forms of urban nature, we can take key aspects of this experimental genre further to consider how race, sexuality, and the ambivalent limits to the human subject are imprinted into the urban landscape. In this section I want to emphasize the significance of the transect as a creative act that transcends the mere application of a botanical method. My argument brings the phenomenological insights of the feminist scholar Sara Ahmed into dialogue with a queered reading of urban ecology.

A transect is a kind of botanical reverie that connects the practice of walking with the space of thought and creativity itself.[61] Every kind of walk can be considered a kind of transect, or at least a "transection," in the earlier etymological sense of the verb *transect*, derived from the Latin *trans* meaning "across" or "beyond" and *secare* meaning "to cut."[62] The botanical walk known as the "transect" has become one of the most familiar methods used in conducting ecological surveys. Unlike the quadrat, which involves closely focusing on a small area and is in essence a stationary method to estimate species diversity, the transect is a mobile form of data collection that involves tracing a line through a designated area, typically on foot, to systematically record what can be found. These static and mobile dimensions to botanical method are complementary and belie the impossibility of knowing everything: there is only so much human time, knowledge, and concentration that can be applied to any kind of ecological survey. And if the objective is to generate longitudinal data, and assess environmental change over time, then there is a kind of epistemological contract with the future, in the hope that perhaps as yet unknown individuals might assist in carrying on with the work.

A botanical transect typically involves walking in a straight line at right angles to a zone of transition produced by natural features such as coastlines, mudflats, marshes, or altitudinal variations. In 1905, for example, the American ecologist Frederic Clements simply described the transect as "a cross section through the vegetation of a station, a formation, or a series of formations."[63] From the outset, however, there has been ambiguity over the appropriate scale for a transect, and by implication how the data derived from this method might relate to broader inferences about the ecological characteristics of a wider area.[64] Hovering between a diagrammatic representation and an analytical tool, it is uncertain whether the purpose of the transect is to reveal identifiable "phytosociological communities" à la Braun-Blanquet and other plant sociologists, or merely serve as a heuristic device for landscape interpretation. Writing in 1954, for example, the American botanist Dorothy Brown considers how the transect might be used for the exploration of "new territory" but refers to undertaking fieldwork by "road cruising," in a tacit admission that most landscapes are simply too vast to be systematically assessed by walking alone.[65] In the movement from a "walking line" to a "road line" there is an implicit acceleration, so that details must be interpreted from more infrequent or distant forms of observation.

Modernity itself can be conceived as a series of lines ranging from the regularization of space to a teleological understanding of its own historiography. For the anthropologist Tim Ingold the act of walking in a straight line is an enactment of modernity that connects with the tracing of plotlines, land measurement, and the elevation of linearity to a mode of rational thought.[66] The idea of "straightening" has multiple connotations extending to questions of racial and sexual difference: an emphasis on linearity as a form of power finds echoes, for instance, in Sara Ahmed's characterization of "the production of whiteness as a straight line."[67] Similarly, the performance of the line as a sociospatial inscription of power is captured in the psychologist Carl Jung's observation, based on his travels in East Africa, that "the white man's idea is to walk straight ahead."[68] The role of walking as a contrary or dogged display of perseverance emerges as a constitutive element in the spatial imaginaries of European modernity. The cartographic stage under modernity emerges as a series of "points or dots," with the establishment of "lines of occupation" serving as the logistical precursors to "settlement and extraction."[69] The historiographic unease generated by a botanical itinerary with a purpose necessitates a careful reflection on the utilitarian dimensions to walking methodologies. Indeed, the use of the transect as an experimental method in its own right, as undertaken by artists and writers, serves as a useful counterfoil to the perpetuation of certain kinds of unreflective methodological empiricism.

Does the appearance of the "urban transect," in all its practical and experimental guises, unsettle or merely overlook its complex historical associations with the utilitarian envisioning of nature? There are a number of botanical studies that simply apply the transect method directly to urban areas in order to examine the plant diversity of specific sites.[70] In other cases a more schematic approach is adopted in order to provide a cross-sectional representation of the urban-rural gradient in terms of variations in the built environment and its characteristic vegetation.[71] The transect has also been adopted within some architectonic formulations as a way of depicting urban topography, or as an urban design tool, yet these approaches are invariably schematic depictions rather than a direct application of botanical method to urban space.[72] More prescient, however, are those botanical surveys that have sought to directly challenge the nativist preoccupations of plant sociology, and its colonial underpinnings, by emphasizing the novel socioecological assemblages to be found in urban space.[73]

Cultural iterations of the urban transect can take several different forms. An urban transect can follow a line produced by infrastructure networks such as roads, canals, or railway lines, as reflected in projects such as the LA-based artist Ed Ruscha's series of photographs entitled *Every building on the Sunset Strip* (1966). The simple practice of walking is also a significant motif within land art, exemplified by Richard Long's performance *A line made by walking* (1967), which records a transitory impression left by the artist walking across a field. In some cases the transect becomes the ecology of the line itself, as in studies of plants growing in the interstitial spaces produced by transport networks such as roadside verges or railway embankments. In other examples a more conventional type of botanical transect is undertaken within a chosen site, as reflected in a variety of works including Gette's 16 mm film entitled *Le transect* (1974).[74]

In some instances, there is an attempt to transect space directly as in the performance art of Simon Faithfull, where, in an homage to Buster Keaton, he scales fences and crawls through windows to follow the meridian line as precisely as possible in *0°00 navigation* (2009). Similarly, with Gordon Matta-Clark's intervention entitled *Splitting* (1974), a recently abandoned house is cut in half so that the transect becomes an active sculptural intervention in urban space. Writers have also used the transect as a way to structure a narrative around movement through space: it can be conceived as a Perecquian literary device to spur heightened forms of observation and creative reflection through the use of an artificial constraint.[75] The idea of the line as a restriction on thought, or a form of conceptual rigidity, is effectively inverted to heighten the experience of space. Examples include the French writer François Bon, who describes the changing landscapes he observes on his regular commute from Nancy to Paris, or François Maspero, who uses the thirty-eight stops along the RER railway encircling Paris as a series of starting points for ethnographic encounters with what he terms the *terrain vague* of the city's outer suburbs.[76] To what extent, however, does Maspero's invocation of the *terrain vague* on the urban periphery belie a certain commonality with the liminal landscapes of colonial governmentality transposed to what the geographer Mustafa Dikeç refers to as the "badlands" of the French republic?[77]

A botanical transect is an embodied methodology par excellence: the systematic recording of plant life involves not just training the eye to notice small details, drawing on sophisticated forms of pattern recognition, but also the use of other sensory clues such as smell to help identify species, haptic interactions with leaves to explore their surface textures, and an awareness of small variations in light and shade, to produce an "incidental sensorium" that is open to the unexpected. The completion of a transect introduces a degree of "slowness" to the navigation of space that echoes Isabelle Stengers's appeal for *ralentissement* (deceleration) in scientific practice.[78] Indeed, the very act of slowing down can be likened to a kind of "ecological loitering" that serves as an entry point into the nuance and complexity of urban space. The point, however, is not that science should simply be done more slowly but should have the imaginative scope to reflect more fully on the practice of research as a relational activity sustained through multiple chains of human and nonhuman interaction.

The transect also resonates with psyschogeographical explorations of the modern city. Yet the situationist excursion or *dérive*, in its classic late 1960s formulation, has little grounding in ecological observation.[79] Methodologically, the *dérive* is a highly impressionistic if not masculinist cultural trope, although we should be cautious in any essentialist form of epistemological critique. For the Paris-based writer Lauren Elkin, it is attentiveness to the shifting and often invisible contours of the "affective landscape" that marks the starting point for urban walking as a form of social critique as well as critical reflection.[80] Similarly, the geographer Morag Rose shows how urban walking can "follow lines of desire, curiosity and coincidence but also invisible threads of power and the whispers of ghosts under the pavement." Rose emphasizes how the use of walking as experimental practice can "disrupt the banal" and reveal "minor epiphanies."[81]

The ethnographic transect is rooted in a slow and often painstaking immersion in urban space that points towards the conceptual orientation of the embodied human subject, in an intellectual maneuver that questions relations between bodies, objects, and others. To queer the transect is to destabilize the walking methodology from a variety of empirical and conceptual vantage points. The relevant insights from queer theory clearly extend beyond the identification of "queer space," or the use of a transect to encounter traces of human sexuality, but rather serve to problematize a series of categorizations, taxonomies, and subjectivities.[82] Queering the transect connects botanical practice with a series of posthuman and other-than-human critical discourses. Following the lead of Sara Ahmed we can introduce the significance of "queer phenomenology" as a matter of orientation that operates on different levels: firstly, the orientation of the human subject towards objects and others (in a variety of configurations that extend to other-than-human nature such as plants); and secondly, Ahmed's play on the word "orient" as a geographical and ideological counterpoint to modernity that connects with the recognition of cosmopolitan urban ecologies.[83] The transect invites a reverie of disorientation as the queered observer becomes "lost" within the line. What appears as simply given, as part of the "taken for granted" world within idealist cosmologies, is rendered off-center or unfamiliar: there is a productive dissonance between the familiar and the unfamiliar. Taking the diasporic dimension to disorientation further, we can characterize cosmopolitan ecologies as a kind of "queer regional imaginary," following the insights of cultural critic Gayatri Gopinath, that stands in explicit contradiction to nativist landscape idioms.[84]

The etymology of the word *transect* also connects with recent developments in trans theory as a further elaboration on the queering of space, materiality, and the human subject. Drawing on the relational rather than nominal import of trans theory, as a matrix of connections rather than a set of categories, we can connect with "proliferating ecologies of embodied difference." The spatial implications of "transing" the world encompass a series of movements where "the lines implied by the very concept of 'trans-' are moving targets, simultaneously composed of multiple determinants."[85] Implicit within the prefix *trans-* is a sense of multiple crossings, not just of walking across but also of interspecies encounters, including an enrichment of the botanical imagination. Trans theory unsettles the reinscription of the (modified) human subject within the burgeoning field of posthumanist studies and widens the potential scope of interspecies interactions.[86] The displacement of heteronormative space opens up the affective potentialities of socioecological assemblages as well as the multiple sexualities that suffuse the nonhuman realm including plant life (to the evident unease of early botanists).[87] The unsettling of existing taxonomies or classificatory schemas holds implications far beyond the confines of urban ecology.

The sense of what is "in place" clearly rests on the operation of repetition, familiarity, and naturalization: a set of ideological coordinates that routinely ignores the complexity of lived space through the interpolation of difference. The socioecological determinants of space comprise a series of lines, both temporal and spatial, that rest on an array of affective and material orientations. If we consider marginal spaces in European cities, and the presence of adventive plants from elsewhere, we can reflect on the ecological lineaments of Ahmed's use of the term "constitutive outside," and the way in which the history of vegetation science in a European context had emerged in the context of anxiety over the perceived inauthenticity of unusual socioecological formations.

The queering of botanical methods connects with a wider set of discourses concerning the human subject, the history of science, and the role of walking methodologies in ecological surveys. I use the term "ecological" in this instance to denote a set of relations that transcend the here and now of site-specific observations. The urban transect emerges as a kind of epistemological and ideological disturbance, not only to the historiography of botany and related fields, but also to the boundaries and categories that inhere within modernity itself. The botanical transect is a means to systematically record material differences but also a spur to thought and reflection in its own right.

3.4 BLACK ECOLOGIES

Questions of race and racism have not been extensively addressed in the field of urban ecology. A shift in the late 1960s can be discerned, especially in North America, where a series of critical interventions sought to redirect aspects of environmental discourse towards the lived experience of urban environments. The conservation biologist Raymond F. Dasmann, for instance, called for cities to be fully integrated into wider deliberations over the protection of nature and biodiversity.[88] Yet the question of race and ethnic difference was rarely placed into articulation with urban ecological discourse or landscape design.[89] The "human environment" was left largely to systems-based ecological models or subsumed within public health concerns. In 1970, for example, the recently founded journal entitled *The Black Scholar* devoted a special issue to the theme "Black cities: colonies or city states" in which the editorial noted that:

> The emergence of the concept of ecology in American life is potentially of momentous relevance to the ultimate liberation of black people. Yet blacks and their environmental interests have been so blatantly omitted that blacks and the ecology movement currently stand in contradiction to each other.[90]

The sociologist Nathan Hare and his colleagues describe how the environmental concerns of wealthy suburbs and poverty-stricken inner cities were not only materially different but also pointed towards contrasting, and to a significant degree irreconcilable, political agendas. Concerns with environmental protection on behalf of the recreation needs of white environmentalists stemmed from a narrow focus on the "chemical and physical or esthetic conditions only," whereas a "black ecology" would raise fundamentally different questions.[91] It is clear, however, that the science and practice of urban ecology have largely remained outside these developments, even as significant strands of work in fields such as environmental justice, toxicology, and access to public space have gathered momentum.

An important exception is the UK-based Black Environment Network (BEN), formed in 1987, which not only addressed issues of access to nature for ethnic minorities but also questioned pervasive xenophobia in environmental discourse. The cofounder of BEN and urban studies scholar Julian Agyeman, who originally studied both geography and botany, made a series of significant early interventions, showing how the UK agencies responsible for improving access to nature were oblivious to social justice issues at either a local or global scale, or to the specific barriers facing inner-city communities of color.[92] Similarly, the sociologist Dorceta E. Taylor carried out research into minority participation in the British environmental movement in 1991. She relates a range of findings, including the degree of hostility and indifference she encountered towards her own work.[93]

In parallel with these early academic contributions we also find some of the first works by British artists to address the intersections between race, landscape, and national identity: the series of photographs taken by Ingrid Pollard in the English Lake District, for example, form part of this emerging discourse. In works such as *Pastoral Interlude* (1987–1988) Pollard explores the disjuncture between the contemplative milieu associated with the nature poetry of William Wordsworth and the sense of unease experienced by ethnic minorities when visiting the same types of "picturesque" rural landscapes. Pollard's work proved a significant source of inspiration for Agyeman, who also draws on an essay by the ecologist James Fenton, published in 1983, that seeks to explore the pervasive hostility towards "alien" plants within nature conservation practice. As we saw in the case of nativist botanical discourses in Germany, the scientific basis for ecological xenophobia is in most cases an ideological response to what is regarded as nonnatural, since most nonnative species do not pose any kind of environmental threat. Fenton suggests that "the scientific reasons put forward against aliens are not the *real* reasons. It is," he continues, "the *concept* 'alien'—or rather 'non-natural'—that is the real reason that introductions are disliked."[94] Interestingly, specific species that have been the focus of conservationist ire such as the evergreen shrub *Rhododendron ponticum* are known to have been part of the British flora in previous interglacial periods. When Agyeman describes groups of conservation volunteers embarking on "rhodo bashing," it is hard to disentangle the ideological and scientific strands to nature conservation practice. Agyeman recalls from his own experience as a conservation volunteer how "the use of a brutal, anthropomorphic language among conservationists in their ecological management work" extended beyond the ecological sphere since it "was not simply related to eradication and removal of nonnatives or alien species, but to 'othering.'"[95] The active targeting of nonnative trees species in Milan, for example, by far-right political activists intent on preventing the "Africanization" of the city, provides a striking illustration of the interplay between racism, xenophobia, and urban nature.[96] For landscape theorists such as Gert Gröning, the organizational dimensions to volunteer-based vegetation clearance programs carry a sinister undercurrent to nature-based forms of political mobilization.[97] The making of more critically reflexive differentiations between specific components of a cultural landscape lies in tension with ideologically driven attempts to create or sustain an imaginary conception of ecological unity and authenticity.

Urban landscapes offer a tiered status of belonging to both human and nonhuman inhabitants. The rendering of the black presence as "out of place" in European or North American cities, for example, rests on a series of exclusionary imaginaries that permeate a set of intersecting environmental discourses in fields such as restoration ecology, conservation biology, and landscape design. The figure of the black birdwatcher, for instance, exemplifies the delicate and sometimes dangerous ecological choreographies of urban space, and the longstanding dissociation between a minority presence and dominant environmental narratives.[98] The articulation of what the geographer Carolyn Finney terms a "black environmental imaginary" must contend with a double essentialization: on the one hand the nonwhite figure is routinely placed outside the European cultural firmament and cast as uninterested in nature or lacking in aesthetic refinement; and on the other hand, the idea of the migrant, irrespective of citizenship status, is allied with conceptions of a "residual rurality" linked to patterns of urban migration and an innate closeness to working landscapes that engenders a more attenuated relation to metropolitan cultures of nature.[99]

The decentering of existing discourses on the urban environment moves beyond the environmental justice lens, with a central focus on differential degrees of corporeal exposure to risk, to encompass a wider set of institutional and ideological articulations surrounding modes of knowledge production and forms of cultural representation.[100] The relative exclusion of both people of color and poorer communities from spaces of urban nature stems from a combination of factors including self-perpetuating patterns of park use that render certain locations inhospitable or threatening. In some cases, the journey to such spaces, or the demographic characteristics of adjacent neighborhoods, can also produce an additional sense of exclusivity. Although most research into "ethno-racially differentiated patterns" of recreation in parks, nature reserves, and other spaces of urban nature has been focused on the North American experience, there are clearly parallels elsewhere.[101] Furthermore, the figure of the lone botanist, birdwatcher, or urban walker connects with the Eurocentric trope of neoromanticist contemplation that is imbued with specific forms of social and cultural stratification.

The intersections between ecology and urban culture reveal the coexistence of different modes of differentiation, spanning both the vernacular and the scientific. For ethnobiologists, the impulse towards the classification of the natural world is regarded as a ubiquitous cultural phenomenon: the point of departure rests on the relationship between direct observation and the deployment of language.[102] Since the 1990s ethnobotanists have become increasingly interested in urban cultures of nature enriched by patterns of migration. Many of these ethnobotanical studies of urban flora serve as a cultural counterpoint to emerging interest in cosmopolitan ecologies.[103] Urban space, as I suggested earlier, can be conceptualized as an "ecological pluriverse" that encompasses myriad cultural, material, and biophysical traces from multiple elsewheres, in contrast with the existing ethnobotanical concern with "traditional," or in some cases premodern, sources of knowledge.[104] The contemporary flora of Detroit, for example, bears the imprint of African-American migration from the South, since many people brought medicinal and culinary plants with them to not only supplement incomes but also sustain botanical traces of collective memory.[105]

Cosmopolitan urban landscapes can be conceived as diasporic spaces of memory that unsettle dominant cultural and environmental narratives: specific traces of nature can connect with "sensual memories" of other spaces and places. These "intimate ecologies," to use the geographer Divya Tolia-Kelly's expression, connect the smells and textures of specific traces of nature such as fruit, herbs, or soil with specific kinds of diasporic longing and belonging.[106] Plants that might be overlooked or reviled under a nativist ecological schema may nonetheless form part of alternative affective geographies of urban space. Some species have acquired radically different connotations for different communities: mugwort (*Artemisia vulgaris*), for instance, originating in Europe and Asia and now naturalized in North America, has been a repeated focus of frenzied removal by so-called "weed warriors" in some American cities, yet it is also highly valued by Korean-Americans for tea making.[107] Practices such as urban foraging reveal the coexistence of multiple cultures of nature within the same metropolitan space. The reconceptualization of urban space as a kind of vernacular botanical garden brings the question of ecological belonging full circle, since spaces of meaning can reside anywhere. In thinking through the implications of cosmopolitan urban ecology we can draw on Stuart Hall's rendition of a "double inscription" emerging from a reconfiguration of relations between the "inside" and "outside" of the colonial project as a global space of flows.[108] An emphasis on ecologies of difference implies not just a more reflexive and historically situated reading of "actually existing" urban ecologies, but also a recognition that cities contain multiple intersecting ecological imaginaries.

FORENSIC ECOLOGIES

The field is not an isolated, distinct, stand-alone object, nor is it the neutral background on or against which human action takes place.

Eyal Weizman[1]

How is the particular fact of *endangerment* produced?

Tim Choy[2]

The new M11 highway that connects Moscow with St. Petersburg has been a focus of bitter controversy. The planned project first came to public attention in 2008, almost by chance, with the discovery of unfamiliar paint markings on trees in the Khimki forest on the northern edge of Moscow. Under discussion since 2004 and shrouded in secrecy, the route of the new motorway would pass directly through the middle of the forest, despite its protected status. Among the habitats under threat were ancient oak woodland, cranberry bogs, and the last undisturbed sections of the Klyazma River floodplain.[3] Studies of the forest revealed a range of rare species listed in the Red Book for the Moscow region, including birds such as the great gray shrike (*Lanius excubitor*) and the spotted nutcracker (*Nucifraga caryocatactes*), along with many endangered plants and insects.[4]

The proposed road immediately provoked intense public resistance, followed by a series of brutal targeted attacks on leading environmental activists. The outspoken local newspaper editor Mikhail Beketov, who published a series of critical articles in *Khimkinskaia Pravda*, was attacked by masked assailants with an iron bar and left for dead in the snow.[5] The leader of the campaign, Yevgeniya Chirikova, who now lives in exile in Estonia after state threats to take away her children, wrote a series of incisive articles highlighting the web of interests behind the project, which extended from local politicians to overseas investors including the French company Vinci, with links to tax havens in Cyprus, Lebanon, and the British Virgin Islands, and the European Bank for Reconstruction and Development (which later withdrew its support under public pressure). The construction project's tentacles reach deep within the Russian state oligarchy: the Prague-based CEE Bankwatch Network revealed that one of the key business magnates behind the scheme, Arkady Rotenberg, is an owner of Stroygazmontazh, a major supplier of gas to Gazprom, and that Rotenberg and his brother were members of Vladimir Putin's martial arts team in the 1960s. The forest had been designated as a site of special scientific interest and formed part of the Moscow greenbelt, but was transformed by presidential decree in 2009 into "vacant land" available for development. The State Forestry Service that had vociferously opposed the scheme was broken up, with conservation responsibilities moved to another ministry.[6] For environmental activists the fight to protect the Khimki forest marks part of a wider struggle to protect vulnerable ecosystems across Russia from commercial exploitation and resist the manipulation of existing forms of legal protection or due process. "It is only a testing ground," warns Chirikova, "for the fine-tuning of methods for the commercial development of conservation areas."[7]

How does the attempt to protect the Khimki forest illuminate contemporary debates over the protection of nonhuman others? For the philosopher Michael Marder, the activists, many of whom used their own bodies to try and save the trees from felling, enacted a form of empathy with vegetal life. "The rationale for their intense commitment," suggests Marder, "is a microcosm of the broader debate surrounding the motivations behind ethical concerns with the environment."[8] Marder argues for an alternative kind of plant-oriented ontology articulated through "an affirmation of the irreducible difference" between vegetal and nonvegetal forms of life.[9] He rejects what he characterizes as a "totalizing vitalism" that stems from an inherently anthropocentric vantage point, and questions the pervasive reliance on degrees of sentience as a marker for subjectivity.[10]

Yet how far can emerging ontologies of the nonhuman illuminate the material dynamics of environmental destruction or identify effective strategies for political resistance? In this chapter I want to explore how specific forms of endangerment become the focus of protection efforts, focusing in particular on different modes of knowledge production within the urban arena. The displacement or decentring of the individual human subject forms a critical starting point for a variety of posthumanist critiques of modernity and its epistemological frames of reference. Yet there are a series of tensions around the explanatory power of these emerging neovitalist formulations, including the need for a more precise reading of agency operating under collective modes of knowledge production. Can the posthumanist emphasis on the "multiplicity" à la Rosi Braidotti and other scholars fully illuminate the state-capital nexus and its associated constellations of cultural, political, and ideological power that underpin the scale of violence against nature and its defenders?

The protection of urban nature oscillates around a series of material, conceptual, and organizational vantage points. Earlier interventions such as the safeguarding of fragments of nature in the face of development pressures to create parks or greenbelts have been supplemented by emphasis on endangered species, fragile ecosystems, and the recognition of novel biotopes. There is a complex matrix of vulnerabilities that has gradually come into view through the efforts of activists, artists, scientists, and others. The evaluation of risk facing nonhuman life includes the use of classification systems such as Red Lists to highlight endangered species or the deployment of more elaborate indices to emphasize vulnerable biotopes.[11] Much of what we know about urban nature comprises a grassroots archive of notebooks, photographs, and field observations, in some cases undertaken over many years. Yet these place-based insights are fragile: many urban sites are under intense threat of damage or erasure, some spaces are relatively inaccessible, and knowledge can easily be lost or dispersed.

The protection of urban nature extends to diverse socioecological assemblages ranging from fragments of preexisting ecosystems to various types of novel biotopes such as postindustrial wastelands. But what kind of arguments can be used to mobilize political support for the protection of urban biodiversity? And who has the power or expertise to interpret ecological data so that it can be related to a wider set of legal or institutional frameworks? Although some organisms may enjoy a degree of protection due to their symbolic status within metropolitan cultures of nature, there are far more species that are largely restricted to the realm of scientific or taxonomic curiosity.

4.1 FORENSIC URBANISM

The intersection between a field of observation and political action relates not only to the adoption of particular data collection strategies but also to the deployment of specific modes of interpretation. In this chapter I introduce the term "forensic ecologies" to bring together two distinctive epistemological frameworks derived from "forensic entomology" and "forensic architecture."[12] Forensic entomology uses the extremely precise environmental data available from the study of insects to help with the reconstruction of crime scenes. The earliest recorded case of a crime solved with the help of insects took place in thirteenth-century China where the lawyer Sung Tźu identified a murder weapon (and the murderer) using the attraction of flies to traces of dried blood. In early modern Europe, advances in observational science included a range of studies on the relationship between insects and corpses: the seventeenth-century Italian biologist and physician Francesco Redi, for example, made some of the first studies of insect eggs, dispelling earlier notions about the spontaneous generation of life (and thereby repudiating aspects of Aristotelian natural history).[13] During the nineteenth century there was increasing interest in the fauna associated with graves and human cadavers. A leading pioneer was the veterinarian and entomologist Jean-Pierre Mégnin who described the flies, beetles, moths, and other organisms associated with different stages of decomposition. In his classic study *Le faune des cadavres*, first published in 1894, Mégnin noted that these "travailleurs de la mort" (workers of death) not only carried out specific ecological roles but always arrived at the scene of death in a precise sequence.[14]

Forensic entomology reduces uncertainty over the timing, and in some cases the whereabouts, of death, yet also illuminates the ways in which the subjective dimensions to human judgment can never be eliminated from scientific method. Unlike much of the natural sciences, the epistemological significance of the human standpoint is widely recognized within critical legal studies.[15] Similarly, a number of prominent statisticians have insisted that probability cannot be simply derived from abstraction but must be rooted in an understanding of human practice, so that the calculation of probabilities can serve as a tool to help refine human judgment in complex situations.[16] With new advances in DNA analysis, however, the scientific status of forensic entomology within legal settings has become less certain.[17]

Insects provide a unique set of insights into the shifting contours of the urban and environmental arena: by scaling up specific indicators or sentinels towards more elaborate multispecies indices, we can build a blinking panel of ecological warning lights. By developing an expanded conception of forensic entomology, the scope and purpose of this scientific practice can be applied to the damaged Earth and its ecosystems. We can use insects to explore the unfolding crime scene of environmental destruction at a global scale, drawing on existing methods used within Quaternary science and other fields to reconstruct past patterns of environmental change based on specific assemblages of insect species.[18] In a contemporary context, the changing behavior, distribution, and population dynamics of insects provide a multidimensional set of insights into climate change, habitat fragmentation, and the effects of neonicotinoids and other toxins. Indeed, mass invertebrate decline has become one of the key markers for what has been dubbed the "sixth mass extinction."[19] These invertebrate sentinels provide an eerie portent of global ecological decomposition that has profound political significance. The pervasive destruction of the web of life represents a form of "slow violence" whose effects ripple through time and space.[20] The decomposing body that was a source of fascination for Mégnin has become that of the Earth itself.

Turning now to the example of "forensic architecture," developed by the architectural theorist Eyal Weizman and his colleagues, we find that the term "forensic" brings together three main elements: first, the "field" as a site of interest or contestation; second, the "laboratory" or studio space within which evidence is analyzed and evaluated; and third, the "forum" or deliberative setting where findings can be presented and scrutinized, such as exhibitions, legal proceedings, and web platforms.[21] Weizman traces the etymological origins of the word *forensic* to the Latin *forensis*, relating to the *forum* or marketplace. Of particular interest, however, is Weizman's emphasis on the Roman orator Quintilian and the use of "prosopopoeia" to give "voice to things that nature has not given a voice."[22] A key concern of forensic architecture is how obscure objects or complex events can be made visible for public scrutiny. In its English usage, "forensic" evolved from the early modern period onwards into a more specific form of legal deliberation over the use of evidence: the more open sense of a public forum in its earlier iteration became successively narrowed to that of the physical space of the courtroom. The term "forensic" became increasingly restricted to the scientific analysis of evidence used in legal deliberations, exemplified by the rise of forensic entomology and related fields during the twentieth century. In contrast, Weizman seeks to recover the earlier meaning of forensic as a form of public deliberation that might enable the inversion of the "forensic gaze" towards rather than from the state apparatus, corporate actors, and their powerful interlocutors. His emphasis on "counterforensics" makes explicit that the tools of forensic analysis are to be used against perpetrators of state-sanctioned violence. Yet the wider connotations of this critically reflexive "forensic turn" become even more complex in situations where legal jurisdictions are unclear, degraded, or simply absent.[23]

At a conceptual level Weizman's project moves in a different direction to the relational ontologies of Latourian actor-network theory, the neovitalist emphasis on the nonhuman, or the neoromanticist aura that surrounds recent interest in object-oriented ontologies.[24] Though Weizman acknowledges the political salience of the animist-inspired extension of legal rights to nature, he recognizes the complexities of navigating between Indigenous knowledge and modern environmentalism.[25] Forensic architecture marks a turn away from the language-oriented poststructuralist legacy in architectural theory towards a postpositivist engagement with evidentiary forms of materialism.[26] There is a sense of dissatisfaction with the existing corpus of architectural theory as predominantly prescriptive, performative, or speculative.[27] The insistence on a radical contextualization of architectural practice holds continuities with the earlier critical traditions elaborated by Aldo Rossi, Manfredo Tafuri, and others yet is radically extended outside the crisis of modernism as the primary field of contestation. The move beyond a reliance on the human sensorium incorporates a wider range of nonhuman and material sensors: the environment itself becomes a kind of recording device or "material witness" to historical events derived from a panoply of different elements.[28] Under this critical lens, material discontinuities do not constitute relativist uncertainties: an acknowledgment of the "confusion of traces," to use the anthropologist Gastón Gordillo's expression, does not preclude a rigorous investigation of how specific geographies have come into being.[29] Weizman's emphasis on the use of buildings or other material traces of human activity marks a transcendence of the classic art historical focus on the individual human subject as an aesthetic observer.[30] The question of agency—distributed or otherwise—is framed in terms of not only causality but also responsibility.

The forensic approach enables an elaborate reconstruction of the socio-ecological characteristics of specific sites that can extend the use of existing indices to the historical and political dynamics underlying the production of space: there is a movement from measures of endangerment towards delineating patterns of causality, agency, and responsibility. Weizman introduces the term "field causality" to encompass "indirect forms of causality" that are "distributed over extended spaces and time durations." The practical and intellectual task is "to reconnect the multiple threads that linear juridical protocols have torn apart."[31] Yet these complex patterns of causality are ultimately matters for political rather than juridical deliberation, since their scope clearly exceeds the actions of a few culpable individuals. Of particular interest is the potential role of nonhuman sensors in reconstructing the environmental impact of historical events. Plants, for example, can serve as a valuable guide to field causality, as illustrated through Weizman's investigation of vegetation patterns left by ruined villages after the genocide of the Ixil people in Guatemala between 1978 and 1983, where the occurrence of "trees such as avocados, papayas, or peach signal the possible former presence of houses and village sites."[32]

Vegetation patterns have long served as sensors or indicators for human activity in archaeology, botany, palynology, and other fields. Archaeological studies of Roman London, for example, have used seeds, pollen, and other botanical traces to enable the reconstruction of everyday life. We can also discern the contours of an urban ecological formation that included adventive plants drawn from across the Roman Empire such as corn buttercup (*Ranunculus arvensis*) and caper spurge (*Euphorbia lathyris*) as well as species cultivated for food such as almonds, peaches, and apple varieties. Abandoned middens—the word is of medieval Scandinavian origin for dung heaps or cess pits—can be spotted by clusters of nettles (*Urtica dioca*) and other plants that thrive on elevated levels of potassium in soil.[33]

Lichens, a composite type of organism that combines fungi with algae or cynanobacteria within one structure, have been used as highly sensitive indicators for air pollution since the nineteenth century (figure 4.1).[34] In a novel study of the Jardin du Luxembourg, Paris, undertaken in the early 1860s, the botanist William Nylander explains how "the lichens give, in their own way, a measure of the *salubrité* [salubrity] of the air, and constitute (so to speak) a kind of very sensitive hygrometer [a kind of scientific measuring device for atmospheric conditions]."[35] The sociologist Jennifer Gabrys notes that these symbiotic assemblages do not comprise a species in the conventional sense but constitute an "ecological microcosm" in and of themselves.[36] For Gabrys, the use of bioindicators provides an opportunity to "go beyond representational modes of politics."[37] But what is implied by "representational" in this context? The impetus behind forensic ecology leads towards a critical representational practice, rather than an elaboration of the nonrepresentational realm. Or, as Weizman notes, in response to epistemological disputes over scientific truth claims, the forensic approach is best conceived as "a mode of problematization that intensifies the research process."[38]

Beyond vegetation patterns or palynological traces we can surely add a wider range of organisms to the roster of human-nonhuman collaborations involved in the biomonitoring of environmental change. The changing migration dynamics of birds, for instance, have implications for biodiversity and epidemiology in the urban arena. Research into the emergence of avian flu and other zoonoses in Hong Kong, for example, has highlighted pathways from wildfowl to humans. Consequently, there is a biosecurity emphasis on detecting avian flu mutations in wild birds before they have a chance to spread via chickens and pigs to the human population. The anthropologist Frédérick Keck uses the term "sentinel" in connection with concerns over biodiversity and biosecurity: the Mai Po wetlands in Hong Kong became not only a focal point for international conservation efforts to monitor and protect migratory pathways but also a

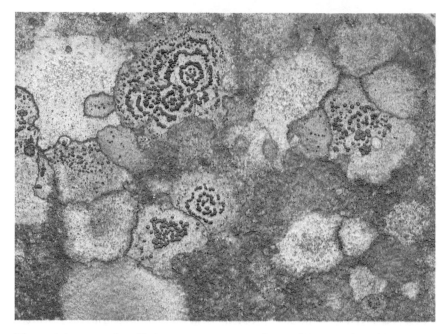

Figure 4.1   Examples of lichens growing on an urban wall (2006). Photo by Margaret Sixsmith.

site for the potential detection of zoonotic threats to human health.[39] Keck describes the transformation of the territory of Hong Kong since the 1970s into a "sentinel post" for the monitoring of avian influenza. For Keck, a pivotal question in relation to the future of zoonoses is the degree of preparedness that spans both human and nonhuman realms, and the development of new forms of collaboration between microbiologists and urban birdwatchers.[40]

The emergence of urban ornithology encompasses changing grassroots cultures of nature set within the evolving dynamics of global nature conservation. Ornithological societies in Hong Kong and Taiwan, for example, gradually morphed into campaigning environmental organizations from the late 1980s onwards, with much larger memberships. In Taiwan, environmental activists were increasingly pitted against landowners and development interests that in some cases deliberately sought to drive birds away. Initial emphasis on "flagship species" such as the fairy pitta (*Pitta nympha*) faltered, however, because of the complexity of the bird's migratory pathways, with the greatest threats to it occurring elsewhere (in this case from deforestation in Borneo). In contrast, the black-faced spoonbill (*Platalea minor*) in the threatened Qigu wetlands in Taiwan provides a clearer focus for urban conservation efforts: this highly endangered bird is now intensely monitored, with some two-thirds of the global population passing through Taiwan as winter migrants. Satellite tagging of individual birds allows the tracking of movements through a series of vulnerable urban wetlands located in China, Korea, Japan, Taiwan, and Vietnam (figure 4.2). "A bird with a satellite tag," notes Keck, "is more than a flagship species: it becomes part of the human collective as it is equipped to send warning signals of extinction."[41] For Keck, the monitoring of birdlife through what he terms "avian reservoirs" lies at the interface of ornithology and virology; it reveals a zone of observation that is concerned with the fragility of the web of life—indicated by measures of endangerment—but also points to the biopolitical and epistemological interface between human and nonhuman occupants of urban space.[42]

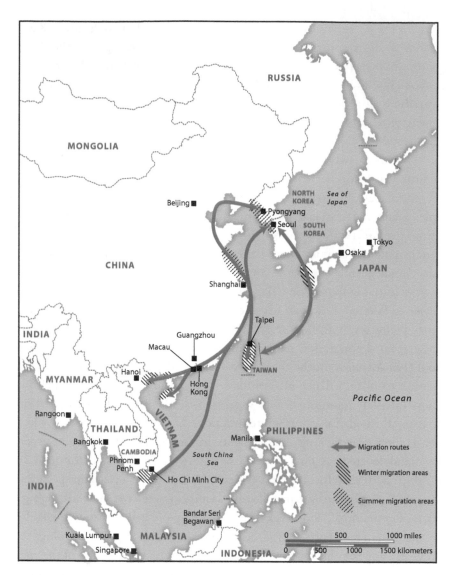

Figure 4.2 Migratory pathways of the black-faced spoonbill (*Platalea minor*). Data derived from various sources including the Partnership for the East Asian-Australasian Flyway. Cartography by Martin Lubikowski.

In many cases, site irreplaceability lies closer than species extinction per se to the dynamics of urban biodiversity as a generator of socioecological novelty or complexity.[43] Yet the presence or absence of a particular organism, or a composite index derived from multiple taxa, can provide invaluable insights into the degree of ecological significance of specific sites. These in-depth studies are more than a matter of scientific curiosity, since they can also play a role in the political arena when grassroots observations become incorporated into legal instruments through the production of what the sociologist Andrew Lakoff terms "biological opinion."[44] But how have comparative indices for levels of vulnerability or endangerment been used (or ignored) in an urban context? The idea of biodiversity as a fragile constellation of different elements involves a distinction among organisms, fragments of ecosystems, or specific biotopes that should be prioritized: it comprises an institutionally framed set of judgments about the relative worth of different elements of nature. Yet at the same time the concept of biodiversity encompasses the functional integrity of the biosphere as a whole. Endangered dimensions to urban nature might include remnant ecosystems such as dunes or estuaries as well as novel biotopes of cultural and scientific significance including ruins or wastelands. The question of scientific value often rests on indices of uniqueness and vulnerability such as measures of the ecological value of woodlands based on the presence of saproxylic invertebrates associated with old trees, or of grasslands using measures based on the diversity of orthoptera (grasshoppers and crickets). Specific species such as the stag beetle (*Lucanus cervus*) may take on symbolic significance, whilst other less charismatic organisms, such as fungus gnats or ground beetles, may have a much more complex or ill-defined relationship with public culture.

The attempt to extend the Linnaean taxonomic system to the classification of ecological zones or distinctive types of vegetation marks a territorialization of the species concept.[45] Yet cities, as we have seen, with their proliferation of novel biotopes, have consistently unsettled existing classificatory schemas for plant communities. How far can we apply ecological concerns with endangerment to urban space? The expression "ecologies of endangerment," as used by the anthropologist Tim Choy, has many possible ramifications in the urban arena. "A different scaling of the problem," notes Choy, "designates a different unit for life's analysis and definition: a population, an ecosystem, a cascade of energy's transformations, a cycle of carbon, nitrogen, or other elements."[46] Similarly, Fernando Vidal and Nélia Dias describe how the status of "endangerment" is marked by a series of incorporations into "archives, catalogues, databases, inventories, and atlases."[47] They evoke an "endangerment sensibility" that suffuses the affective realm of late modernity with the actualities or possibilities of loss.[48] We can trace a relationship between heightened levels of "corporeal affectivity" and the development of more nuanced ethical relations towards nature.[49]

The use of indicator species, and the frequency of the term "indicator" in scientific literature, have grown rapidly since the 1980s, yet many analytical and methodological uncertainties remain.[50] Over what scales or temporalities, for instance, can an indicator species hold its significance? Data on indicator species can be scaled up to produce more elaborate indices marked by gradations in ecological value, from regional to national and ultimately international significance. But is a small or isolated urban population of an ecologically vulnerable species of greater conservation value than other populations in less threatened parts of an organism's range? Cities can serve as refugia for species that have become diminished within agroindustrial deserts beyond the urban fringe: examples include the skylark (*Alauda arvensis*) that now nests in urban grasslands; the presence of water voles (*Arvicola amphibius*) along urban canals; and various species of bumblebees that thrive on flower-rich wastelands.[51]

The protection of biodiversity rests on an uneasy synthesis between the ostensible universality of classificatory schemas and the grounded materialities of environmental change. Knowledge about urban nature represents a bricolage of colonial and postcolonial elements drawing on Eurocentric epistemological and taxonomic framings of the natural world combined with diverse local and vernacular elements. Only certain types of place-specific knowledge appear to resonate with the universalist precepts underlying global environmental discourse. In the case of biodiversity protection in Hong Kong, for example, Choy emphasizes the subtle choreography of expertise in the public arena including the role of gestures, prosody, and the intricacies of simultaneous translation as individual testimonials oscillate between different linguistic and epistemological terrains. Choy notes how political and economic circumstances "have enabled only some people—and what they know—to count as the particular counterpoint to global environmentalism's universality."[52]

Following the insights of the philosopher Miranda Fricker, we can identify forms of "epistemic injustice" that frame different degrees of access to data, interpretive resources, and the opportunity to make credible contributions to policy discourse.[53] Similarly, the philosopher Kristie Dotson draws on Gayatri Spivak's conceptualization of "epistemic violence" to explore processes of knowledge elimination. Dotson identifies two forms of silencing: firstly, that of "testimonial quieting," where "an audience fails to identify a speaker as a knower" through the operation of various kinds of stereotyping; and secondly, "testimonial smothering" arising from various forms of self-silencing in the face of specific situations produced through ignorance on the part of a potential audience.[54] Environmental discourse is riven by epistemological inequalities that shape the interface between science, politics, and public culture. Under what circumstances will the figure of the urban ecologist or grassroots activist be taken seriously? What structural factors might undermine the testimony of an individual contributor to public environmental discourse? And how are different sources of credibility identified or sustained? As Weizman shows, using the lens of forensic architecture, there is a vital role for analytical and representational tactics that can strategically outmaneuver an array of powerful, sophisticated, and well-funded adversaries. The recognition of "tactical epistemologies" brings the biological sciences and legal theory into closer dialogue by highlighting how politics suffuses scientific discourse in the public arena. Making the purpose of research explicit, for example, including sources of funding, adds further nuance to the meaning of objectivity at the interface between science and politics.

The role of legal instruments as potentially counterhegemonic tools, bolstered by grassroots data collection, has gained significance in the context of new forms of authoritarian populism and rising levels of antienvironmental violence. But what counts as truth or evidence in the field of environmental law? How does the segue from grassroots activism to litigation operate in practice? The role of the "ecological citizen" as a litigant has been pivotal to the postwar evolution of environmental law. The increasing significance of grassroots action, especially in North America, is related to a more pluralist interpretation of power that extends to the impact of environmental externalities, systems of political patronage, price fixing, and state subsidies.[55] The potential scope of environmental action has clearly moved beyond the reformist agendas instilled by earlier public health advocates and the evolution of municipal forms of governmentality. Writing in a North American context, the legal scholar Zygmunt Plater notes:

> More than any other area of the modern legal system, environmental law has developed its complex, extended, doctrinal structure in a process dependent upon confrontational, pluralistic citizen activism, operating in every area of governance, but particularly in judicial and administrative litigation.[56]

Plater does not discount the many people within regulatory agencies who have shared environmentalist concerns but suggests that, without external pressure emanating from activists such as Rachel Carson, the full recognition of the complex web of ecological impacts caused by agriculture, forestry, industry, and other sectors would have been ignored to a much greater extent. He suggests that the development of environmental law would have been much more hesitant and piecemeal in its scope without "relentless and sophisticated citizen litigation."[57] A refocusing of environmental discourse towards urban environments from the late 1960s onwards marks a shift of emphasis from a preservationist ethic, and various forms of "ecological zoning," towards a series of structural issues affecting diverse communities such as childhood lead poisoning in major metropolitan areas and cancer clusters associated with petrochemical industries.[58] These campaigns rest on the possibility of combining credibility with causality so that existing sources of scientific knowledge can inform political discourse. A characteristic dimension to this environmental activism is the combination of grassroots observation with insights from peer-reviewed scientific studies in epidemiology, toxicology, and other fields. Yet the political, legal, and epidemiological dimensions to environmental threats are complicated by cumulative and interactive dimensions to risk.[59] Activists must contend with a complex of causalities operating over different scales and temporalities that transcend the strictures of routine scientific studies. Activities such as the monitoring of air quality, looking for signs of tree disease, or involvement in cartographic projects hold wider implications for knowledge, science, and political praxis. The blurring of distinctions between controlled laboratory spaces and various types of field sites can raise tensions over what constitutes legitimate scientific practice.[60] In recent North American campaigns, coordinated by groups such as the National Black Environmental Justice Network, connections have also been made between the companies and utilities that are responsible for polluting communities of color, the sources of banking and finance for these industries, and the extensive donations given to police foundations that support increasingly militarized modes of policing against environmental protests.[61]

The use of legal instruments to protect vulnerable ecosystems has also emerged in the global South. In India, for example, the role of citizen participation in lawmaking has been closely associated with the rise of public-interest litigation (also known as social action litigation) that has been used in a variety of environmental campaigns since the 1980s, and especially in the wake of the Bhopal chemical plant disaster of 1984.[62] Public-interest litigation has developed in part because of concern with the limited legal and regulatory capacities of the Indian state to respond to environmental concerns. These legal instruments have been used to protect vulnerable ecosystems such as lakes or urban wetlands from development pressures (with varying degrees of success), but the meaning of the "public interest" is often ambiguous, with the same legal procedures used to threaten marginalized communities as part of middle-class efforts to "clean up" urban space.[63] The articulation of a "public" discourse in relation to the urban environment is highly fractured and riven by preexisting forms of class, caste, ethnic, and religious difference.[64] In the case of "bourgeois environmentalism" there are situations where middle-class groups who had previously benefited from lax planning controls then reposition themselves as arbiters of environmental protection.[65] In Mumbai's Sanjay Gandhi National Park, for example, public-interest litigation has been used against the poor on the pretext that encroachment is the main cause of environmental damage. The park, which was first established in the early 1940s, now covers an area of about 87 square kilometers, but the fringes are under constant development pressure, including attempts to clear vegetation by the use of fire to create new plots of land. Legal action by the Bombay Environmental Action Group has brought about a series of evictions, but these disproportionately affect informal settlements rather than illegally constructed restaurants, wealthy bungalows, and other commercial developments. Indigenous residents of the forest area, whose presence long predates the growth of Mumbai, have been categorized as "unauthorized" along with slum dwellers and other marginalized groups.[66]

The limits to a public-interest framing for environmental conflict are revealed by more radical challenges to the state-capital nexus. There is a shifting geography of law and its instruments that renders protestors against land grabbing, pollution, or other environmental threats increasingly vulnerable to rising levels of harassment and violence.[67] Globally, the highest levels of violence against environmental campaigners and defenders of nature are concentrated in Latin America, with further epicenters of violence in the Democratic Republic of Congo and the Philippines.[68] The combination of violence, murder, and intimidation with legal chicanery and regulatory neglect contributes to a sense of affective resignation.[69] Environmental law is a dynamic and contested field that is marked by a series of geographical and institutional limits, where the boundaries of legal jurisdiction are exposed through acts of intimidation, violence, and the exercise of alternative sources of power. The interpretation of legal and regulatory geographies, and their ghostly interstices, is suffused by what the geographer Nick Blomley has termed "compounded opacity" in which law and space cannot be conceptualized as autonomous fields.[70] Similarly, the legal theorist Boaventura de Sousa Santos identifies "a symbolic cartography of law" that marks the emerging significance of space within critical legal studies. For Santos, legal space draws on a series of pluralistic, relational, and to some degree metaphorical modes of thinking since "the conception of different legal spaces [is] superimposed, interpenetrated and mixed in our minds as much as in our actions."[71] Intersecting spheres of "juridical capital" operate at different scales of intervention, ranging through diverse local contexts to the international arena, marked by multiple tensions between doctrinal and procedural forms of legal deliberation.[72]

4.3 THE URBAN FIELD

By the nineteenth century the distinction between "natural history," still dominated by amateur enthusiasts, and the institutional framing of the "natural sciences" was becoming more apparent.[73] The professionalization of botany, entomology, and other fields, partly in response to the strategic needs of the modern state in relation to agriculture and pest control, led to a shift of emphasis away from curiosity-driven taxonomic research.[74] Yet the field of natural history, as a grassroots form of scientific activity, has continued to flourish through the latter half of the twentieth century and beyond, even as its institutional support structures and scientific status have been in decline.[75] Interest in urban natural history—especially birds—appears to be thriving, spurred by technological advances in digital media and the sharing of information. A fascination with urban birds has become an increasingly global phenomenon spreading beyond established centers such as London and New York.[76] Hundreds of metropolitan areas now have active ornithological societies sharing information on urban birds, including Chennai, Kigali, Manila, and many other cities in the global South.

In Chennai, for instance, the flourishing interest in urban birds is not just a middle-class phenomenon but extends to a thriving Tamil public sphere.[77] Natural history societies that were once the preserve of colonial-era elites have had an influx of new members, and have in some cases widened their remit to raise public awareness of threats to biodiversity such as the destruction of vulnerable ecosystems including urban wetlands. The resurgence of grassroots interest in urban nature could also be interpreted as a partial reprise of its preinstitutionalized phase, in the sense that the separation between "field-" and "laboratory-" based modes of environmental knowledge production, instituted from the late nineteenth century until the last quarter of the twentieth, marks something of a scientific interregnum within a longer history of human engagement with the natural world. The flourishing of natural history across much of Europe from the late eighteenth century into the first half of the nineteenth century can be read as a grassroots scientific dimension to the egalitarian political foment of the era, as observational science became popularized through a series of widely disseminated publications and the local sharing of knowledge. In its early phase, natural history had to contend with advances in taxonomy and questions of nomenclature, as well as the evolving relationship between lay forms of existing knowledge, such as the identification of medicinal plants, and the gradual evolution of institutionally framed scientific practices.[78]

Almost any urban site can become a focus of curiosity. In 1941, for example, an entomology curator for the American Museum of Natural

History in New York, Frank E. Lutz, published a book detailing a series of "expeditions" just a few meters from his home.[79] Another striking example is Jennifer Owen's thirty-year investigation of her back garden in the city of Leicester in the English midlands, begun in the early 1970s. Owen's meticulous study recorded over 2,200 species of insects, including 20 that had never been found in Britain and four that were new to science.[80] The biodiversity of this single garden compares favorably with one of the UK's most intensively studied nature reserves at Monks Wood, Huntingdonshire, an ancient woodland site that is around 1,000 times larger.[81] This is not to argue that localities such as Monks Wood are less important than ecologists might claim: the question is one of openness towards different types of study sites and novel forms of scientific practice. What is especially striking from Owen's long-term study of her garden is a significant decline in invertebrate biodiversity that mirrors the results of other long-term surveys elsewhere.[82] In a contemporary context, with growing interest in mass invertebrate decline, the existence of independent longitudinal studies in "urban field stations," such as that undertaken by Owen, has suddenly acquired critical significance as an additional source of evidence for environmental change.

The Linnaean classification system is a specific kind of scientific language that enables the representation of nature. It is a cultural tool that furthers precise forms of human communication, even if its origins lie shrouded in forms of European cultural hegemony including the use of slave routes to acquire scientific specimens.[83] The formalized binomial language of natural history is a Janus-faced cultural artifact that emerges out of the violent origins of global modernity yet also serves as a rich source of shared knowledge about the web of life. The ability to recognize different species, and place them within this cultural framework, can be likened to the experience of learning a new language. The urban landscape can be recast as a complex, dynamic, and continually surprising kind of textual interface. The interpretation of urban space as a collage of words is alluded to by André Breton, René Char, and other enthusiasts for experimental walking in the 1920s and 1930s.[84] Similarly, the idea of the city as text connects with an emphasis on signs and symbols in the urban landscape and concerns with the legibility (or otherwise) of urban space. For some readers of the street, such as Franz Hessel, the role of the flâneur or urban observer is relatively straightforward: various human and nonhuman elements such as trees and human faces are simply characterized as letters of the alphabet.[85] In contrast, his contemporary Siegfried Kracauer conceived of the city as a mix of intentional and unintentional landscapes, in a conceptual formation that prefigures more recent interest in the intersections between the independent agency of nature and urban design.[86] In methodological terms, the idea of the city as an archive owes much to the cultural archaeology of Walter Benjamin and the interweaving of different stratigraphic forms rooted in history, memory, and the imaginative scope of modernity.[87] For Benjamin, nature serves as a kind of historical archive, with fragments and traces scattered all around. But if we loosen our ties to the Linnaean classification systems, to the set of taxonomic constellations by which Benjamin and others have navigated the landscapes of urban nature, what are the wider implications for the field of urban ecology? Can a postcolonial taxonomic sensibility hold in play these tensions between past and present sources of knowledge?

The city can be envisaged as a kind of laboratory in its own right; or rather, as a series of overlapping or intersecting fields of curiosity ranging from the microecology of pavements or balconies to more complex multisite investigations. The practice of urban ecology unsettles the methodological contours of scientific knowledge and existing distinctions between amateurs and experts (see figure 4.3). The term "amateur," for instance, can be misleading, since some self-taught specialists have acquired an international reputation for their taxonomic expertise and local activists often have far greater knowledge of specific sites than regulatory agencies.[88] The grassroots study of urban nature can supplement the small numbers of professional staff employed by museums and nature conservation organizations. The elaborate data collection programs organized by the Cornell

Figure 4.3　One of the displays at the annual exhibition of mushrooms organized by the Société Mycologique de France in the Parc Floral de Paris (2009). Events of this kind blur the distinction between experts and amateurs through the enrichment of public cultures of science and taxonomy. Photo by Matthew Gandy.

Lab of Ornithology, the British Trust for Ornithology, and the Botanical Society of Britain and Ireland rely on many thousands of volunteers to help build up their databases. The UK-based Big Garden Bird Watch, with some 9 million observers, is among the largest citizen science projects in the world, including the widespread participation of children.

These shifting hierarchies and practices of knowledge production are in part necessitated by lack of research funding but also through a reliance on vast numbers of volunteers to provide the necessary data. In some cases we find the development of "urban parataxonomy," especially for more obscure or "difficult" groups of organisms, where the scope for public participation is more circumscribed and knowledge hierarchies more persistent.[89] The Cornell Lab of Ornithology runs several types of citizen science projects that combine pedagogic and scientific dimensions to data collection. In scientific terms, however, the evaluation of achievement is often skewed towards conventional metrics whilst the pedagogic dimension is rooted in "public scientific literacy" that is very much driven by goals such as "improved participant understanding of science."[90] The idea of science itself as a contested field, rooted in specific institutional contexts, is routinely occluded in these centrally directed surveys, along with the potential role of science in public policy deliberation: what constitutes science, the scientific process, or the evolving interface between science and politics lies outside the field of reference. Participants in data collection schemes typically have little control over how their data is used or what questions are posed. Grassroots ecologists are sometimes wary about sharing information such as cherished local lists of birds, plants, or other organisms, often compiled over many years, with centrally controlled databases that can be "mined" by environmental consultants or private companies.[91] Indeed, as the geographer Rebecca Lave has shown, there is an emerging "horizontality" to knowledge claims, as forms of scientific expertise, commercial knowledge production (or acquisition), and grassroots data gathering begin to fracture the existing hierarchies between science, citizenship, and environmental discourse.[92]

Citizen science is a complex epistemological terrain that spans grassroots ecological survey work as well as the campaign-oriented legacy of environmental justice activism: the politics of urban biodiversity brings these two strands into closer articulation through the figure of the activist urban ecologist who combines taxonomic knowledge with the use of legal instruments. The study of urban nature is caught between the centrifugal dynamics of greater democratization of knowledge through practices such as "popular epidemiology" and "street science" (to use Lave's terms) and the growing commercialization of environmental expertise, including new institutional configurations operating outside established networks of universities, museums, and research centers. In some fields, such as architecture and landscape design, the commercial efficacy of ecological knowledge is at least as significant as established peer-review procedures within scientific journals: the appropriation of ecological rhetoric provides multiple examples of what the philosopher Jean-François Lyotard has referred to as the supplanting of scientific "truth" with market worth.[93] Equally, we might ask what it is worth *not* to produce knowledge, as in the absence or even suppression of ecological data collection that might disrupt land use planning decisions (there are clearly parallels here with the uneven regulatory landscapes for many forms of environmental pollution). The outcome of these developments remains mired in a degree of epistemological uncertainty since the efficacy of environmental knowledge in the political arena, as we have seen, extends to multiple forms of rhetoric, performance, and structural inequality. What the sociologist Thomas Gieryn refers to as the "cultural cartographies of science-in-culture" highlight how different degrees of epistemic authority are located in a diverse set of local and historical circumstances.[94] Since the study of urban nature remains a predominantly field-based endeavor, which is richly informed by grassroots scientific practice, this adds further complexity to the wider cultural resonance of different forms of epistemological veracity.

Grassroots ecology holds parallels with citizen science, but not necessarily in a normative sense. Citizen science is often promoted as a means to enrich "scientific literacy" and contribute towards greater public participation in policymaking.[95] Yet these wider objectives are far from straightforward: there are competing definitions of "environmental democracy," and the question of scientific literacy is routinely framed within narrow epistemological or taxonomic parameters. Hidden within state-led enthusiasm for citizen participation we can often find a strategic emphasis on reducing public expenditure. Grassroots data collection can shift regulatory responsibilities from the fading capacities of the expert state onto civil society. Less certain, however, are situations where the purpose of grassroots data collection begins to diverge from externally set objectives, or where the focus moves beyond the monitoring of environmental change towards systemic forms of critique. The very notion of "environmental citizenship" is complicated by an emphasis on the individualized human subject under modernity, which looks increasingly tenuous in the face of emerging challenges posed by a variety of posthumanist, postcolonial, and Indigenous vantage points.

Grassroots data collection can operate independently of state agencies, mainstream science, or established environmental organizations, thereby exposing regulatory voids marked by varying degrees of official indifference.[96] Examples include the work of the so-called "bucket brigades" in North America to monitor air quality (the moniker is derived from improvised ensembles of affordable everyday items with scientific equipment). Beginning in the predominantly African-American community of Diamond in Louisiana, and then spreading elsewhere, these grassroots monitoring campaigns have led to a series of successful legal actions against major polluters.[97] A recent example is the work of residents in the city of Denton, Texas, located just north of the Dallas-Fort Worth metropolitan region, in response to the impact of nearby fracking activities. The failure of regulatory agencies to measure the impact of gas extraction led to crowdfunded forms of monitoring, deploying the same equipment and procedures as those used by state authorities.[98] The enactment of what we might term "regulatory mirroring" holds similarities to the counterhegemonic role of forensic ecologies. This is not merely a matter of following verifiable scientific procedures but also a means to counter the inadequacies of available expertise, including the political capture of regulatory agencies.

How do broader concerns with environmental justice and threats to human health connect with the more specific aim of protecting urban biodiversity? In addition to regulatory voids, we encounter taxonomic voids created by the lack of resources to carry out research into biodiversity.[99] The limits to the cataloguing of the natural world can be considered an example of what the philosopher Andrew Feenberg refers to as "epistemological finitude" in recognition of practical constraints on advancing scientific knowledge.[100] Funding for taxonomic research, so vital to understanding biodiversity, has been in sharp decline almost everywhere, along with reduced numbers of specialists in museums to look after collections or even sort through existing materials. A combination of neglect and underinvestment contributed to the devastating fire at the Museu Nacional in Rio de Janeiro, in September 2018, in which millions of specimens and cultural artifacts were lost, mirroring earlier fires that destroyed the Instituto Butantan in São Paulo in 2010 and the National Museum of Natural History in Delhi in 2016.[101] These conflagrations have left major scientific archives in ruins. The future of taxonomic research must somehow emerge out of the debris of its colonial past.

Cooperation between institutionally based scientific research teams and grassroots data gathering can have wider political implications. The sociologist Jens Lachmund describes the emergence of a "biotope-protection regime" in West Berlin during the 1970s and 1980s that brought scientific insights from urban ecology into the heart of land use planning.[102] For Lachmund, this distinctive "nature regime," which he models loosely on the typology developed by Arturo Escobar, brings together three main elements: first, the presence of specific types of urban nature that are considered worthy of protection; second, the innovative combination of taxonomic expertise and methodological practices that can recognize these vulnerable elements of urban nature; and third, the presence of "individuals, collectives, and institutions that assemble around and actively sponsor these claims."[103] West Berlin emerged as a global epicenter for urban ecology in the 1970s and 1980s: the urban enclave was unusual in many ways, with relatively low demand for land set within a progressive social and political milieu, including the presence of one of the first radical environmentalist political parties.[104] During the same period, a series of *Bürgerinitiativen* (citizens' initiatives) had mobilized to defend existing green spaces against the construction of new motorways and other developments.[105] In some cases neighborhood activists linked up with urban ecologists to protect specific sites on the basis of accumulated scientific evidence backed by grassroots public support (as we saw in chapter 3).[106] Yet the longer-term success of this phase of urban environmental discourse, especially after German reunification, has rested on an uneasy interplay between development pressures and a series of "compensation landscapes" that link the enhancement of public space in one location to the loss of sites elsewhere in the city.

In other cases "civic ecology practices" involve the combination of ecological enhancement—rather than ecological restoration—to allow biodiversity to flourish in an otherwise inhospitable urban context and also to serve as a focal point for children's experience of the natural world.[107] An interesting example from the 1990s is the transformation of a large concrete pond located in a public park in Yokohama, Japan, which formerly had only three species of dragonflies, into a vibrant aquatic ecosystem supporting nearly thirty species in close proximity to local schools. Since the completion of this project there have been many more "dragonfly ponds" created in the Yokohama metropolitan region, including elaborate "natural laboratories in which ponds are the centerpiece of an outdoor classroom," where an observation platform "provides a space to set up desks and chairs with microscopes and other laboratory equipment within the artificial habitat."[108] To what extent, however, can these types of pedagogic projects be regarded as counterhegemonic ecological practices? For leading advocates, such as the environmental studies specialist Hiromi Kabori, the creation of these intricate urban wetlands mimics a largely lost type of traditional cultural landscape known as *satoyama* (里山), at the interface between mountains and plains, where there is a small-scale mosaic of forestry, rice cultivation, and irrigation ponds. The strong emphasis on the protection or enhancement of native species under urban *satoyama* has wider ideological resonance beyond conservation biology (as we saw in the last chapter) and also with elements of ecological design that can embellish the underlying dynamics of urban development.

The relationship between urban ecology and land use planning places the question of biodiversity at the heart of capitalist urbanization. Civil society is relatively powerless to protect sites from development in the absence of specific configurations between science, public culture, legal instruments, and the state apparatus. More commonly, however, the state acts as a facilitator for speculative urban development, sometimes through violent forms of enclosure or land clearance, so that the state's monopoly over the use of force is effectively mobilized on behalf of capital. In the case of Istanbul's Gezi Park, for instance, political demands to protect this green enclave from development in 2013 involved not just environmental concerns but a wider contestation of authoritarian populist rule. The Gezi Park movement made connections with other protests across Turkey against land grabbing, mega-infrastructure projects, and the destruction of vulnerable ecosystems in the service of "crony capitalism."[109] The complexities of social mobilization in defense of urban nature reflect varying degrees of interaction between aesthetic, social, and ecological aims, so that movements can be highly heterogeneous in terms of their structure and organization.[110] As we saw in the case of India, some urban environmental campaigns contribute towards social inequality, with agendas based on "environmental improvement" or "cleaner cities" that have little to do with ecological concerns and certainly no connections with social justice.

Interest in citizen science has grown out of heightened levels of public disenchantment with remote and technocratic forms of expertise. The sociologist Alan Irwin, writing in the early 1990s, draws on the work of Ulrich Beck and others to emphasize the sense of a modernity project that is increasingly shrouded in doubt.[111] In contrast to Beck, however, Irwin presents science as a heterogeneous field marked by the intersection between different and sometimes competing bodies of knowledge; rather than an overarching crisis of modernity we are faced with a series of interrelated zones of contestation. We can also draw on the work of Andrew Feenberg, who outlines a "critical theory of technology" to illuminate a range of possible sociotechnical pathways through modernity in contrast with teleological or reductionist accounts. Drawing on the insights of the Frankfurt School, notably via Theodor Adorno and Herbert Marcuse, Feenberg develops a critique of the ideological elision between political and technological reason.[112] Feenberg contrasts this position with two other approaches: first, the pervasive dominance of instrumental perspectives underpinned by assumptions about ethical neutrality; and second, a more pessimistic critique of modernity presented as a series of inescapable constraints, emerging through an intellectual lineage that spans figures such as Max Weber, Martin Heidegger, and Jacques Ellul. Ellul, for example, decried an "inhuman atmosphere" produced through the transformation of the nineteenth-century machine into the full-scale "mechanization" of society.[113] In the final analysis, however, Ellul's technological determinism is so complete and overwhelming that it leaves little or no scope to create alternative futures. For Feenberg, much contemporary environmental discourse lies trapped between variants of these two positions: on the one hand a dominant technomanagerialist paradigm in the service of global capital, and increasingly allied with geoengineering and resilience discourses; and on the other hand, a reactionary counterdiscourse that is rooted in essentialist conceptions of nature and technology, operating in the service of different constellations of power, and often allied with explicitly reactionary ideologies.[114]

4.4 GROUND TRUTHS

The mapping of nature has often been a prelude to its control or destruction. The French geographer Yves Lacoste, for instance, showed how the detailed vegetation maps used in the Vietnam war were instruments of military power to better enable the annihilation of nature in acts of genocidal aggression.[115] In a similar vein, the anthropologist Erik Harms describes how the mapping of marginal spaces on the periphery of contemporary Ho Chi Minh City has served as a precursor to their elimination through a process of "knowing into oblivion." Harms notes a continuity between the colonial era French *mise en valeur* and the postcolonial dynamics of *khai phá* (clearing the wasteland).[116] In contrast to the insights of Lacoste and Harms, however, can we also conceive of the cartographic imagination as a means to protect specific elements of nature? Can cartography serve as a tool that enables hidden or overlooked elements of nature to enter urban consciousness and experience?

The inevitable distance between any form of cartographic representation and a putative "ground truth" is alluded to by the Argentinian writer Jorge Luis Borges. In a single paragraph entitled "On exactitude in Science," published in 1946, Borges explains how "the Art of Cartography attained such Perfection that the map of a single Province occupied the entirety of a City." Borges describes a map "whose size was that of the Empire, and which coincided point for point with it."[117] Perhaps the closest we have to a Borgesian map in the field of urban ecology is the attempt to create a complete botanical map of West Berlin that we encountered in the last chapter, prepared by Herbert Sukopp and his colleagues, which involved charting the city's flora street by street. The intricate outcome of this work, which was completed in the 1980s, divides the whole city into a series of precise vegetation zones, and represents the outer limits of what a dedicated team of botanists can achieve given the inevitable constraints on human time and expertise, and the fact that their object of study is in a state of constant flux (figure 4.4).[118] Although Sukopp's map clearly does not match Berlin "point for point" à la Borges, it nevertheless gives a sense of the cartographic limits to the ecological imagination at a citywide scale.

Other elaborate cartographic depictions of urban nature include studies of the effects of meteorological phenomena such as the "urban heat island." From the middle decades of the nineteenth century we find growing

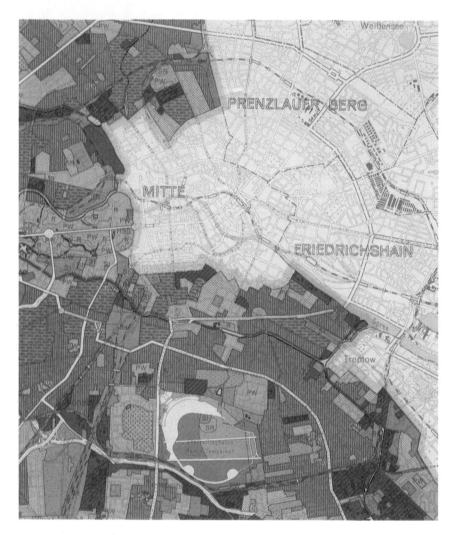

Figure 4.4 Detail from the biotope map for West Berlin produced by the Technical University's Institute of Ecology in the mid-1980s. Note the grayed-out botanical terra incognita of East Berlin, including the central district of Mitte. Source: Institute of Ecology, Technical University, Berlin.

scientific interest in the climate of cities in France, Germany, and elsewhere, focusing on features such as temperature, visibility, and atmospheric conditions.[119] In the early decades of the twentieth century a number of studies were undertaken on the urban heat island to explore the effects of higher temperatures on the phenology of vegetation. By the 1970s, research on Berlin, Brussels, Utrecht, and other cities used a combination of botanical research with cartographic techniques to delineate variations in temperature.[120] In Berlin, for example, the botanist Frank Zacharias demonstrated the effects of urban climatic variations on lime trees, which had been extensively planted along city streets, to show that trees came into bud up to ten days earlier in warmer parts of the city.[121] Similarly, in the case of Brussels, maps prepared by Paul Duvigneaud and his colleagues showed springtime differences in the date on which horse chestnut (*Aesculus hippocastanum*) trees came into bud (figure 4.5).[122] In these examples, trees play the role of environmental sensors to enable an isothermic representation of urban space.

There are a number of design-oriented maps of imaginary urban natures that owe their inspiration to a combination of ecological models with advances in the technical sophistication of data representation (popularly referred to as geographic information systems). The landscape ecologist Eric W. Sanderson's Mannahatta project, initiated in the late 1990s, creates an intricate reconstruction of Manhattan island at its moment of first encounter with Europeans in 1609. The simulations of a pre-European landscape present a mosaic of forests and swamps with only occasional wisps of smoke alluding to a human presence. Sanderson uses a variety of models, including what he terms "Muir webs," to depict the many different ecosystems that flourished under the human ecology of the original Lenape people, who occupied an extensive area stretching from Delaware to New Jersey.[123] Whilst Sanderson uses cartography as a form of didactic experimentation, the earlier use of maps, under British colonial rule, was an indispensable tool for the erasure of "first nature," including the topographic flattening of the island to leave just a few rocky outcrops to be

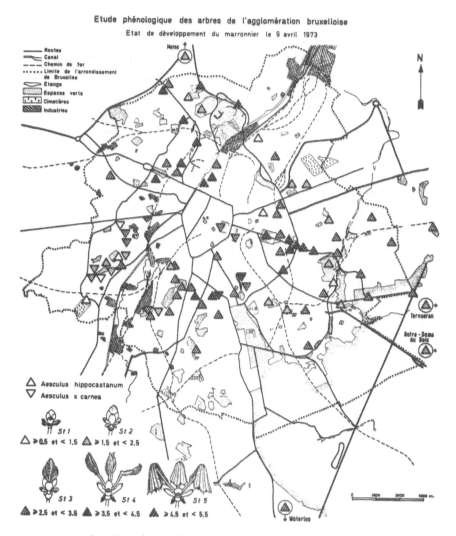

Figure 4.5 Phenological map of Brussels showing differences in the rate of bud development of the horse chestnut (*Aesculus hippocastanum* and *Aesculus x carnea*) in response to the urban heat island effect on 27 April 1973. Source: Paul Duvigneaud Centre, CIVA archive, Brussels.

woven into naturalistic park design. Nevertheless, in Sanderson's conceptualization of contemporary New York as an ecosystem, there is a peculiar elision between economic and environmental processes that mirrors the epistemological reductionism of the Chicago school. "Like economic fundamentals," writes Sanderson, "if these basics of nature are functioning free from market manipulation, life will thrive."[124]

Cartographic projections have also been used to characterize contemporary environmental change as a design opportunity. A cluster of interventions that might be gathered under the aegis of the "adaptive Anthropocene" have presented rapid environmental change as a kind of technical puzzle to be solved.[125] Tellingly, many of these representations rest on an implicit interchangeability of species that belies a limited grasp of ecological processes: urban nature is often represented in a generic form that can deliver specific kinds of "ecological services." Examples of maps within this idiom include the *Nature atlas* (2017) developed by the urban designers Ruchika Lodha and Timon McPhearson to explore the socioecological characteristics of the Gowanus Canal in Brooklyn, New York. Similarly, researchers at ETH Zurich have developed new cartographic projections of Switzerland as an urbanized cultural landscape amenable to the "orchestration" of environmental change based on new forms of territorial experimentation.[126] At issue here is a radical divergence between urban ecology as a distinctive field of scientific knowledge and newly emerging digital genres in ecological design.

Other cartographic representations of urban nature closer to Sukopp and the Berlin school seek to present actually occurring nature in immense detail. There is a shift from a simple range of land use types towards a much finer differentiation of ground-level ecologies, including the highlighting of specific sites of interest. Examples include the recent urban nature maps that have been produced for Amsterdam, Berlin, Rotterdam, and other cities. These highly detailed representations serve as an encouragement to discover spaces of nature, typically on foot or by bicycle. They are invitations to orient oneself "slowly" through urban space, with implicit links to heightened levels of what might be termed "ecoliteracy." Similarly, a series of new urban floras have been produced, notably for Birmingham and Hamburg, that are supplemented by suggested itineraries for botanical exploration. The recent guide prepared by the Hamburg Botanical Society, founded in 1891 and still going strong, contains 95 suggested walks covering every kind of habitat in and around the city. The urban botanist needs no expensive or specialist equipment but just a "preparedness to keep their eyes open" and should maybe bring a few items along such as a plastic bag for leaves or other fragments of vegetation that might need closer inspection at home, a hand lens and tape measure, a notebook and pen, a digital camera (though a smart phone will do fine), and two books: an "excursion flora" with specialist keys and a "picture flora" to enable "swift orientation."[127]

The limits to conventional forms of cartographic representation become especially apparent with excursions through various marginal zones of the city, where we encounter sites that are too complex or uncategorizable for conventional types of maps. In the case of Paris, for instance, Philippe Vasset explores the "blank spaces" that elude cartographic representation or administrative categorization.[128] In Portugal, an interdisciplinary project that combines de Solà-Morales's concept of *terrain vague* with the "heterotopia" of Michel Foucault has undertaken a series of explorations of vacant land, often dotted with ruins, to emphasize how ostensibly empty sites can serve as zones of cultural and ecological experimentation (figure 4.6). The project uses a series of multisensory approaches to the material complexities of marginal spaces in Lisbon, Guimarães, Barriero, and Vizela, cataloguing street art and abandoned objects as well as animals, plants, and other organisms.[129] Site surveys include recordings of birdsong to help identify which species are present as well as a series of podcasts to bring these soundscapes to a wider audience.[130] There is something especially significant about the use of methodologies that involve the human body interacting with, and moving through, urban space. The appropriation of the botanical transect, for example, as a distinctive kind of walking methodology, framed by a

Figure 4.6   The Matinha gasworks site, Lisbon (2018), is one of the sites investigated as part of the NoVOID project. The pampas grass (*Cortaderia selloana*) is regarded as invasive under Portuguese law whilst the black poplar trees (*Populus nigra*) in the foreground are considered to be an archeophyte (ancient introduction) to the region. Photo courtesy of the NoVOID project.

Perecquian set of constraints, can produce new and unexpected insights (as we saw in the last chapter).

4.5 ECOLOGICAL VOIDS

Urban nature lies at the intersection between the universalist aspirations of conservation biology, with its comparative indices of endangerment, and a variety of cultural and material complexities on the ground. The identification of regulatory voids in relation to the protection of biodiversity should not necessarily be interpreted as evidence of an absent state, but rather one of calculative reconfiguration. The state no longer has a monopoly over the production of environmental knowledge, if indeed it ever did, leaving a more dispersed set of sources in its wake, ranging from counterforensic collaborations and citizen science initiatives to neoliberal data mining on behalf of environmental consultancies. A significant challenge for grassroots activists is the production of scientifically robust data that can play a role in legal or political deliberation.[131] Established scientific practices for cross-checking observations can be emulated to some degree through online sharing of data to help determine the identification of plants, birds, insects, and other potential bioindicators. The corroboration of biological records enables grassroots data to become more politically effective so that the use of shared observations can bolster degrees of scientific certainty and credibility. The shift towards more collaborative forms of knowledge production, and insights gained from "evidentiary materialism," marks a reorientation of urban analysis towards socioecological processes operating across different scales and temporalities.

What might a decolonized conception of urban biodiversity look like? Would it involve dispensing entirely with existing classification schemas or comparative indices? Or would it signal a tactical combination of different discourses as part of counterhegemonic knowledge formation? Critical legal studies emphasize that truth is a contested field rather than the self-evident outcome of scientific research. This is not, of course, a relativist or social constructivist position but rather an explicit recognition that the veracity of scientific knowledge can never be fully disentangled from its social and historical context. Legal instruments are often marked by a degree of ambiguity and fragility. As we saw in the case of Moscow's Khimki forest, legal protections for nature can simply be ignored or overturned. If environmental campaigns are supplemented by insights from other fields, such as forensic accounting, as is strikingly revealed in the Russian example, we can uncover who benefits from specific instances of environmental destruction. The role of law has had a complex relationship with progressive environmental discourse, with cross currents concerning intellectual property rights (and bioprospecting), rights of assembly, and the prohibitive cost of undertaking many forms of legal redress.

The urban arena has witnessed many instances of large-scale mobilizations to protect fragments of urban nature. A striking example is the *Almbråket* (elm fight) that took place in the Kungsträdgården park in central Stockholm in the spring of 1971. On the night of 12 May a group of workmen with chainsaws attempted to remove a stand of old elm trees in the city center during the hours of darkness to avoid protestors, but their activities nevertheless drew the attention of thousands of people, some of whom climbed into the trees as they were being felled so that the operation had to be suspended. The *Almbråket* protest, involving violent clashes with mounted police, attracted international media attention and signaled a wider sense of unease with a technocratic planning ethos that appeared to shatter connections with the past, with nature, and with the public realm.[132] More recently, the city of Sheffield in northern England has been the focus of a bitter dispute since 2012 over the removal of thousands of street trees, including fine avenues of elm, lime, and cherry trees, some of which harbor rare insects. The conflict was sparked by the city's award of a 25-year contract worth over £2 billion for street maintenance to a private contractor, marking a shift away from expensive forms of aboricultural care towards the creation of "simpler streets" that are cheaper to manage.[133] These two examples of efforts to protect street trees mark part of a wider set of demands to prevent what the architect Nicolas Soulier has referred to as "street sterilization" and the elimination of "unruly" elements of urban space.[134]

The plight of urban trees raises important questions about urban ecologies in a state of transition. With many older trees facing a combination of disease, neglect, and the threat of removal, the neoliberal "urban forest" of the future is likely to consist of smaller and younger trees that support far less biodiversity. Furthermore, planting schemes for urban trees appear to be polarizing between the maintenance of perceived cultural landscapes, which often have little connection to actually existing urban ecologies, and the attempt to grow "resilient forests" through the use of "climate-proof" species that offer little input to urban biodiversity. If we step back from the politics of urban trees, as viewed through the technocratic lens of municipal policymaking, it is clear that urban space is kept in a form of active temporal suspension from the generalized drift towards some form of ecological entropy. The cultural trope of the "abandoned city" that I consider in the next chapter shows how any city is actually a latent urban forest in the making: the logical end point for urban rewilding, without human interference, would be a dense and ultimately impenetrable tangle of vegetation.

If we shift our focus from individual street trees or wild animals in the city towards the ecological relations between species, how does this alter the conceptual parameters of ethical and legal theory? An emerging theme within critical legal discourse is the contested intersection between the field of animal rights oriented towards the life of an individual animal (though rarely a lower-order organism) and the strategic parameters of conservation biology focused on the protection of more complex assemblages of different species such as ecosystems. Can concerns with environmental justice be extended to the web of life as a whole? Clearly a tension exists between a neo-Benthamite emphasis on degrees of sentience and attempts to extend legal standing to invertebrates, plants, and other organisms. A recurring point of contention is whether elements of nature, from individual organisms to whole ecosystems, might be conferred some type of legal standing or personhood. When we consider the intensity of political discourse surrounding individual street trees, for example, it is apparent that the question of other-than-human protections is being radically recast. An increasing number of philosophical interventions have sought to delineate how nonhuman nature might be granted some kind of constitutional rights or protections. Michel Serres, for instance, has sought to elaborate on the possibility of a "natural contract" that reframes the political scope of the Enlightenment.[135] Serres is careful, however, to suggest that the identification of legal subjects in nature should not stem from an elaboration of existing forms of personhood but rather emerge from the articulation of new kinds of collective subjectivity that combine human and nonhuman elements.[136]

Viewed through the prism of an idealized nature threatened by the "new extractivism," the idea of a natural contract has implications for a reconceptualization of urban nature that extends beyond the utilitarian logic of ecosystem services.[137] Innovative legal approaches to the protection of river systems in New Zealand, for instance, have influenced regional deliberations in Colombia, India, and elsewhere, yet there are already fierce counter responses in terms of the defense of existing legal precedents, emanating from within the field of jurisprudence, or outright skepticism towards extending rights to nature as articulated by Alain Badiou, Jacques Rancière, Slavoj Žižek, and other (generally male) social theorists who suspect an ideological cover in play for an unchanged polity.[138] A fundamental break with anthropocentric conceptions of legal subjectivism would involve a movement towards relational and more context-specific conceptions of rights.[139] At issue here is the possibility of articulating a posthuman variant of legal theory which does not dispense with an epistemological framework that can guide meaningful forms of political action. For the sociologist and critical legal scholar Gunther Teubner, the gradual extension of legal personhood towards nonhuman entities, both organic and inorganic, marks an "ecologization" of law and a proliferation of potential actors. By synthesizing a Latourian elaboration of agency to include the recalcitrance of nonhuman "actants" with a Luhmannian emphasis on the specific modes of communication that can sustain multiple intersecting kinds of collective actors, Teubner offers a historically grounded conceptualization of the elaboration of legal practice.[140]

The paradox, however, is that the realm of the nonhuman is becoming increasingly incorporated into human societies in order to provide greater protection from human action.[141] Furthermore, the extension of legal systems of protection to different spatial scales, as elaborated by calls for "global environmental governance," rests on a shaky foundation of pre-existing political ontologies rooted in a technomanagerial conceptualization of environmental problems.[142] To what extent would an alternative conceptualization of legal theory remain mired in existing epistemological and regulatory regimes? Can the exercise of environmental law within the urban arena serve as the starting point for a new conceptualization of science, citizenship, and democracy? Or, alternatively, would a shift in ethical relations towards the nonhuman prefigure these other changes so that a different kind of political ontology could begin to emerge? In the next chapter we turn to the question of periodicities in relation to environmental change in order to explore contrasting urban ecological imaginaries.

5

TEMPORALITIES

Time, like a film reel running through a faulty projector, was moving at an erratic pace, at moments backing up and almost coming to a halt, then speeding on again. One day it would stop, freeze forever on one frame.

J. G. Ballard[1]

The climate events of this era, then, are distillations of all of human history: they express the entirety of our being over time.

Amitav Ghosh[2]

It is shortly after dawn and I am standing next to a group of amateur ornithologists with binoculars directed towards the remnants of a lake, surrounded by signs of construction activity. This small expanse of water is a fragment of the extensive wetlands that once dominated what is now the metropolitan landscape of Chennai in southern India. Just a few meters in front of me a pied kingfisher (*Ceryle rudis*) hovers midair over a pool of water strewn with plastic bags, pieces of polystyrene, and other detritus.

In a matter of minutes we have seen over sixty species of birds including the black-headed ibis (*Threskiornis melanocephalus*) and the painted stork (*Mycteria leucocephala*), both of which are designated as "near threatened" by the International Union for Conservation of Nature (IUCN), an organization founded in 1948 and now the principal global arbiter for degrees of endangerment through its compilation of Red Lists for both species and ecosystems.

Chennai, a city of over seven million people according to the 2011 census, lies under the Central Asian Flyway, one of the most important, yet least studied, migratory pathways for birds, extending from eastern Europe through Siberia and East Asia, before reaching southern India.[3] Along their migratory routes birds face multiple threats including habitat loss, hunting, and the disruption of climatic zones.[4] Around 90 of the 400 or so bird species recorded in the state of Tamil Nadu are migratory, with evidence of sharp declines in many species over recent decades.[5] By some estimates over 90 percent of the once plentiful lakes in the Chennai metropolitan region have been lost in less than thirty years, and the disappearance of creeks and water bodies has also contributed to a series of devasting floods.[6]

Chennai's degraded system of lakes and watercourses illustrates an intersection between different temporalities. For the historian Dipesh Chakrabarty, the contemporary epoch is marked by three timescales—the human, the evolutionary, and the geological—as the conventional parameters of the human sciences become overwhelmed by the speed and scale of environmental change.[7] Chakrabarty's call for a new kind of critical interdisciplinary synthesis rests on a reprise of universalism: in this case, however, the impetus for an overarching explanatory framework emanates from the field of history rather than the biophysical sciences. In the place of naturalism, or other types of reductionist schema, we find a more open set of causal interactions between the geological and historical realms of human experience. There are evident continuities with earlier attempts to build nonpositivist forms of transdisciplinary historical analysis in the work of Roy Bhaskar and others. Bhaskar's rejection of any narrow emphasis on

Figure 5.1 Perumbakkam, Chennai (2016). One of the last remaining wetlands in the metropolitan region. The sign indicates the importance of the site for migratory birds. Photo by Matthew Gandy.

"empirical regularities" in the face of experimental practice, his inter-weaving of causalities, and his sense of the shifting contours of scientific knowledge remain a useful starting point for an expanded conception of environmental change.[8] We should note Bhaskar's insistence on different forms of philosophical incommensurability: contra recent developments in urban ecology, and the attempt to delineate emergent socioecological sys-tems, it is simply not possible to combine all conceptual schemas into one overarching explanatory framework. Yet significant uncertainties remain over the distribution of agency (and responsibility) within this expanded conceptual field.

The global environmental crisis is often conceptualized in terms of the twin threats of climate change and loss of biodiversity. Yet an emphasis on climate change points towards the malleability of human environments rather than forms of extinction and irreversible loss. A greater emphasis on climate change, rather than on the destruction of biodiversity, in contem-porary environmental discourse is reflected in the relative strength of geo-constructivist perspectives, various resilience paradigms, and an emerging faith in terraforming, extending even to the creation of "off-base" worlds. The emerging interest in ecological reconstruction and "novel ecosystems" falls at the interface of the paradigms of "malleability" and "irreversibil-ity" through a sense that nature is in a state of perpetual change that can be steered, to some degree at least, towards human purpose.

The question of temporalities touches on the material dynamics of actu-ally existing ecologies but also extends to imaginary future scenarios. My use of the term "urban ecological imaginary" resonates with the existing elaboration of the "social imaginary" à la Cornelius Castoriadis that has emerged in response to more deterministic interpretations of collective cultural forms.[9] The imaginary in this context is not simply a matter of heightened cultural experience or new modes of representation, although these can often reveal fascinating insights into alternative worlds, but is in itself a source for the articulation of alternative socioecological pathways. By invoking an ecological imaginary I am seeking to highlight a cultural mediation of collective life, both existing and projected, that extends to the realm of the nonhuman. Where should we locate an urban ecological imaginary in space and time? Can the cultural idiom of alternative ecological futures be disentangled from a narrowly Eurocentric context in which future pathways remain largely constrained by the ideological tentacles of the past? My critical engagement with cultural forms such as science fiction literature highlights an oscillation between neoromanticist and technologically determined tropes. Much less frequent are posthuman accounts that reflect on new forms of sociability between human and nonhuman life within the more mundane settings of what the novelist J. G. Ballard once referred to as the "near future."[10] A future ecological imaginary is simultaneously a topographic imaginary; it is a spatial rearrangement of material elements that denotes far more than a scientific metaphor.

But what is the precise role of cities within what has been termed the "sixth mass extinction" facing the history of the earth? The rapid growth of Chennai and other cities signals the direct contribution of urbanization to ecological disturbance, but as we move our analytical lens towards the regional or global scale the picture becomes more complex. Are cities to be subsumed within a broader environmentalist critique of modernity, or can they serve as the focal point for alternative socioecological pathways? Cities and biodiversity are conventionally conceived to lie in an antagonistic relationship to one another, although global patterns of mobility and land use (including agricultural intensification) serve to unsettle existing assumptions. Urban space is presented as the antithesis of ostensibly "pristine" ecosystems, especially through a Euro-American cultural prism, that have long been the focus of ecological analysis and global conservation efforts. More recently, however, the idea of nature has been extended to encompass a spectrum of socioecological formations, which I have been exploring throughout this book. If the idea of nature itself is opened up to critical reflection, and is no longer considered separately from modernity, then a more nuanced understanding of urbanization can emerge as a set of processes intersecting with distinctive material topographies.

5.1 CITIES IN DEEP TIME

It is striking that most of the significant theoretical work in relation to the Anthropocene has only engaged indirectly with the urban arena beyond the identification of specific empirical parameters or material traces. Interest in the Anthropocene, as a putative new geological epoch, has thus far been driven primarily by contributions from the biophysical sciences, and especially by variants of "earth systems science" although the balance of disciplinary contributions is shifting. Emerging calls for "a sustainable Anthropocene" based around a resilient "techno-biosphere" rest on an uneasy combination of behavioral change, large-scale geoengineering, and other kinds of technical interventions.[11] Emerging typologies of urban nature clearly resonate with the Anthropocene debate: the cultural critic Steffen Richter, for example, traces the term "third nature" to emerging interest in the complexity of nature as a kind of cultural synthesis under the Anthropocene.[12] There are parallels here with the anthropologist Anna Tsing's characterization of precarious forms of multispecies coexistence as a kind of "third nature" that persists in spite of, or in the midst of, global capitalism.[13] If Richter's framing of the term "third nature" is oriented towards the periodization of the Anthropocene, then Tsing's use points towards an acknowledgment of socioecological complexity.

The case for a revised geological periodization draws on two interrelated sets of arguments: first, the accelerating momentum in human sources of environmental change; and second, a neocatastrophist interpretation of the earth's history that emphasizes not just the scale of previous geoenvironmental transitions but also their rapidity. If we locate the emergence of cities within "deep time" we can begin to address some of the assumptions made about the connections between modernity, capitalist urbanization, and environmental change. The idea of deep time, first articulated by the eighteenth-century geologist and polymath James Hutton, connects with an extended historical horizon towards the origins of a habitable Earth. Hutton's conception of time proved influential for the geologist Charles

Lyell and the doctrine of uniformitarianism which contends that the surface of the earth has been subject to unvarying natural processes over a very long period.[14] In contrast, the emphasis on sudden change, often referred to as "catastrophism," can be traced in particular to the French zoologist and early paleontologist Georges Cuvier who showed that past worlds were radically different.[15] Cuvier's *Essay on the theory of the earth*, first published in 1813, presented extensive evidence for extinct species, whose dissapearance he attributed to some kind of traumatic past event. For Cuvier, this lost world was one of "scattered and mutilated fragments" that might somehow be gathered under a system of understanding that could "burst the limits of time."[16] The excavatory impetus of modernity contributed to the emerging paleontological imagination, with its new perception of a diminishing human presence within an enlarged sense of time. The limestone quarries of Montmartre, for example, required for the reconstruction of nineteenth-century French cities, generated a wealth of new and unfamiliar fossils.[17] Over time, the idea of catastrophism, and the impact of what Cuvier termed "great events," have ultimately served as a more prescient marker of geological time for the contemporary epoch, and resonate with emerging evidence for an acceleration in the rate of environmental change.[18]

Several different starting points have been suggested for the Anthropocene: the late Pleistocene megafauna extinctions; the emergence of agriculture, including the specific methane spike associated with the expansion of rice production; historical traumas such as disease and violence, exemplified by the European impact on the New World; the rise of global capitalism; the switch from water to steam power in late eighteenth century England; rising mercury concentrations within the global food chain; radionuclide traces left by the detonation of the first nuclear weapons; and more recently, the extraction of rare earths such as indium and gallium in the service of digital capitalism and the transition to postcarbon futures. Running through this range of potential beginnings, however, is a tension between the stratigraphic emphasis on a globally synchronous geological marker—the so-called "golden spike"—and an alternative focus on developments in human history that mark a step change in patterns of global environmental change. Every suggested starting point serves simultaneously as an epistemological vantage point from which any putative environmental history must be articulated in relation to rival temporalities.

If the start of the Anthropocene is marked by the first synchronous human impact at a global scale, then there is a case for using the changes in vegetation cover following the late Quaternary megafauna extinctions as the initial instance of human-induced changes in the composition of the atmosphere. Later traumatic events such as plague mortality, peaking in the fourteenth century, have also left measurable traces in the composition of the atmosphere.[19] Recent research in the geosciences connects the early sixteenth-century temporary depression in levels of atmospheric carbon dioxide, referred to as the Orbis spike, to the environmental impact of the European "discovery" of the New World, whereby a combination of disease, genocide, and the abandonment of agricultural land created a vast carbon sink through the extensive regrowth of forests.[20] The Orbis spike provides an analytical bridge between the search for stratigraphic markers and the emerging dynamics of European imperialism and global capitalism.[21]

The more critical literature on the Anthropocene has sought to link causality with periodicity. The patenting of steam power in 1784, for example, has also emerged as a convenient focal point within a broader history of increasing human impact on the biosphere. The historian Andreas Malm, following Anthony E. Wrigley's insights from a thermodynamic Ricardian perspective, highlights the transition in fossil fuel use from basic thermal needs for human survival to the scaled-up generation of mechanical power as the decisive break from the strictures of the organic economy.[22] The switch from water to steam power increased the mobility of capital and enabled the transfer of production from predominantly rural areas to fast-growing industrial towns.[23] The presence of large concentrations of potential workers tilted the balance of power in favor of capital and freed production from the constraints of nature, location, and potential labor shortages. The increasing use of fossil fuels, and in particular the transition from wood to coal, greatly expanded the potential scope and scale of industrial production, thereby adding further impetus to capitalist urbanization. With the rising dependence on oil in the twentieth century, a further set of transformations rippled through the global economy, including new patterns of urban growth and infrastructure provision.[24]

The recent emphasis on the "Great Acceleration," as articulated by J. R. McNeil and Peter Engelke, stems from a synthesis between mainstream geological perspectives and environmental history.[25] There is now a range of literature that points to isotopic traces of the nuclear age coinciding with the postwar increases in energy use, urbanization, and other environmental markers. Indeed, the growth of cities has even been referred to itself as a potential "golden spike" within the future stratigraphic record, marked by a distinctive mélange of material traces.[26] McNeil and Engelke note an underlying paradox in that cities have had a huge environmental impact, both directly and indirectly, but have also played a key role in reducing environmental damage as "centers of creativity and innovation."[27] Cities, in other words, have long served as cultural and political crossroads between different strands of modernity.

In some metahistorical accounts of environmental change there is a blurring of human agency under geohistorical modes of interpretation. The historian Yuval Noah Harari presents human history "as the next stage in the continuum of physics to chemistry to biology."[28] There is an implicit teleology to his view of human history operating at a broad scale of cultural evolution from the Neolithic revolution onwards. Yet a species-based account of historical change leads towards a reductionist and undifferentiated reading of human agency. "While providing important and indispensable insights," notes the historian Sebastian Conrad, "the category of 'species,' and of large time frames alone, does not enable us to address questions of responsibility, either historically or in the present."[29] The emphasis on human history as an extended "species history" marks a further narrowing of agency to a teleological rather than relational reading of social and environmental change.

Even if agreement may ultimately emerge over geological nomenclature, and its evidentiary and temporal parameters, much of the current discussion lacks a sense of the historicity of the Anthropocene debate in terms of its conceptual antecedents, emerging incongruities, and contemporary cultural resonance. The history of the geological sciences, for example, is widely occluded within these emerging interdisciplinary formations. Alternative conceptualizations of the Anthropocene have extended the parameters of what the geographer Kathryn Yusoff refers to as "geologic subjectivity" encompassing ontologies of both human and nonhuman agency.[30] Furthermore, emerging perspectives from the global South question the imposition of the Anthropocene as a "closed scientific category" that precludes alternative futures, and have insisted on the need for a new ethics of interdependence that acknowledges modernity as a "precarious ecological achievement."[31] Recent contributions from Chile, India, and elsewhere are moving beyond a compendium of perceived challenges for environmental governance towards the analysis of ethical and epistemological issues connected with capitalist extractivism and the neocolonial production of toxic zones.

The paleontological imagination provides us with a portal into the current scale of environmental change, and in particular the extent of biodiversity loss under past mass extinction events. Yet the reemergence of life after these previous catastrophic episodes holds ambivalent implications for conservation biology. If life forms are in a constant state of flux, then why should mass extinctions be such a cause for concern? The evolutionary biologist Chris Thomas recasts the current wave of extinctions as the "sixth genesis" driven by the evolutionary pressures of human modified environments.[32] Thomas suggests that the long-term human impact under the Anthropocene might even "increase the number of species on the Earth's land surface" within "the human-dominated Pangean Archipelago."[33] For Thomas, the "New Pangaea," a term that first emerged in the early 2000s, serves as "an apt metaphor for the accelerated connections of the modern world."[34] He questions the pervasive "neophobia" across the ecological sciences and promotes the rise of "novel ecosystems that contain mixtures of species never seen before."[35] His characterization of the earth as "a biological park" resonates with the "rambunctious garden" described by the ecologist Emma Marris, who is another prominent exponent of "novel ecosystems." In the writing of Marris, for example, we find an explicit acknowledgment that intact ecosystems no longer exist and that there are practical limits to what conservation biology can hope to achieve.[36] The articulation of normative temporalities derived from an arbitrary baseline is replaced with an emphasis on hybrid temporalities and novel socioecological assemblages. The vibrancy of urban nature forms part of a reorientation of ecology towards new socioecological assemblages and the recognition that biodiversity is a culturally and historically specific phenomenon, rather than a fixed marker or benchmark. A tension remains, however, over what species loss actually means: is it an irreversible dynamic that portends wider forms of ecological disturbance, including zoonotic dimensions to urbanization, or does it signal part of an inherent process of change within a larger geohistorical timeframe?

The invocation of the world as a garden, or especially a park, connects with a wider impetus towards the "planification of the future" as elaborated in the writings of Manfredo Tafuri and Massimo Cacciari.[37] The term "planification" brings colonial and neocolonial variants of the plantation system, and its high-tech successors, into dialogue with global discourses of spatial control. There are parallels here with emerging interest in the Plantationocene as an alternative conceptualization of global environmental change that emphasizes the centrality of racial difference to the construction of extractive frontiers.[38] The rendering of global ecology as a series of plantation systems or biopolitical zones holds very different connotations to the ostensibly more benign construct of a garden or park. An emerging characteristic of the Anthropocene is the steady replacement of an array of intricate and interdependent socioecological systems with a series of "vulnerable saturated monocultures."[39] The "plantation ecologies" of late modernity are at odds with the remaining fragments of earlier types of cultural landscapes. For the literary critic Jeremy Davies, a focal point for environmental justice under the Anthropocene will be countering the "simplifying tendencies of the Holocene's final phase."[40] These tensions between simplicity and complexity operate on a number of cultural and material levels, including the protection of both social and ecological forms of difference.

5.2 Capital, ecology, and urban space

How might we relate the production of the built environment, and its associated ecologies, to the circulatory dynamics of capital? Neo-Marxian analysis of the secondary circuit of capital, notably advanced by David Harvey, traces the flow of excess capital out of production—the primary circuit of capital—into different facets of the built environment—the secondary circuit.[41] The production of metropolitan nature has been extensively entrained in this process through the construction of infrastructure, landscape design, and other elements in the shaping of urban space. The state operates as an enabler at multiple levels, from the provision of land (by force if necessary) to the fostering of growth coalitions that attempt to steer the urban process. And capital itself is by no means a singular actor in this process, as evidenced by various forms of "interfactional conflict."[42] Finance capital in particular, especially from the early 1970s onwards, has extensively shaped patterns of investment in urban space. Furthermore, perturbations within the finance sector, including periodic banking and credit crises, have directly impacted the built environment through the specific temporalities of investment cycles.[43] These cycles of investment and disinvestment have produced their own forms of ecological succession as abandoned sites or buildings acquire distinctive assemblages of fauna and flora, so that the periodicities of capital intersect with the temporalities of nonhuman nature. Conversely, waves of investment in urban space, and the production of metropolitan nature, have produced new ecologies associated with parks, gardens, and infrastructure systems.

The Belgian economist Ernest Mandel points out, following Lefebvre, that the problematic dimensions to urbanization are an outcome of the structural and political characteristics of capital rather than inherent features of the modern city:

> The blatant deformation of urban development since the industrial revolution, has been the unequivocal product of social conditions: private ownership of land; real-estate speculation; systematic subordination of town planning to the development of "growth sectors" of private industry; general underdevelopment of socialized services.[44]

Although Mandel deploys the phrase "blatant deformation" to imply a historical divergence of urbanization from an idealized form, as expressed through a long history of architectonic experimentation, we might extend the notion of deformation to evolving relations between human societies and the realm of the nonhuman. The history of urban planning has been marked by diverse attempts to embellish or incorporate specific cultures of nature within urban space: these socioecological incorporations mirror structural dimensions to human societies, with varying degrees of autonomy from existing power structures. For Mandel, the substitution of a capitalist logic with an ideological attachment to "technical rationality" fails to address the underlying contradictions to urbanization:

> These societal conditions, far from being suspended or neutralized by any technical logic, in their turn determined technological development—for example, the backwardness of industrial methods in the construction industry—and aberrant development (high-rise blocks, dormitory cities, and so on).[45]

Urban form is thus a reflection of the political dynamics of urban space rather than an innate dimension to modernization (a critique that could be extended to state socialist models of urbanization). Mandel's observations can be read as an indictment of a particular kind of twentieth-century technocratic modernity that reflects the structural dimensions of political contestation in the urban arena. His argument is also apposite in relation to the emerging technomanagerial logic that marks the contemporary resilience paradigm and its geoconstructivist ambitions. The frontiers of commodification have moved inexorably into every dimension of ecological relations and other-than-human life.[46] Can the existing neo-Marxian emphasis on circuits of capital capture this evolving dynamic? Or do we need to recalibrate circulatory metaphors to take account of distributed conceptions of agency and the work of nonhuman actors in the shaping of urban space?

The conceptual terrain of the Anthropocene is marked by a tension between an "adaptive Anthropocene," an optimistic view in which rapid change is presented as an opportunity to reconstruct human environments, and a "dystopian Anthropocene" in which destructive, unpredictable, and irreversible dimensions to environmental change threaten to overwhelm human capacities to respond.[47] Most of the adaptive variants of the Anthropocene mark a radical elaboration of existing discourses within the sustainability literature, whereby a combination of technological innovation, market dynamism, and behavioral change takes precedence over structural forms of social and political transformation. Geoconstructivist responses to the Anthropocene favor large-scale technological fixes such as nuclear-powered desalination plants to enable the limitless expansion of vast littoral agglomerations.[48] In a similar fashion, the rapid development of biotechnologies enables greater control over food production, enhanced forms of domestication, and increased possibilities to manipulate genetic resources.[49] Under the adaptive Anthropocene, cities are presented as concentrated nodes of resource efficiency in a symbolic inversion of earlier urban-rural antinomies within the environmental literature. The recently articulated "ecomodernist manifesto," for example, promoted by the California-based Breakthrough Institute, notes that cities occupy less than three percent of the global land surface area. In this spatial logic, urbanization protects the global environment from more damaging patterns of land use through a process of "decoupling" modernity from nature.[50]

Advocates of the adaptive Anthropocene use evidence from archaeology and paleoecology to contrast the agricultural efficiency of modernity with the denuded subsistence landscapes of past human societies. It is a global history within which ecological systems are inherently dynamic, biodiversity is in a state of constant flux, and new natures are continuously emerging. There is, however, a conflation between land use and biodiversity, so that phenomena such as "large-scale forest recoveries" become part of a wider argument for global market integration and the "displacement of production needs" to more efficient (i.e., poorer) locations.[51] The direct effects of landscape disturbance, such as ecological endangerment or zoonotic disequilibria, are absent from the adaptive paradigm. For critics of the "good Anthropocene," with its emphasis on new modes of "Earth system governance" and "planetary stewardship," this political agenda is dismissed as little more than a high-tech Promethean fantasy. The geographers Erik Swyngedouw and Henrik Ernstson, for instance, drawing on the insights of Frédéric Neyrat, cast the Anthropocene as "a peculiar periodization that splits modernity in two."[52] The question of starting points is thus recast in terms of enhanced forms of technologically enhanced human agency rather than specific geochemical traces. The Anthropocene can be characterized as a historically specific cultural construct that combines capital, technocratic modernity, and a belief in the malleability of nature into a new kind of ideological synthesis. There are parallels between the messy omnipotence of the Anthropocene and the contours of postmodernism in the 1980s, as a culturally and historically specific kind of ideological formation.[53]

In contrast to the techno-optimism of the adaptive Anthropocene, a range of more critical perspectives have sought to elucidate the emerging patterns of environmental destruction under "digital capitalism" and its associated material articulations. The idea of the "stack," for instance, as elaborated by the sociologist and design theorist Benjamin H. Bratton, denotes a complex interplay between new digital infrastructures and landscapes of resource extraction. Bratton's use of the term "third nature," which he deploys to describe the digital realm, does not denote a process of dematerialization but a new set of digital-material interfaces emerging under the technological aegis of late capital. The "earth layer" for Bratton is simply the starting point for a "planetary-scale computation" that "disembowels geological resources."[54] Media theorist Jussi Parikka similarly emphasizes "the unsustainable, politically dubious, and ethically suspicious practices that maintain technological culture and its corporate networks."[55] For Parikka, these concerns fold into a reading of "deep time" that moves beyond the existing use of the term by Siegfried Zielinski and others and encapsulates the geological realm, its associated nonhuman processes, as well as the contemporary global ecological crisis.[56] Parikka's alternative neologism "Anthrobscene" forms part of an emerging critical literature that questions geocapitalist models of environmental adaptation.[57] In a similar fashion, the literary theorist Frédéric Neyrat offers a critique of technocratic omniscience or "hypermodernity" in relation to the irreducible and ultimately uncontrollable dimensions to nature. In contrast to the geoconstructivist faith in the earth as little more than a stage set for human manipulation, Neyrat emphasizes the fragility, interconnectedness, and unknowability of nature as an alternate conceptualization to neo-organicist readings of earth systems.[58] From the geoconstructivist perspective nothing lies fundamentally "outside" the scope of the technological apparatus of modernity; ontologies of flux and uncertainty have been co-opted into the geoconstructivist schema to underline a sense of the material realm as a space of infinite malleability.

Panglossian interpretations of the Anthropocene tend to assume that novel combinations of capital, technology, and human innovation will carry the earth through its moment of danger. In some variants of this hypothesis the future planet will be inhabited by an enhanced human subject—a technoeuphoric version of the posthuman. Such dreams of omnipotence mark a variant of twentieth-century futurism reprised for the twenty-first century and beyond. In contrast, the term "Capitalocene," as deployed by Jason Moore, Andreas Malm, and others, emphasizes the inherent ecological tensions underlying global capital. For Moore, the "Capitalocene" denotes "a way of organizing nature—as a multispecies, situated, capitalist world-ecology."[59] The periodization of environmental change is centered on the history of capital rather than involving a more vaguely framed emphasis on the human impact and by extension a Eurocentric conception of the human subject.

The question of irreversibility is captured in the term "metabolic rift" which frames modernity as a form of radical environmental disjuncture. With its origins in Karl Marx's critique of capitalist agriculture, the term "metabolic rift" denotes a different emphasis from that of "urban metabolism" (when conceived as a flow-based model of urban space or as a neo-organicist conceptualization of cities as a concentrated zone of life supporting infrastructure systems).[60] One of the first scientific elaborations of irreversible environmental damage is captured in the agricultural research of the nineteenth-century German chemist Justus von Liebig, who described how the complexities of natural cycles that replenish soil fertility stood in stark contrast with the dominance of short-term speculation within the capitalist economy.[61] The idea of metabolic rift can be used in a double sense here to evoke the fragility of soils that take a very long time to form and the discontinuous stratigraphies of human history carved into the surface of the earth. Liebig's emphasis on protecting the fertility of the soil found echoes in the emerging emphasis on the "organic city" in the prebacteriological era, exemplified by Baron Haussmann's plans for Second Empire Paris.[62] In particular, Liebig's analysis of damage to the "soil cycle" influenced Marx's critique of the destructive impact of capitalist agriculture and his identification of an "irreparable rift" whereby large landed property reduces the agricultural population to an ever decreasing minimum and confronts it with an ever growing industrial population crammed together in large towns; in this way it produces conditions that provoke an irreparable rift in the interdependent process of the social metabolism, a metabolism prescribed by the natural laws of life itself. The result of this is a squandering of the vitality of the soil, which is carried by trade far beyond the bounds of a single country.[63]

By "natural laws" Marx emphasizes the question of irreversibility within the web of life rather than the ideological use of nature to legitimate historically specific kinds of social relations. Marx notes how "progress" in the field of capitalist agriculture "is a progress in the art, not only of robbing the labourer, but of robbing the soil," in an early appreciation of the ecological contradictions of capitalist abstraction.[64] Marx's reading of Liebig has recently been elaborated to provide a contemporary analytical tool for understanding destructive relations between society and nature. Jason Moore, for example, emphasizes how the expanding "frontiers of appropriation" that have shaped the "world-ecology of capital" are based on a fundamental tension between a finite nature and a capitalist imperative that is "premised on the infinite."[65] Though the idea of metabolic rift has its origins in the critique of capitalist agriculture, it holds wider implications for the intersection of capital with nature, including the conceptualization of urban metabolic pathways.

Neo-Marxian readings of urban metabolism mark a divergence from industrial ecology and other flow-based models that we encountered in the introduction to this book, by extending the analytical frame from the original emphasis on soil to include infrastructure, technological networks, and other functional components of urban space. A modified conception of metabolism, developed under the aegis of urban political ecology, emphasizes the intersection between the circulation of capital and the production of the built environment. For the geographer Erik Swyngedouw, the process of urbanization is founded on "the perpetual metabolic transformations of nature."[66] Swyngedouw uses the term "metabolism" to delineate the biophysical inputs required for the production (and maintenance) of urban space but also, in a metaphorical sense, to show how flows of capital become entrained within cultural artifacts, including the immense complexity of the built environment.[67] In the work of Swyngedouw and others the secondary circuit of capital becomes analytically aligned with a neo-Marxian elaboration of the concept of urban metabolism, yet this flow-based model of urban space is very different from systems-based conceptualizations of urban ecology. In the case of systems-based ecological models the difference between human and nonhuman components of urban space are not made analytically distinct: indeed, these circulatory dynamics are effectively elided in the quest for some form of epistemological unity. In contrast, the neo-Marxian conception of circulation places capital at its center, as a historically specific driver for the urban process.

The classic "spatial fix," as articulated by Harvey, Swyngedouw, and others, moves towards the more generalized metabolic dynamics of a "socioecological fix" that extends to energy transitions, postcarbon futures, and other strategic diversions of capital.[68] An expanded reading of urban metabolism holds parallels with recent developments in critical theory that draw on a synthesis between neo-Lefebvrian analysis of urban space and the field of neo-Marxian cultural analysis à la Fredric Jameson.[69] Jameson, for instance, is particularly interested in the "peculiar and specialized abstractions" associated with the rise of finance capital and their inflection with both cultural and material forms through land speculation.[70] Following Harvey, he emphasizes that the value of land originates in forms of "fictitious capital" that are "oriented towards the expectation of future value."[71] The reconstruction of urban nature forms part of a speculative strategy to leverage a notional future value into the capitalization of the material present. There is an intensified shift towards the externalization of the temporalities of capital circulation through the search for new investment opportunities in urban space.[72] What Jameson refers to as the "colonization of the future" is fundamental to the emerging interface between capital, ecology, and global patterns of indebtedness.[73] By 2019, for example, global levels of government debt had reached some 66 trillion dollars, equal to 80 percent of global GDP, around double the level witnessed in 2007 before the global banking crisis.[74]

A renewed emphasis on metabolic rift connects with an earlier phase of political economy that occurred before the rise of marginalist economics, the Kuznets curve, and the "disembedding of the economy from natural constraints."[75] The recent elaboration of an environmental Kuznets curve, for instance, posits that economic growth initially worsens environmental problems before eventually leading to widespread improvements. Yet this reprise of the Kuznets curve rests on the notion of "reversible" externalities, along with misplaced assumptions about the relationship between economic productivity located in one place and facets of environmental destruction displaced to another.[76] Urbanization involves the continuous redistribution of environmental externalities: just as nineteenth-century cities banished abattoirs or cemeteries to the urban periphery, we find that many contemporary cities have successfully displaced environmentally damaging processes to increasingly distant localities or less regulated metropolitan areas. It therefore makes little sense to focus on the environmental achievements of individual cities in isolation since they form part of an elaborate set of interconnected processes and transformations.

5.3 URBAN REFUGIA

Cities form part of what has been termed the "new Pangaea" comprising an increasingly integrated global space for ecological processes and evolutionary dynamics. For some commentators the drift towards "biotic homogenization" marks a deleterious step towards a degraded global ecology.[77] For skeptics towards the environmental value of urban biodiversity, cities provide an example of high alpha diversity but low beta diversity, to use the American ecologist Robert Whittaker's distinction, in which a large concentration of species in one place masks a declining number of endemic or regionally distinctive organisms.[78] In contrast, many urban ecologists, and especially urban botanists, have made these novel socioecological assemblages the focus of sustained analysis as part of a new kind of urban ecological imaginary. I would argue that we can be concerned by global biodiversity loss but also fascinated by the cosmopolitan intricacies of urban nature. In the case of plants, for example, many urban settlements have a flora that is now at least one third nonnative so that existing botanical typologies become much harder to sustain. As we saw in chapter 3, the flora of modern cities has become steadily more cosmopolitan as successive waves of species from all over the world have become established, sometimes as garden escapes or other naturalized additions, but most often simply through chance via trade routes, transport networks, and other means. A recent study of European cities found that around 40 percent of the spontaneously growing plant species were introduced, with the figure approaching 60 percent in some cases.[79] A study of Kiev in the early 2000s reveals the presence of over 500 nonnative plant species, derived mainly from North America and the Mediterranean.[80] In the city of Chonju in South Korea a survey found that some 17 percent of the flora was nonnative, with most of these species originating in North America, Europe, China, and elsewhere in Asia, but a number also from Africa and Latin America.[81]

This drift towards a more intensely connected global ecological formation has been under way for millennia, with a sharp acceleration under modernity. Significant developments include the environmental impact of European imperialism, marking an unprecedented degree of interaction between hitherto separate ecological systems.[82] Although many, perhaps even most, of the species that constitute the "new Pangaea" have no deleterious or disruptive consequences for existing ecosystems, there are some species that have become the focus of intense ecological or epidemiological concern. The global impact of species exchange has been highly uneven, with islands or relatively isolated ecosystems often worst affected.[83] Biodiversity hotspots with high concentrations of endemic species are also vulnerable to forms of ecological disturbance, and where these zones intersect with urban areas we can encounter conflicting ecological imaginaries spanning generic forms of "urban greening" as well as more locally specific kinds of environmental discourse. In the case of Cape Town, for example, the metropolitan area coincides with the Cape Floristic Region which has a high concentration of endemic species of global conservation concern. The geographer Henrik Ernstson has shown how the complexities of a "settler public sphere" in relation to botanical discourse in Cape Town are laid bare through tensions between generic conceptions of nature such as the need for shade-providing trees and the so-called "fynbos fanatics" devoted to the protection of endemic species and distinctive ecological biomes (the word *fynbos* is of Afrikaans origin and refers to the species-rich shrubby vegetation of the Cape mountain region).[84] In this instance a nativist botanical discourse that is itself rooted within a colonial legacy of taxonomic ordering and landscape appraisal is simultaneously mobilized *against* the ecological impact of European imperialism. The implications of a decolonial analytical lens for the elucidation of postcolonial urban ecologies adds an additional layer of complexity to the question of botanical symbolism and cultural belonging. In the Delhi metropolitan region, for example, the sociologist Amita Baviskar has highlighted how some invasive species such as the mesquite (*Prosopis juliflora*), a drought-resistant shrub originally from Central and Latin America, owe their origins to colonial-era afforestation programs and are now dominant elements of ecologically degraded landscapes.[85] The provincializing of urban ecological discourse unsettles existing understandings of relations among conservation biology, novel ecosystems, and variants of cosmopolitan environmentalism.

Biodiversity discourse has evolved to make distinctions between plants that pose little threat, and may even provide specific benefits such as roosting opportunities for birds, and species that negatively impact existing ecosystems, including the increased risk of fire beyond the levels that might be naturally associated with specific types of forest ecosystems. Whilst there is a degree of public support for this more nuanced approach to invasive species, at least where the outcome does not effectively diminish access to nature, there is much less consensus over attempts to eradicate birds or mammals such as the Himalayan thar (*Hemitragus jemlahicus*) on Cape Town's Table Mountain.[86] The attempt to remove invasive mammals has often proved controversial, especially where species have been adopted within contrasting cultures of nature, as illustrated by the violent eradication program directed towards the brushtail possum (*Trichosurus vulpecula*) in New Zealand, where some schools had even encouraged children to mock dead animals.[87] Here we find an illustration of Clare Palmer's relational approach to animal ethics: that whilst humane control of sentient animals might be justifiable under specific circumstances, the underlying cause of the ecological disturbance in question should also form part of the analytical frame. In particular, Palmer seeks to navigate between an emphasis on the individual rights of animals—the unwanted possum in this case—and larger-scale concerns with species, populations, and ecosystems.[88] The urban context provides further complexity to these ethical dilemmas, since a city is clearly not an intact ecosystem and many of these "problem animals" occupy the liminal "contact zone," to use Palmer's expression, that lies somewhere between wild and domesticated forms of nature.

The diverse socioecological assemblages encountered in cities problematize the meaning of ecological authenticity: the rediscovery of urban rivers, for instance, or the creation of "natural floodplains" raises questions about the temporal specificity of any putative baseline ecology. In the case of the UK, for example, the Turkey oak (*Quercus cerris*), which is ordinarily regarded as a nonnative tree, was in fact present during the previous interglacial period some 130,000 to 115,000 years ago. Having been introduced over the last 300 years it has brought with it a distinctive insect assemblage that is now serving as a food source for birds whose usual diet has been disrupted by seasonal shifts induced by climate change.[89] The example of

217

the Turkey oak illuminates three specific issues: the relatively arbitrary choice of ecological baselines; the presence of harmless novel ecological assemblages; and the potential advantages of new socioecological interactions under changing environmental conditions. Yet this tree is routinely categorized as an invasive species: the organization Plantlife, for instance, which promotes the protection of native wild flowers and their habitats, has been lobbying the UK government to have this species added to Schedule 9 of the Wildlife and Countryside Act that would prohibit its sale or planting. There are now several vegetation management programs under way in the UK "aimed at eliminating Turkey oak and other alien tree species."[90] Yet climate change is now unsettling any meaningful distinction between "native" and "nonnative" species: it is predicted, for example, that walnuts and other southern European trees may soon become self-sustaining in the parks and woodlands of northern Europe. Furthermore, some cities are being reconceptualized as "climate change forests," thereby connecting the idea of urban space as a laboratory for future ecologies with emerging resilience paradigms in landscape design.[91]

As biodiversity has declined in many nonurban ecosystems, the role of cities in nature conservation has become more significant. The idea of the ecological refugium has tended to be applied to isolated or relict populations in relation to Quaternary environmental change, but there is now growing interest in the idea of describing a wider range of habitats as "safe havens" under the Anthropocene.[92] The French horticulturalist and landscape designer Gilles Clément, for example, has explored the possibility of creating ecological refugia in urban landscapes. Clément links the idea of urban refugia to what he terms *le tiers paysage* or the "third landscape" characterized by its higher biotic diversity than surrounding areas. For Clément, this has three forms: the abandoned or *délaissé* space associated with past agricultural or industrial uses; the *réserve* spaces that have been scarcely modified by human activity, whether by chance or inaccessibility; and designated nature reserves or *ensembles primaires* which enjoy some form of legal protection.[93] It is the idea of *délaissé* space, however, that is of particular significance for marginal or relatively overlooked urban sites with high levels of biodiversity. In his design for Parc Henri Matisse in Lille, for instance, Clément created an inaccessible raised island for spontaneous ecological processes that he hopes will serve as an ecological refugium, or seed bank, to allow more vulnerable species to survive and recolonize the surrounding area (figure 5.2).[94]

The cultural and ecological significance of urban seeds is also an inspiration for the Brazilian artist Maria Thereza Alves, who has examined exposed earth after construction activity to search for dormant seeds (some species can germinate many years or even decades after being buried). For Alves, these residues of life serve as a kind of ecological archive which she has further explored in her "ballast flora" projects in Bristol, Liverpool, Marseilles, and other port cities.[95] The unusual plants found on ballast hills near harbors became a recognized component of urban floras during the nineteenth century, with detailed lists of species associated with sites where ships offloaded their ballast rocks and other refuse.[96] In her ongoing project *Seeds of Change* (1999–), Alves uses the sites of ballast waste from ships as a means to examine interconnections between colonialism, migration, and trade networks, as a radical starting point for the historical exploration of urban flora (figures 5.3 and 5.4). In the case of Liverpool, for instance, an extensive ballast flora had developed in association with the transatlantic slave trade, producing what Alves terms a "borderless history" that she excavates through a combination of historical botanical surveys and contemporary site visits. In these and other instances the collection of urban seeds serves to sustain collective memory through the "spectral materialism" of past violence that is reflected in global patterns of biodiversity.[97]

The role of cities, or at least of designated zones or enclaves within the urban fabric, has gained elevated significance as part of wider conservation strategies. The literary critic Ursula Heise, for instance, refers to these refugia as "urban arks" where vulnerable species can be protected from wider threats. For Heise, these ecological sanctuaries imply a "multispecies

Figure 5.2   Gilles Clément's design for Parc Henri Matisse in Lille includes an ecological refugium created out of construction debris (2011). Photo by Matthew Gandy.

Figure 5.3 Maria Thereza Alves, *Seeds of Change: New York—A Botany of Colonization* (2017). Installation view, Vera List Center/Sheila C. Johnson Design Center, The New School, November 2017. Photo by David Sundberg, courtesy Vera List Center for Art and Politics.

Figure 5.4 Maria Thereza Alves, *Seeds of Change: Liverpool—A Botany of Colonization* (2017). Courtesy of Maria Thereza Alves.

justice" that rests on the shaping of urban space for "both biological and cultural diversity."[98] Cities, and their diverse biotopes, are now emerging as an unexpected sanctuary for many species under severe pressure within their natural range. She describes how the red-crowned parrot (*Amazona viridigenalis*) has spread since the 1980s along the San Gabriel Valley to the northeast of downtown Los Angeles. The LA metropolitan area is now a sanctuary for this species because it has become endangered in its native range in northeast Mexico, where a combination of logging activities and poaching for the international wildlife trade has dramatically reduced its numbers: it is thought that the Los Angeles population of the red-crowned parrot may now exceed that of Mexico.[99]

The attempt to connect biodiversity with landscape design is now an increasingly significant element in urban environmental discourse, reflected in the work of Kate Orff, Maike van Stiphout, and others. "To design nature inclusively is to work with complex information," notes van Stiphout, "together with diverse disciplines and on multiple scales."[100] The move beyond ecological rhetoric towards more precise taxonomies marks a more systematic set of engagements between ecological science and landscape design. The cohabitation of urban space with nonhuman others is taken as a starting point for conceptualizing urban form. For Orff, an expanded conceptualization of urban ecology lies at the intersection between the Anthropocene and its acceleration into an "unknowable complexity," on the one hand, and the "new systemic challenges of the petrochemical era," marked by multiple and proliferating environmental impacts, on the other.[101] In some cases, however, designing with nature can go badly awry: the award-winning Bosco Verticale (vertical forest) project by Studio Boeri for two "forest-clad" residential towers in Milan, completed in 2014, was later replicated for the Qiyi City Forest Garden in Chengdu, China. Following completion in 2018, however, the eight vegetation-covered towers of the Chengdu project became plagued by swarms of dengue-carrying mosquitoes. By the autumn of 2020 only a few residents had dared to move in, with Beijing-based Agence Presse-France reporting that "plants have almost entirely swallowed up some neglected balconies, with branches hanging over the railings all over the towers."[102] It is ironic perhaps that these vertical urban topographies have provided breeding grounds for disease vectors that have evolved in relation to the ecological niches provided by the complex physiognomy of forest trees. These and other examples illustrate how generic urbanism, including various types of ecological façades, lies in tension with the maintenance complexities of infrastructure systems and the specificities and temporal dynamics of microecologies in the urban environment.

If the focus of an ecological imaginary shifts from the protection or reconstruction of "pristine" ecosystems, or fragments of "first nature," towards the production of new natures by experimental forms of rewilding, then there is an interplay between predetermined forms of intervention and nonintervention from the outset. The grazing regime used on the Oostvaardersplassen urban rewilding project east of Amsterdam, for instance, has proved highly controversial: the sight of horses and cattle in a nature reserve left to die from starvation under the controlled conditions of an ecological experiment has provoked public disquiet, raising questions about the purpose and limits of human intervention in relation to "new natures."[103] Urban rewilding experiments such as the Oostvaardersplassen are rooted in a putative form of Pleistocene ecological authenticity, albeit at variance with existing ecological imaginaries, through an emphasis on more open rather than densely forested types of landscapes, in settings that place greater emphasis on the role of grazing herbivores. For the Dutch biologist Frans Vera, the creation of experimental spaces of rewilding such as the Oostvaardersplassen connects with a "concern for the European nature of tomorrow," yet this is ultimately a synthesis between two different kinds of ecological imaginary.[104] Vera wonders what kind of nature should exist for nature conservation if "untouched nature" no longer exists. In an urban context the ideologically framed distinction between natural ecosystems and various forms of human modification has little relevance. The application of controlled forms of "noncontrol," as part of an experimental field of urban ecology, holds parallels with interest in the use of "nondesign" for the creation of urban landscapes, except that the ethical issues become more complicated when extended to animals.

5.4 LABORATORIES OF THE FUTURE

Cities can serve as laboratories for the study of future ecological scenarios: urban biotopes can be recast as experimental zones for the production of scientific knowledge, as evidenced by ecological research programs under way in several cities. As early as the 1970s, the urban botanist Herbert Sukopp and his colleagues, based on research in West Berlin, were suggesting that cities could play the role of laboratories to study "the prevailing ecosystems of the future."[105] Similarly, the Rotterdam-based ecologist Jelle Reumer notes that the city will be "the dominant ecosystem of the future" so that scientists will be concerned with understanding how cities function "in a biological, ecological and evolutionary sense."[106] A study of urban ants in Brazil, for instance, notes that cities can "serve as excellent natural experiments for quantifying the impact of climate change on organisms."[107] Specific species such as the tree of heaven (*Ailanthus altissima*) have served as experimental prototypes for future urban forests on account of their capacity to withstand pollution, poor soils, lack of water, and higher temperatures.[108] The tree of heaven has served not just as an ecological indicator for the urban heat island effect but also as a cultural and political symbol for spontaneous forms of urban nature, as I outlined in chapter 3. A queered or postcolonial reading of the presence of the tree of heaven in marginal landscapes provides a nonhuman elaboration of the violent displacements and differentiations that mark the structuring of urban space.[109] The twin processes of ecological erasure and "green gentrification" surrounding the New York High Line project, for instance, emphasize the presence of "queer memory and mourning for a space of successional ecological emergence and for counter-normative social relations."[110] In this sense, as we saw in chapter 2, the strange ecologies of void spaces also enable the flourishing of alternative social and cultural life, creating experimental zones for the human and nonhuman realms alike.

Writing in the late 1990s, in the wake of the appearance of coyotes and other wild animals in Central Park, the nature writer Anne Matthews reflects on "this new and baffling wilderness" emerging in American cities.[111] Matthews evokes a neo-Ballardian vision for a future New York by the mid-twenty-first century:

> A dusty, battered Central Park will be inhabited chiefly by junk trees and weedy species. Oaks, wildflowers, and songbirds will be gone, or nearly so, from the new, overheated New York, though other, more adaptable varieties of urban wildlife should spread and thrive, from the harbor herons to the midtown coyotes. Under city floods will drive millions of New York rats to the surface, but rats are eminently adaptable, and should enjoy the incessant sewage backups along the avenues of midtown. And New York in 2050 will have become a fine haven for warm-climate insects of every sort, from yellow jackets and killer bees to Formosan termites, fire ants, and a great many mosquitoes, many of them carrying viruses, some familiar, others emerging.[112]

At about the same time that Matthews was completing her manuscript, the mosquito-borne West Nile virus did reach Central Park. The park was consequently closed in the summer of 2000, following the discovery of infected birds at a series of East Coast locations, marking the arrival of a previously unknown pathogen to North America.[113] Future ecological imaginaries can pose questions that are extensively occluded in systems-based metabolic scenarios. The denuded and weedy future that Matthews describes is reminiscent of the kind of "trash forest" that Emma Marris evokes in her survey of spaces that have long played only a marginal presence in ecological theory and conservation biology.[114] In a "trash ecosystem" we find that matter is decidedly out of place in relation to existing ecological paradigms.

Cities have become the domain for a new wave of nature writing, as well as novel forms of ecocriticism. There is renewed interest in the imaginative insights of Robert Browning, Richard Jefferies, and other nineteenth-century accounts of modern cities overrun by nature: we find a degree of continuity, for example, between early articulations of a "ruin aesthetic" and more recent neoromanticist or science fiction depictions of an urban ecological imaginary.[115] Franz Hohler's parable *Die Rückeroberung* (The recapture), first published in 1982, describes the return of wild animals to Zurich. The first indications are multiple sightings of eagles, wolves, deer, and bears, followed by the aggressive growth of plants (ivy and butterbur in particular) that quickly begin to engulf the entire city.[116] This power of plants to tear cities apart has become a source of fascination for "speculative ecologies" that weave scientific insights with imaginary future scenarios. In Alan Weisman's *The world without us*, published in 2007, he explores what would happen to cities and other human landscapes in the absence of people. Weisman draws inspiration from spaces of abandonment that serve as large-scale ecological experiments, such as deserted zones of geopolitical exclusion: he describes, for example, the settlement of Varosha, abandoned after the partition of Cyprus in 1974, where within just six years a carpet of white cyclamen flowers had broken through tarmac roads.[117] The removal of people does not, however, lead to a return of previous ecosystems but to the emergence of new ecologies of abandonment: in the case of the Cyprus buffer zone, a strange ecology has emerged in which packs of feral dogs have become the top predator.[118]

These dog ecologies are reminiscent of the posthuman "dog civilization" described by the American science fiction writer Clifford D. Simak, who published his *City* series between 1952 and 1973. Simak describes how dwindling human populations have retreated into rural isolation and are gradually replaced by intelligent dogs, who reflect back on the oddities of human civilization including the existence of cities.[119] The city form is presented as an archaeological curio within his futurist canine pastoral (the dogs themselves are ultimately displaced by mutant ants in the final book in the series).[120] More recently, in Emily St. John Mandel's novel *Station Eleven*, published in 2014, a postpandemic Toronto is described as the "silent metropolis" after a variant of swine flu has killed off most of the human population. In this quiet zone of devastation set in the Great Lakes region, human society has been relegated to a relatively minor presence within the abandoned landscapes. But does this emphasis on nature's resilience provide ideological sustenance for the "adaptive Anthropocene"? It is notable that elements of "auto rewilding," to use Anna Tsing's expression, such as the return of fish to urban rivers or the arrival of animals on the urban fringe, have enabled postindustrial ecologies to serve as a wider marker for the innate resistance of nature.[121] A distinctive right-wing strand of environmental thought that we might term "biodiversity decline skepticism" has emphasized the presence of coyotes, foxes, kites, and other motifs for a resurgent urban nature to insist, along with Erle Ellis and others, on the wider benefits of "sustainable intensification" within wider agricultural and industrial landscapes. Instances of land abandonment and the regrowth of forests are enlisted into an ideological schema for the active role of nature in remaking landscapes under the "adaptive Anthropocene."[122]

Emerging patterns of global environmental change have increasingly challenged what the Indian writer Amitav Ghosh refers to as the "literary forms and conventions" that developed in parallel with the modern accumulation of carbon dioxide in the atmosphere.[123] In the wake of the Great Acceleration, however, there has been an increasing disjuncture between the imaginative scope of fiction and the scale of environmental disruption. For Ghosh, the subject matter of rapid and uncontrollable environmental change has been routinely relegated to the genre of science fiction, in a creative disavowal that marks "an aspect of the broader imaginative and cultural failure that lies at the heart of the climate crisis."[124] The cultural resonance of the Anthropocene has been marked by a shift of the literary imagination from the fragmentary genre of the postmodern novel towards a more expansive global sense of scale. The postmodern emphasis on irony, identity, and self-reflexivity has been partially displaced by less individualized or inchoate narrative structures in which the human subject becomes decentered. The classic urban-rural antinomy of the nineteenth-century novel has been widely superseded by a new set of geographical and temporal displacements between the local and the global, and between the material and the immaterial.[125]

A contrasting trope to the depiction of abandoned cities overrun by nature is that of future cities where it is the animals that have disappeared or been deliberately extirpated. In the Argentinian author Agustina Bazterrica's extraordinary novel *Cadáver exquisito*, first published in 2017, there is a deadly virus that has rendered all nonhuman life both dangerous and inedible.[126] Under what Bazterrica terms "the Transition," the protein needs of society are to be provided by an organized shift towards state-sanctioned cannibalism, accompanied by the mass culling of animals. The few animals remaining are regarded with a combination of fear and disgust: umbrellas are widely used to avoid the lethal threat of bird droppings or any other traces of wild nature. Meanwhile, the eating of human flesh, referred to as "special meat," is gradually advanced through the use of slick advertising campaigns and the use of hidden industrial facilities to raise people purely for food. At first, the only shoppers to request human meat are "the maids of the rich" purchasing choice cuts for their employers, whilst a parallel black market evolves for poorer-quality human meat diverted from hospitals, mortuaries, and other sources. For the very rich, the trophy hunting of humans is allowed, followed by the elaborate culinary preparation of their quarry for exclusive banquets. In one passage the main protagonist, who works at a state-of-the-art human slaughterhouse, reflects on the city now emptied of its nonhuman inhabitants:

> He drives slowly through the city. There are people around but it seems deserted. It's not just because the population has been reduced. Ever since animals were eliminated, there's been a silence that nobody hears, and yet it's there, always, resounding throughout the city.[127]

Tellingly, one of the only traces of nonhuman nature in the novel is a litter of stray dog puppies living in the city's abandoned zoo, but they are brutally beaten to death when discovered by human intruders. Bazterrica transposes the technosphere of modern protein production to a near future where human society has transitioned to extreme forms of social stratification driven by the politics of food, fear, and a nihilistic modernity (being sent to a human abattoir is a routine punishment for any form of social dissidence). If this is indeed a novel for the Anthropocene then it explores the ramifications of a world that is emptied out of both nature *and* humanity. As Claude Lévi-Strauss wrote in a 1993 essay for *La Repubblica*, entitled "Siamo tutti cannibali" [We are all cannibals], the commercial exploitation of the human body, including the utilization of tissue and body parts, has entered a new and technologically sophisticated phase with multiple and uncertain consequences.[128]

A pervasive sense of separation from nature has suffused emerging interest in "weird ecologies," articulated in particular by the philosopher Timothy Morton as part of a new corpus of literature at the boundaries of "speculative realism" and object-oriented ontologies within the human sciences.[129] These theoretical developments are also discernible within science fiction literature and especially in the literary imagination of Jeff Vander-Meer, whose work has become a focal point for explorations of the "new weird" in ecocriticism.[130] In VanderMeer's *Southern Reach* trilogy, published in 2014, the question of scale is reworked as a series of spatial monstrosities such as the mysterious and forbidding "Area X" that clearly transcends the kind of urban-rural antinomies associated with the modern novel. Implicit within recent developments in ecocriticism is the sense that the Anthropocene marks a disjuncture in terms of not only scale but also levels of biophysical and cultural complexity.[131] VanderMeer's waterlogged zone of ruination could also be interpreted as a reprise of the "dismal swamp" from the plantation-era landscapes of the Deep South transposed to a late modern setting. In this sense the novel might be better read as a portal into the political monstrosities of racism within the American landscape rather than specific forms of ecological disturbance associated with the Anthropocene.

If we were to characterize this genre as a literary counterpart to "speculative realism," then, it marks a distinctive kind of synthesis between contemporary literature and developments within the biophysical sciences that exchange the technological fascination of Cold War science fiction and the cyberpunk genre with a late modern turn towards the biological. Under the conceptual aegis of presenting "weird natures" as a fundamentally unknowable domain à la Morton, however, we find a transposition of historical agency into the more elusive and politically ambiguous plane of autopoiesis and biomimicry. Pivotal to the genre of the weird is a posthuman bricolage of genetic code with nonhuman matter and elements. Here it is worth recalling that Roger Caillois's classic essay on mimicry, first published in 1935, questions Darwinian evolution with an interest in neovitalism and epiphenomenal forms of creativity in nature.[132] In ontological terms, therefore, Caillois refuses the narrow instrumentalism of biological reductionism with its emphasis on the clearly defined organism in favor of an extended conceptualization of its relation to an external environment. Following the insights of the psychiatrist Eugène Minkowksi, who sought to combine the fields of phenomenology and psychopathology, Caillois considers how "dark space" is experienced as a blurring of the "distinction between the milieu and the organism" rooted in a pantheistic urge to become matter itself.[133] The body, in such a view, becomes an organic camera that reflects its surroundings. The aestheticization of damaged landscapes in contemporary fiction lies in a state of tension with the analytical scope of critical cultural and environmental discourse. Missing from the new fascination with the weird, however, is any sense of the "violent foundational structures" that have brought these unsettling material and imaginary landscapes into being.[134] When proponents of the Anthropocene invite their readers to "love your monsters," as advanced within the postenvironmentalist thesis, they are not referring to the political monstrosities of violence and injustice.[135]

What are the implications of radical estrangement from nature for cultural representations of urban space? What kind of socioecological formations might inhabit future cities? Around the 1980s, although it is difficult to be precise, the symbolic coordinates of the science fiction city begin to shift as part of a deeper listing of the modernity project. Future cities become increasingly represented in the form of material bricolage, where the detritus of modernity serves as a kind of stage set for partially realized alternate worlds.[136] In Lauren Beukes's 2010 novel *Zoo city*, set in a kind of cyberpunk Johannesburg, any human character who has killed someone is thereafter accompanied by the animal that appeared at the scene of the crime.[137] The main protagonist Zinzi December must carry a sloth around at all times after she is held responsible for her brother's death. The animals that accompany Zinzi and the other outcast characters serve as a kind of creaturely stigmata for past violence: implicit within the novel is the devastating timeline for the global spread of HIV, with its zoonotic origins and its high death toll in the cities of sub-Saharan Africa. Indeed, the permanent acquisition of animals in the wake of criminal acts is described by Beukes as Acquired Aposymbiotic Familiarism (AAF), but is referred to more colloquially by her characters simply as the Zoo Plague. Here we encounter an eco-noir urban imaginary haunted by the presence of animals.

The literary critic Ursula Heise calls for a "different architecture of narrative" that marks a decisive break with existing narrative forms.[138] The scale and complexity of environmental change evokes a neo-Kantian sense of the sublime derived from the proliferation of new data sources, the advent of ever more elaborate models, and increasingly sophisticated modes of digital representation. The novelist William Gibson's classic exposition of cyberspace as a form of "unthinkable complexity" has proved highly prescient. Yet the cyberpunk city, nestling within what Jameson refers to as "the seamless Moebius strip of late capitalism," is an urban landscape that is largely devoid of nature.[139] In Gibson's *Neuromancer*, published in 1984, we encounter a series of emerging dislocations between "meat space" (note the implicit posthuman conceptualization of corporeality) and an emerging skein of dematerialized spaces and networks. Yet there is a tension within the novel between the organizational nexus of modernity and the "outlaw zones" that comprise the material and political dimensions of a constitutive outside that includes the extractive economies of resource depletion.[140] With its neo-noir constellation of concrete and neon, the future city is presented as a kind of dense technological palimpsest with few traces of nature. This emphasis on scale reflects the shifting epicenter of urbanization towards the global South, and especially East Asia, so that earlier leitmotifs of modernity such as London, Paris, and New York, or even postmodernity in the case of Los Angeles, have been replaced by cities such as Chongqing, Shenzhen, and Tianjin.

But where should we locate agency—both human and nonhuman—within these evocations of the future city? Gibson, for instance, has recently turned his attention to the questions of artificial intelligence as well as global forces such as the circulation of capital, rising levels of xenophobia, and environmental uncertainties. In a recent interview, he noted the steady merging of the "offline" and "online" domains, so that the world is effectively undergoing a kind of "spatial eversion" through the materialization of the digital realm.[141] This process is reflected in an expanding landscape of stack infrastructures, terraforming, and extractive capitalism, along with the emerging digital-robot interfaces of urban warfare. In the face of such pressures, Gibson fears for the collapse of the "only liveable planetary ecosystem we know of anywhere," since the material basis of life remains inescapable. In the final instance, his urban ecology is ineluctably bound up with an analog mode of theorization that counterposes the fragility and finitude of life with the eerie abstractions of binary code. Might it be possible to envisage a different kind of ecological interface between the digital and the material that could bring noncapitalist worlds into being?

The increasing sophistication of digital representations of nature can produce heightened levels of verisimilitude, yet how do these visual projections relate to the complexities of material spaces on the ground? Are existing power relations already inscribed in these emerging representational practices? Or can the shifting sociotechnical dimensions to spatial representation reveal new kinds of posthuman environments that explore intersections between agency, sentience, and the human subject? Generic forms of landscape representation mark a sense of estrangement from material spaces, whether these be design models or other kinds of cultural projection. In the digital dystopias of videogames such as *Death stranding* (2019), for instance, created by Hideo Kojima, we encounter a series of postapocalyptic landscapes set amid the ruins of modernity. In Kojima's earlier breakthrough project *Metal Gear Solid* (1994), the origins of these destroyed landscapes are made more explicit through connections with the cultural memory of wartime destruction and the postwar flourishing of black markets and organized crime syndicates. Kojima's exploration of themes such as bioethics, nanotechnology, and the menacing reach of the "military-industrial-nuclear complex" is given further poignancy through the use of archival documentary sources depicting the aftermath of Hiroshima and sites for nuclear waste disposal.[142] Can critically reflexive dystopian genres, as articulated by Kojima and other avant-garde designers, enable alternative worlds to become part of a new kind of social and ecological imaginary? Or does this process of radical cultural "dematerialization" mark a digital precursor to material annihilation? Emerging strands of cultural criticism are now grappling with the disjuncture between narrative form and the scope of environmental threats. Yet environmental discourse, especially that surrounding the adaptive Anthropocene, remains committed to a technologically mediated form of spatial and political retrenchment.

The impetus towards environmental control ranges from an emphasis on specific organisms to the geoengineering of earth systems. In the process, we are witnessing an elaborate technological extension of the human subject, including cryopreservation after death, the construction of elaborate postapocalyptic bunkers, or even the development of "off-world" space settlements.[143] Under more fantastical scenarios, technological change is seen as a race against time to enable artificial intelligence to serve as an escape portal for humanity from a degraded material reality into a purely digital realm, enabled by the mass download of human brains.[144] Yet the replacement of corporeal materiality with a form of digital existence can never be complete: there remains a need for vast stacks of humming servers and other types of material infrastructure. The "artificial infinite" can be contrasted with the inescapable materialities of the human body, ecological processes, and earth systems.

Some of the most interesting questions about the future of modernity are posed through forms of cultural experimentation rather than scientific interventions. The visual arts, for example, can generate theory rather than simply serving as a cultural prism through which existing philosophical or ethical ideas are refracted.[145] The intensity of the emerging Anthropocene discourse within the visual arts has forced an expansion, or even disorientation, of the scientific imagination. "The Anthropocene is to natural science," suggests cultural critic Jennifer Fay, "what cinema, especially early cinema, has been to human culture."[146] "Thus, not only is cinema like the Anthropocene in its uncanny aesthetic effects," continues Fay, "but also, insofar as cinema has encouraged the production of artificial worlds and simulated, wholly artificial anthropogenic weather, it is the aesthetic practice of the Anthropocene."[147] The emphasis by Fay and others on the "artificial" dimension to Anthropocene aesthetics is complicated, however, by the fact that climate change and other kinds of environmental disturbance are real phenomena. Visual culture can highlight the role of the simulacrum in public culture through the "reality effect" and the construction of artificial worlds, but this remains a suspension of disbelief that is also culturally and materially grounded. The environmental simulacrum creates its own category of the real through its efficacy in the material present. The divergence from what Fay terms the "temperate norms of the Holocene epoch" is producing a pervasive manifestation of the ecological uncanny in which the familiar is rendered unfamiliar through a form of environmental estrangement.[148]

5.5 ANALOG ECOLOGIES

A focus on deep time clearly places cities outside the conventional frame of historical analysis, including the *longue durée* as articulated by Giovanni Arrighi, Fernand Braudel, and others. Cities have been widely characterized as markers, or even progenitors, of the environmental crisis. The first cities, emerging in the wake of the Neolithic period some 12,000 years ago, mark the advent of agriculture, new forms of domestication, and sharper forms of social stratification. For some commentators the Neolithic revolution marks a fateful teleology that has led inexorably towards the contemporary environmental crisis.[149] Subsequent waves of city building, spurred by the expanding reach of empires and trade networks, led to distinctive patterns of environmental destruction including extensive deforestation in China, the Mediterranean, and elsewhere.[150] Later waves of urbanization, including the growth of industrial cities driven by the circulation of capital, also produced characteristic patterns of environmental degradation, including new landscapes of extraction and vast environmental sinks.

The emerging space-time geographies of modernity have been superimposed on the existing temporalities of the natural world, including circadian rhythms, migratory pathways, and the dynamics of evolutionary change. How might we combine these different temporalities under an integrated analytical framework? Whilst the concept of deep time helps to contextualize capitalist urbanization, it does not resolve the tensions between different types of human and other-than-human temporalities. And, as Chakrabarty notes, "the Anthropocene would not have been possible, even as a theory, without the history of industrialization" emerging under an "energy intensive" set of unbounded human capacities.[151]

What of the geography of environmental politics itself? How, for example, does the Anthropocene debate, and its associated global premise, interact with alternative perspectives in the global South, or in the fast-growing cities of East Asia? If we return to the example of Chennai, how does Anthropocene discourse help to illuminate the underlying dynamics of environmental change, and in particular the destruction of wetlands on the urban fringe? There is a thriving Tamil public sphere that is engaging with environmental threats, both local and global, marked by a synthesis between vernacular and global discourses.[152] The tension between the universal and the postcolonial that Chakrabarty, Choy, and other scholars have explored is being played out at ground level in Chennai and elsewhere.

The more technophile readings of the Anthropocene share ground with emerging geoengineering paradigms and increasingly elaborate resilience scenarios. Yet geoconstructivist conceptions of environmental change do not address irreversible damage to the biosphere. Indeed, many of the proposed technological fixes to climate change would exacerbate other forms of ecological damage. Furthermore, if such schemes were abandoned or interrupted, then the underlying dynamics of climate change would simply resume. In contrast to geoconstructivist interchangeability, exemplified by various forms of terraforming, we find that biodiversity discourse emphasizes the irreplaceability of both species and ecosystems. In an urban context this generates a tension between various forms of generic urbanism and the specificities of actually existing ecosystems. An emphasis on "knowing what things are," as stressed within alternative kinds of ecological practice, illustrates the uniqueness of different species and ecological formations. In contrast, emerging design-based discourses such as "ecological urbanism" mark a radical elaboration of the longstanding emphasis on the technical modification of environmental systems under capitalist urbanization.[153]

The Linnaean classificatory system can serve as a language of complexity: the horticulturalist Gilles Clément, for instance, decries the pervasive lack of species-level knowledge amongst landscape designers. Yet there is a tension between the appropriation of scientific rhetoric and different forms of epistemic uncertainty as facets of nature become incorporated into environmental discourse. A shift of emphasis from knowing relational ecologies through the interactions between named species towards a post-Linnaean concern with assemblages of living matter may ultimately recast biodiversity as little more than a series of replaceable biophysical configurations. The literary critic Derek Woods has suggested that the "subject of the Anthropocene is not the human species but modern terraforming assemblages," in a critique of both the "species history" invoked by Chakrabarty and also the atomistic underpinnings of earth systems science. The replacement of the species with what Woods terms "assemblage x" provides an intriguing parallel to VanderMeer's "area X" that we encountered earlier, as part of a radical and multiscalar fusion between disparate bodies of knowledge.[154] Yet the political implications of an ecological discourse without species remains unclear: how are precise measures of endangerment to be articulated for a shifting matrix of configurations for living and nonliving matter? And if the human subject is lost as a conscious agent of historical change, then what kind of autonomous or algorithmic fields of agency are left in its wake?

The city—inasmuch as there exists an agreed definition—is widely portrayed as a powerful agent in environmental destruction through its concentrated demands on land and resources, either directly through habitat loss (especially on the urban fringe) or indirectly through the production of food, energy, and other needs. We should be cautious, however, in ascribing some form of "urban agency" to cities that operates independently of the wider ecological contradictions of capital.[155] It is the extractive realm of "cheap nature," to use Moore's term, and its associated environmental sinks, including the atmosphere and oceans, that are pivotal to ecological destruction, rather than the development of cities per se.

For the literary critic Jeremy Davies, a richer reading of environmental change over the long term must encompass a history of what it means to be human, of the rights and protections that have been won for individuals and communities, and of the characteristic forms of social inequality that are embedded within the specific socioecological relations that have accompanied fossil fuel extraction, the expansion of industrialized agriculture, and the destruction of vulnerable biotopes.[156] For Davies, the principal lesson to be drawn from an extended historical timeframe is that there have been many different kinds of human societies and that environmental vulnerabilities such as disease or food scarcity can produce intensified forms of social stratification.

Even if the scientific validity of the term "Anthropocene" depends on an agreed start date, the neologism has nevertheless quickly diffused into public culture as a focal point for wider debates about the rate and scale of environmental change. The term has thus become a divergent cultural motif that connects with posthuman (or superhuman) hubris as well as anxiety over an accelerated and potentially perilous phase in global environmental history. It may be that the contemporary emphasis on a new periodization will ultimately fade from cultural consciousness in a similar fashion to that of postmodernism in the late twentieth century; the novel insights may simply become incorporated into new variants of environmental discourse as matters of political exigency supersede concerns with nomenclature or periodicity. In the context of emerging concerns with urban epidemiology in the wake of the Covid-19 pandemic, however, there are already signs that relational rather than temporal conceptualizations of environmental change hold greater potential for the interpretation of the nonhuman material realm.

During the emergence of the global coronavirus pandemic in 2020 many cities under lockdown found their streets deserted. The absence of people created an eerie atmosphere reminiscent of a de Chirico painting, yet the tranquil emptiness was now filled with birdsong.[1] Animals of every kind began to venture into urban areas along quiet roads, often nonchalantly grazing in front gardens or roadside verges. Newspapers contained many accounts of people rediscovering their affective affinities with the nonhuman realm. A number of writers drew attention to the "healing" powers of contact with nature as a form of respite from the isolation and uncertainty. Yet this newfound sense of affinity with nature belies the brutal relations with nonhuman lives that engendered the public health crisis, along with the uneven epidemiological landscapes of race, class, and age framed by different levels of exposure to risk.

Over the last twenty years interest in the ecology and biodiversity of metropolitan areas has steadily grown. "Almost without anyone noticing," suggests the geographer James Evans, "urban ecology has become a foundational ingredient of governance in the Anthropocene."[2] The interdisciplinary scope of urban ecology has become the paradigmatic case for its enhanced status within a realignment of the environmental sciences under the Anthropocene and a shift of emphasis away from "non-anthropogenic nature."[3] Furthermore, there has been a growing sense of expectation that cities might be capable of providing new forms of policy innovation that can contribute towards global environmental challenges.[4] Key swathes of environmental thinking have shifted from a broadly antiurban position towards a focus on the potential role of cities in promoting new modes of living, more efficient resource use, and cutting-edge contributions to ecological design.[5] But what does this recent elevation of urban ecological thinking within wider environmental and political deliberations portend?

I have argued that the field of urban ecology remains dominated by systems-based conceptions of space—notwithstanding the presence of other approaches derived from botany, natural history, and further field-based observational traditions—and it is this more abstract epistemological framing that has had the greatest influence on emerging environmental concerns with resilience, green infrastructure, and attempts to "ecologize" policy discourse. But can a systems-based urban ecology be scaled up to take account of operational landscapes for resource extraction, the multiple pathways of the global technosphere, and other kinds of dispersed interconnections between nature and society? Is there a potential zone of conceptual engagement with other intellectual traditions, such as urban political ecology or multispecies ethnographies within the ecological pluriverse? In a recent overview of the field of systems-based urban ecology we find a passing mention of "political ecology" in a list of 23 different "disciplines and theories" enlisted into the task of building an integrated approach to urban environmental research. Yet the authors note that "there is no unifying conceptual or methodological approach for investigating complex urban systems throughout the globe." Despite the search for "common elements," they observe that the desire for a consensus moves inexorably in a normative rather than a conceptual direction: it is easier to agree on policy intentions than find an elusive unity within the underlying intellectual landscape.[6] If we accept that some kind of epistemological merger is neither possible nor desirable, then the intellectual challenge becomes framed differently in terms of a more pluralist analytical framework that can accommodate diverse insights, including the decentering of the bounded human subject.

In the introduction I introduced four vantage points for urban ecological thinking: systems-based models, observational tropes, urban political ecology, and the ecological pluriverse. Each of these modes of engagement with urban nature holds specific implications in terms of ecological practice and ways of intervening (or not intervening) in urban space. The long-established and highly influential systems-based idiom clearly connects powerfully with the development of urban sustainability discourse and its associated engagement with the adaptive Anthropocene. The observational legacy of nature study, and affective delight in various manifestations of spontaneous urban biodiversity, connects with the gradual articulation of new idioms in conservation practice and landscape design that seek to valorize novel biotopes. The critical impetus of urban political ecology, with its neo-Marxian underpinnings, points to structural inequalities in the urban arena that impact health, human well-being, and access to metropolitan nature. And the ecological pluriverse seeks to find new ways of incorporating the nonhuman inhabitants of urban space into architecture, design, and everyday life. Although there are tensions between these perspectives, and my argument has sought to navigate between some of these, there is an underlying emphasis within the broader field of urban ecology on the centrality of the urban experience to a richer kind of global environmental discourse.

In isolation, the systems-based emphasis on the scaling up of interconnected metabolic units presupposes the possibility of some kind of interconnected skein of urban governance that might challenge the power of nation-states in the environmental arena. Yet this technomanagerial imaginary lies in tension with the weakness of cities as distinctive actors in politicolegal terms, as critical legal scholars have long observed.[7] As we saw in chapter 4, the status of "legal instruments" in relation to environmental protection raises a series of questions about the institutional and political context for scientific epistemologies. This brings us back to the question of the location of "the political" within urban environmental discourse in the face of the recent upsurge in authoritarian, antienvironmental, and anti-metropolitan ideological constellations.

Many cities could lay claim to being the inspiration for new ways of looking at urban nature: Baltimore, Cape Town, Paris, Rotterdam, and Los Angeles among them. However, it is the city of Berlin, and especially the geopolitical enclave of West Berlin, that has featured most prominently in my own thinking about urban ecology, for two interrelated reasons: first, the postwar landscapes of Berlin presented a unique set of ecological assemblages that became the focus for arguably the most intensive program of study in the world at this time; and second, the cultural and political milieu of the island city, and especially the Institute for Ecology at Berlin's Technical University, fostered one of the most prominent international programs for the study of actually existing urban ecologies on the ground. My own opportunity to spend time at the center in early 2013 allowed me to directly observe the work of urban ecologists, including the pedagogic dimensions to a vibrant field science operating in an urban context. One of the themes that have driven my own project is an attempt to delineate the distinctive contribution of urban ecology to wider developments in urban and environmental thought. Beyond the ecological specificities of urban space, for instance, what kind of unique cultural or political configurations can be observed?

In this book I have adopted a mix of entry points, both conceptually and methodologically, ranging from grassroots encounters with specific sites to the analysis of cultural representations of urban nature in films and novels. At a theoretical level, I have tried to hold a productive tension in play between the local and the global, between the intricate ecologies of urban space and attempts to delineate a broader set of geographical and historical processes. But where does this leave a theory of urban nature? What boundaries or zones of transition, we might ask, lie outside the legacy of systems-based urban ecology? We have seen, for example, that interstitial ecologies can occur in any part of the city, so that the concept of the "urban fringe" or periurban landscapes only operates effectively in relation to specific socioecological formations. In writing this book I have sought to articulate an alternative conceptual synthesis to systems-based models, building on alternative fields such as urban political ecology, along with a range of embodied, ethnographic, and in some cases ecstatic encounters with urban nature. I have sought to bring existing forms of materialist analysis into dialogue with feminist, queer, and postcolonial insights to build a richer sense of the scope for urban ecology.

How should we characterize a politics of urban nature? Can there be a politics of nature that seeks simply to "let live" outside of an ecocentrist framework, which is itself rooted in forms of human intentionality, even if philosophically couched in various rights-based discourses directed towards the nonhuman? In this book I have sought to navigate away from more obscurantist or romanticist ontologies to emphasize how cultures of nature are context-specific, emanating from precise delineations of both human and nonhuman forms of agency and subjectivity. I find contemporary interest in new modes of cohabiting appealing since this shifts the emphasis of environmental discourse towards a wider spectrum of possibilities, from the role of novel or spontaneous ecological assemblages in landscape design to emerging recognition of the recuperative properties of affective encounters with plants and animals in the urban arena. There is an evident tension between, on the one hand, continuities in environmental discourses of manipulation and control, including landscape design traditions of staging metropolitan nature, and, on the other hand, a longstanding countertradition of cultural and scientific fascination with the unexpected outcomes of actually existing urban ecological formations. Whilst aesthetic claims for the protection of marginal spaces of urban nature can often fail in isolation, there are nonetheless a range of additional arguments that can be brought into play such as enhanced protection from asthma, floods, or even existential anxiety. Of course, we should be careful, when enlisting urban nature in mental health discourse, not to drift towards essentialist tropes derived from variants of environmental psychology, yet there seems little doubt that many plants and animals perform affective work.

What are the wider cultural and political implications of the recent surge of enthusiasm for urban nature? Comparative studies of urban biodiversity reveal a paradoxical set of developments: on the one hand, cities often exhibit elevated levels of biodiversity on account of their many biotopes and traces of global connectivity; on the other hand, cities are bound up with broader processes that mark a step change in the destructive momentum of global environmental change. There is, nonetheless, another dimension to this dynamic in terms of the role of cities as generators of new cultures of nature, as reflected in fields such as grassroots scientific practice, ethical debates over relations with the nonhuman, and experimental aspects to landscape design. Sensory contact with nature holds the potential to counter a pervasive sense of affective disengagement from the natural world. These emerging affective and sensory affinities with urban nature can also extend to an awareness of the brutality of the global technosphere, with wider implications for ethical relations with the nonhuman. Yet can the metropolitan cultural sphere be envisaged as more than a cosseted realm of consumption? Ethical relations towards nonhuman others are complicated by questions of context, proximity, and awareness: if I notice a snail or caterpillar on the pavement, I usually move it to a place of safety. The killing or mistreatment of sentient beings in the urban arena has long been a point of political contention. The exhausted working horse of the preautomobile metropolis may have disappeared, but load-bearing animals are still a frequent sight in the cities of the global South or parts of East Asia. The sale of live animals for food in urban markets has become an unsettling spectacle in the context of rising concerns over the zoonotic city. The mass killing of urban animals has often provoked unease or outright opposition, whether this extends to nonnative possums in Christchurch or stray dogs in Bengaluru. The "urban cull" remains a matter of cultural and political contestation.

A recurring theme in this book has been the connections between enhanced environmental knowledge and the political salience of biodiversity. High-profile ecologists such as Edward O. Wilson have been keen to promote a scientifically informed environmentalism. Wilson contends that there is an innate love of nature that he terms "biophilia" rooted in a deepening of knowledge about individual species and their precise role in the web of life.[8] Implicit in such formulations is the sense that modernity has broken a preexisting bond with the natural world. The difficulty with Wilson's position, however, is that it elides the understanding of science with the power of nature as a symbolic dimension within political consciousness. The idea of nature, as Lorraine Daston has shown, cannot serve as the basis for human reason, since it "displays so many kinds of order that it is a beckoning resource with which to instantiate any particular one imagined by humans."[9] The idea of ecoliteracy, as articulated by Wilson and others, implies some kind of familiarity with a taxonomic canon that is itself culturally and historically constituted, whether this be the European legacy of classification systems or the appropriation of other kinds of local or vernacular sources of knowledge.

Can the valorization of urban nature offer an alternative to the violent exclusions that have characterized aspects of biodiversity discourse at a global scale? Existing strands of conservation discourse have often been framed around externally imposed forms of control over land use, sometimes involving the forcible removal of existing human populations or the use of violent measures against intruders.[10] The recent interest in Wilson's "half-earth" proposal, for example, to set aside vast areas of occupied land on behalf of global biodiversity, could be read as a radical elaboration of neocolonial conceptions of global cultural heritage. These forms of "scientific enclosure" can be contrasted with the accidental or vernacular spaces of nature that have formed the focus of my book. Can more participatory dimensions to biodiversity discourse within the urban arena be extended to a global scale as part of wider concerns with mass extinction that might build on grassroots forms of environmentalism? By bringing the cultural and scientific aspects to urban biodiversity into closer dialogue, I have tried to open up the relatively closed field of conservation biology but at the same time work towards new ways of imagining urban nature. In this sense the relational dynamics of urban ecology constitute far more than a set of biophysical flows or interactions, extending to the reimagining of urban space as an experiential realm of cohabitation.

More recently, the rise of what the urbanist Henri Lefebvre referred to as "complete urbanization" underlines a sense that modernity, globalization, and urbanization have become increasingly synonymous, even if the relations between urban and nonurban spaces are more complex than some neo-Lefebvrian formulations allow. The recent articulation of the "planetary urbanization" thesis by Neil Brenner and Christian Schmid has engendered criticism in terms of those voices, ideas, and spaces that are perceived to lie "outside" this overarching vision of an urban planet.[11] In this book I have sought to articulate a productive tension with the neo-Lefebvrian position: on the one hand, the extent and scale of global urbanization are unprecedented, so that very few spaces can be considered to lie fully outside of an urban world; yet on the other hand, the category of the urban requires a finer degree of conceptual and material disaggregation. It is useful, for example, to make a distinction between the city, as a particular kind of social and political arena, and urbanization, as a broader set of ecological and sociotechnical relations. The analytical task is not to subsume the city within the urban or vice versa but to trace connections between diverse cultural and material entanglements, spanning both human and other-than-human realms. The distinction *between* cities and broader processes of urbanization remains significant for a more critically engaged reading of the politics of the biosphere. The implicit redundancy of existing spatial categories such as "city" or "region" raises difficulties in terms of the scale of nature as a cultural and political discourse. Urban biodiversity is not a generic or universal phenomenon, even if the advent of "cosmopolitan ecologies" is indicative of global patterns of interaction and change. Indeed, an overemphasis on "methodological globalism," to use the historian Sebastian Conrad's expression, risks obscuring the differences that matter in the articulation of alternative modernities.[12] Cities lie at the forefront of the development of new ecological imaginaries. The growing salience of the metropolitan-versus-nonmetropolitan political divide reinforces the significance of the urban arena as an experimental field within which new conceptualizations of nature, biodiversity, and the human subject are taking shape.

I have long been fascinated by the language of nature. It is clear that utilitarian terms such as "natural capital" or "ecosystem services" fail to capture the cultural and scientific richness of the web of life. As I suggested in chapter 2, many aspects of urban nature, and especially marginal spaces, confront ambiguities in language and meaning. The beguiling words "ecology" and "nature" are fraught with difficulty, but I am not advocating that we dispense with either of these terms, as some authors have suggested, since they are so richly connected with urban environmental discourse. Furthermore, while the word "ecology" is closely bound up with systems-based perspectives on urban nature, it is increasingly encountered in its plural form, "ecologies," as if to underline the transition from a positivist scientific construct to a range of cultural and political metaphors. As a result, we no longer tend to speak of a singular ecology or a generic ecosystem in relation to cities.[13] I am suggesting instead that we look towards an expanded and more critically reflexive urban lexicon that extends to words or phrases such as *Brache*, *terrain vague*, *poramboke*, and *Stadtwildnis*, and self-consciously look beyond the confines of the Anglo-American linguistic realm to recognize the subtleties and specificities of other languages. The question of communication moves beyond one of agreed terminologies to encompass distinctive cultural histories of urban nature that may hold wider analytical resonance. In some cases, we encounter the tactical appropriation of terms such as "wasteland" or *terrain vague* to signal a different kind of socioecological constellation. The protection of marginal spaces suggests the figure of the "botanist killjoy" (to adapt Sara Ahmed's term "feminist killjoy"), who resists the incessant commodification and erasure

249

of fragments of nature that have not yet been entrained within the speculative dynamics of urban development.[14] How annoying that a human interlocutor might once in a while enable a rare plant or beetle to delay or even prevent the planification of the urban realm.

How should we interpret the marginal or interstitial spaces of urban nature that have been a source of cultural and scientific fascination? Do we need an alternative conceptual lexicon to the jaded dualistic registers of "utopian" and "dystopian," or "beautiful" and "sublime," that are often used in relation to urban and industrial landscapes? In my previous writings on cinematic and postindustrial spaces, I have used neo-Kantian terms such as the "technological sublime," which remain couched in the romanticist aura of the lone human observer. By moving towards a postphenomenological frame of reference, however, in which the bounded human subject is no longer the focal point of analysis, we can delineate a more multisensory engagement with the material dimensions to urban nature. We can acknowledge the presence of "composite subjectivities" that emerge out of affective associations between human and nonhuman inhabitants of urban space, such as a kitchen spider that is noticed everyday but simply left alone, a fox that knows which people might offer food, or a small patch of weeds that are watered from time to time.

The presence of "disordered" urban ecologies represents a kind of queering of the modernity project with its attachment to regularity and simplification. I have drawn on insights from queer theory to highlight forms of socioecological complexity that move beyond the boundaries of urban ecology as conventionally conceived. Spaces of spontaneous nature are more than just unusual biotopes; they serve as portals into alternative conceptualizations of agency, identity, and modernity. I have highlighted at various points a certain playfulness towards the use of methodologies and a commitment to forms of empirical and conceptual experimentation. The role of artists, for example, is notable in posing unexpected questions or subverting existing forms of knowledge. Similarly, vernacular appropriations of marginal spaces reveal lived possibilities beyond the disciplinary reach of architects, planners, and other types of technical intervention.

Alternatives to generic urbanism can be provided by place-specific adaptations to ecological uncertainty such as "cloudburst plans" for surface water runoff, xerophytic planting schemes for semiarid landscapes, and larger-scale reconceptualizations of the biophysical complexity of urban deltas. The construction of bioswales, for example, to capture surface water from sidewalks, car parks, and other impervious surfaces has become a significant element in green infrastructure schemes aimed at reducing flood risk and enhancing biodiversity. These distinctive strips of vegetation in otherwise built-up areas have gradually evolved from a focus of curiosity, or even hostility, towards a familiar element of urban nature in "living streets." In the place of generic conceptualizations of nature there is scope for complexity and nuance in landscape design: indeed, the meaning of design itself is brought into question through the acknowledgment of nonhuman agency and the serendipitous qualities of cohabitation in urban space.[15] Innovative interventions such as Gilles Clément's interest in the "garden in movement" mark an explicit acknowledgment of the temporal aspects to landscape complexity.[16] An other-than-human approach to ecological design can be contrasted with geoconstructivist variants of terraforming, in which relational ethics are displaced by the utilitarian ethos of market-driven landscape urbanism.

Cities can play a dual role in the protection of biodiversity: first, through the provision of a kind of ecological haven for flora and fauna; and second, by enabling the exploration of different socioecological pathways through modernity. Recent patterns of political turbulence have necessitated a partial reprise of the historical role of cities as sanctuaries for human and nonhuman nature alike. Growing threats to biodiversity at a global scale have prompted calls to extend legal rights to nature as an elaboration of existing humanist doctrines. Yet there is a tension here between the extension of a formal rights-based elaboration of existing models of citizenship and the articulation of posthuman conceptions of subjectivity that can extend to multiple configurations of human and nonhuman life. There are clearly limits to a neo-Benthamite ethical and legal framework for the nonhuman à la Martha Nussbaum and others, which emanates from a recognition of putative similarities between human and nonhuman creatures in terms of their degrees of sentience, capacity to experience pain or emotion, or other attributes of life that are worthy of protection.[17] A recognition of multiple forms of subjectivity, both singular and plural, human and nonhuman, in an infinite variety of configurations, requires a very different kind of theoretical basis for its articulation.

How should we navigate questions of agency and subjectivity through the contemporary urban landscape? Jennifer Wolch draws on feminist epistemologies of difference, including Haraway's figure of the cyborg, to help develop a more nuanced ontological landscape for the contemporary city.[18] The emphasis is not on the technologically augmented human subject, or even the fragile interdependencies of urban space, but rather on new socioecological configurations of nature that transcend the human subject as conventionally understood. Wolch uses the idea of zoöpolis, as we saw in the first chapter, to develop a more fluid understanding of relations between society and nature and as a means to counter the latent antiurban sentiment that is still prevalent in much environmental discourse. The embrace of the zoöpolis is not just a celebration of cosmopolitan urban ecologies but also a philosophical and ethical challenge regarding the treatment of nonhuman others that extends to distant sites and locales far beyond the conventional boundaries of the urban. Yet we should be cautious of any straightforward distinction between human and nonhuman forms of agency: reflexive modes of historical agency now extend to multiple kinds of interactions with the nonhuman realm, including distributed modes of cognition that are increasingly associated with algorithmic modes of governmentality. In this respect, various kinds of "smart landscapes," such as the ecological deployment of technological monitoring systems, intersect with material spaces. These emerging ecotechnological surveillance configurations serve as an ambiguous counterpoint to forensic ecologies and the production of counterhegemonic forms of knowledge. The question of the posthuman is thus not simply a matter of transcending the bounded human subject but also of acknowledging new kinds of material interfaces with binary code, information gathering, and the latest iterations of technomanagerial ecological control. What kind of urban ecological imaginaries might be invoked by these new configurations of agency and technology? Do these new types of landscapes begin to blur the boundaries between the material and immaterial realm, or actually reinforce a sense of separation between ecological intricacies on the ground and new modes of governmentality?

Since the watershed UN convention in Rio de Janeiro in 1992, the salience of the term "biodiversity" has been extended to include the preparation of "biodiversity action plans" by urban municipalities. Interest in the urban biosphere is a related but more recent phenomenon, emerging in particular from the second international conference on biosphere reserves held in Seville in November 1995, and marks increasing concern with the intersection between expanding cities and vulnerable biotopes.[19] An increasing number of urban municipalities have sought recognition as "biosphere reserves," "national parks," or "national forests" as an emerging facet to ecologically inflected forms of strategic planning. At one level, these initiatives are indicative of a topographic reimagining of urban space, recasting the modern city as a mosaic of biotopes, including a series of distinctive "urban woodlands" encompassing streets, parks, spontaneously regenerating zones of semiwild nature, and a myriad of other types of land use. London, for example, was classified as a "national forest" in the early 2000s on account of its extensive tree cover, and a campaign has recently emerged to designate the entire city as a national park. This latest initiative, dating from 2015, is rooted in the idea that nature is everywhere. "It prescribes no boundary for nature," notes the geographer Cara Clancy, "nor does it specify what counts as nature."[20] Similarly, in Bogotá there have been plans since the early 2010s to create the largest urban ecological park in Latin America, serving as a kind of green corridor between bio-diversity hotspots.[21] Support for new kinds of urban parks devoted to the enjoyment of nature stems also from the recognition that many nonmet-ropolitan cultural landscapes are spaces of danger or exclusion: backing for London's national park initiative from the Black Environment Network, for example, stems precisely from this awareness that communities of color often have least access to nature within cities and restricted contact with spaces beyond.

My book has sought to capture a series of intersections between urban ecology, as an observational scientific practice, and the parallel develop-ment of diverse cultures of nature. I have emphasized how cosmopolitan urban ecologies emerge from a series of articulations between the local and the global. Implicit here is an extension of prevailing cosmopolitan ideals towards the inclusion of nonhuman others: the existing distinction between political and scientific spheres of cosmopolitan discourse, exem-plified by interventions from Ulrich Beck and Bruno Latour respectively, is undergoing a further set of reconfigurations in the light of postcolonial and posthuman theoretical insights.[22] There are clearly some commonalities between my interest in cosmopolitan urban ecologies and the term "eco-cosmopolitanism" elaborated by the literary critic Ursula Heise. Whereas Heise directs her emphasis towards the articulation of some kind of global citizenship, however, I take as my starting point the characteristics of urban ecologies on the ground.[23] As we saw in chapter 3, these actually existing forms of urban nature reflect colonial and postcolonial histories, constitut-ing a kind of living archive on city streets. Early twentieth-century observ-ers of urban nature sometimes referred to animals such as rats and mice as "cosmopolitan pests" to emphasize how their global distribution repre-sented a particular kind of ecological disturbance.[24] Over time, however, more subtle differentiations have emerged over nature, place, and belong-ing so that cosmopolitan ecologies have become a source of cultural and scientific curiosity. Yet the idea of "cosmopolitan space" operates in rela-tion to a universalist geographical imaginary that nestles uneasily within a postcolonial present.[25] The identification of cosmopolitan urban ecologies clearly problematizes a narrower reading of the cosmopolitan ideal as the elaboration of a specifically European political project. But can a cosmopolitan urban ecological imaginary point towards a postcolonial political sensibility? This interface between ecological analogies and political ideals is fraught with difficulty, since ecological relations are context-specific whereas cosmopolitanism aspires to some form of rights-based universalism. The tension arises from an articulation of cosmopolitanism without history. What I am arguing here is that an engagement with urban ecological complexity provides additional nuance to our understanding of cosmopolitan ideals that are derived from specific historical and epistemological legacies. The reframing of urban space as a kind of spontaneous global garden blurs the distinctions between intentional and unintentional, between private and public, and between "here" and "there" so that new patterns of cohabitation can be recognized. A project of this kind resonates with what the philosopher Isabelle Stengers refers to as "possible nonhierarchical modes of coexistence" and "an ecology of practices that would bring together our cities."[26] A postcolonial reading of urban ecologies reveals how the cultural and political resonance of socioecological assemblages in London or Berlin, for example, will differ from the experience of Chennai, Cape Town, or other cities in the global South.

The recognition of cosmopolitan urban ecologies sits uneasily with the colonial legacies of the natural history archive. Urban environmental discourse operates within the ongoing classificatory impetus of the Linnaean binomial system, so that new understandings of ecological endangerment must contend with inherited taxonomic and institutional frameworks for the interpretation of nature. If we loosen our ties with the Linnaean classification system, and its associated Red Lists and other indices of endangerment, what are the implications for urban ecology? One of the difficulties facing urban nature conservation has been the need to articulate the value of "artificial ecologies" in the face of a wilderness-oriented nature discourse within which cultural landscapes play an uncertain ideological role.

A perceived choice between the protection of fragments of preexisting ecologies and the valorization of novel socioecological assemblages raises specific questions in terms of the global comparability of endangerment measures. What I am arguing for here is a greater appreciation of an ecological pluriverse that can encompass both brownfields and rainforests, along with anything and everything interesting that we might find in between. As the Euro-American hold over biodiversity discourse loosens, what kinds of new cultures of nature might evolve? The center of gravity for urban ecology research has already begun to shift slightly, with increasing input from Brazil, China, India, South Africa, and elsewhere.

By emphasizing a variety of metropolitan cultures of nature, I have outlined an argument against the arbitrary baselines and blueprints that suffuse both conventional approaches to conservation biology and also the search for nature-based ontologies in environmental discourse. I have suggested that spontaneous urban ecologies hold cultural and political implications far beyond the confines of the subdisciplinary domain of urban ecology itself. I began my book with an example of how a marginal site in a marginal place, ironically referred to as a "brownfield," could nonetheless be transformed into a source of cultural inspiration and community engagement. Postindustrial spaces, and other types of transitional landscapes, offer creative possibilities for grassroots ecology that run counter to the generic cultures of nature that accompany the speculative and utilitarian impulse of capitalist urbanization. How might alternative sociotechnical pathways, including a postcarbon transition, intersect with the full diversity of urban ecologies on the ground? Are there synergies to be found between alternative environmental futures and new cultures of urban nature?

The question of agency, and the independent dynamics of urban ecology, run through my interpretation of urban nature. I have used the term "other-than-human geographies," however, to introduce an element of agential nuance into the discussion. Contra some of the new materialist literature, I wish to retain an emphasis on the specificities of human agency in combination with other forms of vitality. I think this leaves the door open for a conceptual dialogue between elements of historical materialism and the multiple forms of human and nonhuman subjectivity that shape urban space. The neo-Marxian analytical lens via Jameson, Swyngedouw, and others remains a critical yet not determining element of my interpretative framework. We need an ecological analysis of capitalism because economic perturbations produce their own distinctive ecologies of investment and abandonment before we even move to an examination of the role of capital in shaping the political dynamics of the urban arena.

In the introduction I noted that urban political ecology, a body of ideas with which my own work has been closely associated, had paid relatively little attention to the subdisciplinary field of urban ecology itself, including new developments in the biophysical sciences, and the material intricacies of urban ecologies on the ground. At different points in my book I have sought to address these lacunae by emphasizing the need to take questions of causality and evidentiary materialism seriously, extending our analytical lens to fields such as epigenetics, evolutionary biology, and ethological dimensions to affective encounters with the nonhuman. If political ecology is taking tentative steps towards a more thorough engagement with the biophysical sciences, there remain further intellectual challenges in relation to ecocriticism, multispecies ethnographies, and evolving approaches to environmental research that seek to unsettle the bounded human subject. These postphenomenological developments have sometimes been bundled together under the loosely defined category of posthumanism, but I have urged caution here to emphasize the need for a relational and context-specific ethics towards nonhuman others. In methodological terms, I have been especially fascinated with those studies that emphasize "slowness" and embodied interactions with urban space, ranging from ethnographic or long-term studies of specific sites to moments of reverie in which time appears to slow down. Discourses of endangerment and collective memory can be fostered through painstaking efforts to record and understand the intricacies of socioecological assemblages, including the use of diagrammatic and cartographic modes of representation.

In writing this book I have tried to capture something of my enthusiasm for the endlessly fascinating realm of urban nature. I have suggested that there is a productive tension between neo-Lefebvrian accounts of complete urbanization and the intricate realm of actually existing ecologies on the ground; I have sought to expand on the wider political significance of metropolitan cultures of nature as a potential portal into alternative modernities; I have questioned the merits of an adaptive Anthropocene that renders biodiversity generic, malleable, and interchangeable; I have argued for a degree of agential nuance that holds onto the contingencies and specificities of historical agency; I have explored novel socioecological assemblages in relation to the interpretation of ecological endangerment at different spatial and temporal scales and ongoing tensions with nativist conceptions of landscape authenticity; I have engaged directly with insights from the biophysical sciences as part of a commitment to evidentiary materialism and forensic ecologies; I have drawn inspiration from more embodied and longer-term methodological insights, including the work of artists, to build a richer sense of urban ecologies as a multisensory domain; I have made a tentative case for an affirmative biopolitical paradigm in which strategic interventions in the web of life can be justified on ethical grounds, extending to epidemiological aspects of human well-being, as well as attempts to enable new modes of cohabitation through urban rewilding or innovative approaches to landscape design; I have reflected on the limits to existing modes of interpretation for cultural representations of urban nature, including recent developments such as ecocriticism; and, above all, I have sought to identify the contours of a conceptual synthesis that is capable of combining elements of historical and materialist analysis with a range of newly emerging fields that question the centrality of the bounded human subject. In short, I have tried to articulate an alternative perspective to that envisaged by systems-based ecological models or the grand sweep of a totalizing urban theory.

NOTES

INTRODUCTION

1. "The Brownfield Research Centre" ran from 15 June to 22 July 2018 at the AirSpace Gallery in Stoke-on-Trent. In addition to the gallery space, the program of work also featured six artists in residence (Rodrigo Arteaga, Edward Chell, Rebecca Chesney, Anna Francis, Lucy McLauchlan, and Victoria Sharples). The exhibition notes also provide an alternative definition of the brownfield as "vacant or derelict land which, through wilful human neglect, and barricading, has been successfully reclaimed by nature, developing thriving and important natural ecosystems, and acting as a model for future urban land use."

2. Anthony Barnett, *The lure of greatness: England's Brexit and America's Trump* (London: Unbound, 2017). On the resurgence of far-right politics in postwar liberal democracies see also the timely republication of Theodor W. Adorno's lecture given in 1967, *Aspects of the new right-wing extremism*, trans. Wieland Hoban (Cambridge: Polity, 2020 [2019]).

3. On the etymology and history of the garden as a kind of sanctuary see, for example, Vittorio Magnago Lampugnani, Konstanze Sylva Domhardt, and Rainer Schützeichel, eds., *Enzyklopädie zum gestalteten Raum: Im Spannungsfeld zwischen Stadt und Landschaft* (Zurich: gta Verlag, 2014).

4. On the distinction between forms of nature in Marxist thought see in particular Alfred Schmidt, *The concept of nature in Marx*, trans. Ben Fowkes (London: Verso, 2014 [1962]). The question of a putative boundary between "first" and "second" nature certainly varies within the literature. The geographer and historian William Cronon, for example, tends to use the term "second nature" in relation to the specific landscape transformations

associated with capitalist commodification. On the original Hegelian sense of second nature see, for example, Georg W. Bertram, "Two conceptions of second nature," *Open Philosophy* 3 (1) (2020): 68–80.

5. See also Neil Smith, *Uneven development: nature, capital, and the production of space* (Oxford: Blackwell, 1984).

6. See Theodor W. Adorno, "The idea of natural history," *Telos* 1984 [1932] (60): 111–124.

7. The idea of an originary nature retains a powerful ideological hold for those whom the anthropologist Philippe Descola terms "partisans of nature." Philippe Descola, *The ecology of others*, trans. Geneviève Godbout and Benjamin P. Luley (Chicago: Prickly Paradigm Press, 2013 [2007]).

8. Lorraine Daston, *Against nature* (Cambridge, MA: MIT Press, 2019).

9. Christa Davis Acampora, *Contesting Nietzsche* (Chicago: University of Chicago Press, 2013), 15.

10. Hartmut Böhme, *Aussichten der Natur: Naturästhetik in Wechselwirkung von Natur und Kultur* (Berlin: Matthes & Seitz, 2017).

11. See, for example, Julie Goodness and Pippin M. L. Anderson, "Local assessment of Cape Town: navigating the management complexities of urbanization, biodiversity, and ecosystem services in the Cape Floristic Region," in *Urbanization, biodiversity and ecosystem services: challenges and opportunities: a global assessment*, ed. Thomas Elmqvist et al. (Dordrecht: Springer, 2013), 461–484; and Tania Katzschner, "Cape flats nature: rethinking urban ecologies," in *Contested ecologies: dialogues in the South on nature and knowledge*, ed. Lesley Green (Cape Town: HSRC Press, 2013), 202–206. On the interface between cities and biodiversity "hotspots" see also Richard Weller, Zuzanna Drozdz, and Sara Padgett Kjaersgaard, "Hotspot cities: identifying peri-urban conflict zones," *Journal of Landscape Architecture* 14 (1) (2019): 8–19.

12. Feike De Jong, "Which is the world's most biodiverse city?," *Guardian* (3 July 2017).

13. See, for example, Cord Riechelmann, *Wilde Tiere in der Großstadt* (Berlin: Nicolai, 2004); Wolfram Kunick, "Zonierung des Stadtgebietes von Berlin (West). Ergebnisse floristischer Untersuchungen," *Landschaftsentwicklung und Umweltforschung* 14 (1982): 1–164; Stefan Zerbe, Ute Maurer, Solveig Schmitz, and Herbert Sukopp, "Biodiversity in Berlin and its potential for nature conservation," *Landscape and Urban Planning* 62 (3) (2003): 139–148.

14. See Ingo Kowarik, "Novel urban ecosystems, biodiversity, and conservation," *Environmental Pollution* 159 (8–9) (2011): 1974–1983.

15. Matthew F. Vessel and Herbert H. Wong, *Natural history of vacant lots* (Berkeley: University of California Press, 1987).

16. John Vidal, "A bleak corner of England's Essex is being hailed as England's rainforest," *Guardian* (3 May 2003).

17. W. H. Hudson, *Birds in London* (London: Longmans, Green, 1898), 5.

18. Hudson, *Birds in London*, 5.

19. Richard S. R. Fitter, *London's natural history* (London: Collins, 1945); and John Kieran, *Natural history of New York City: a personal report after fifty years of study and enjoyment of wildlife within the boundaries of Greater New York* (New York: Fordham University Press, 1982 [1959]). Another contemporary account of urban nature in a North American context is Leonard Dubkin's report of encounters with fragments of nature in Chicago in *The natural history of a yard* (Chicago: Henry Regnery, 1955). See also Michael A. Bryson, "Empty lots and secret places: Leonard Dubkin's exploration of urban nature in Chicago," *Interdisciplinary Studies in Literature and Environment* 18 (1) (2011): 47–66.

20. John Kieran, interview with Warren Bower, the NYPR archive collections (29 July 1968).

21. An installation by the Chinese artist Zheng Bo in 2015 entitled *Survival Manual I* (2015) reproduced these illustrations from the original Shanghai guide. At first glance these images seem little different from the kind of intricate botanical drawings to be found in a field guide or a naturalist's sketchbook.

22. See, for example, Marcus Nyman, "Food, meaning-making and ontological uncertainty: exploring 'urban foraging' and productive landscapes in London," *Geoforum* 99 (2019): 170–180; and Melissa R. Poe, Joyce LeCompte, Rebecca McLain, and Patrick Hurley, "Urban foraging and the relational ecologies of belonging," *Social and Cultural Geography* 15 (8) (2014): 901–919.

23. See Alex Morss, "'Not just weeds': how rebel botanists are using graffiti to name forgotten flora," *Guardian* (1 May 2020).

24. Samuel John and Vena Kapoor, "Nature underfoot: wild flora and fauna on city pavements and waysides," *Roundglass | Sustain* (accessed 10 May 2021).

25. David Rothenberg, *Nightingales in Berlin: searching for the perfect sound* (Chicago: University of Chicago Press, 2019). On birds and urban soundscapes see also Joeri Bruyninckx, *Listening in the field: recording and the science of birdsong* (Cambridge, MA: MIT Press, 2018).

26. See, for example, Anna Chiesura, "The role of urban parks for the sustainable city," *Landscape and Urban Planning* 68 (1) (2004): 129–138; and Danielle F. Shanahan, Richard A. Fuller, Robert Bush, Brenda B. Lin, and Kevin J. Gaston, "The health benefits of urban nature: how much do we need?," *BioScience* 65 (5) (2015): 476–485.

27. Listen, for example, to Lee Patterson's *Diptych for Sint-Katelijnesplein*, featuring aquatic insects recorded in Brussels on the CD that accompanies Matthew Gandy and B. J. Nilsen, eds., *The acoustic city* (Berlin: jovis, 2014).

28. These sidewalk ecologies also include the base of urban trees where dog urine produces a "canine zone" that is conducive to specific species of algae. See Fabio Rindi and Michael D. Guiry, "Composition and distribution of subaerial algal assemblages in Galway City, western Ireland," *Cryptogamie. Algologie* 24 (3) (2003): 245–267; and Fabio Rindi, "Diversity, distribution and ecology of green algae and cyanobacteria in urban habitats," in *Algae and cyanobacteria in extreme environments*, ed. Joseph Seckbach (Dordrecht: Springer, 2007), 619–638. Fabio Rindi notes that the diversity of cyanobacteria and microalgae in urban habitats has been generally underestimated. Personal communication with the author (4 September 2018).

29. See, for example, Heather K. Allen, Justin Donato, Helena Huimi Wang, Karen A. Cloud-Hansen, Julian Davies, and Jo Handelsman, "Call of the wild: antibiotic resistance genes in natural environments," *Nature Reviews Microbiology* 8 (4) (2010): 251–259; Lihong Gao, Yali Shi, Wenhui Li, Jiemin Liu, and Yaqi Cai, "Occurrence and distribution of antibiotics in urban soil in Beijing and Shanghai, China," *Environmental Science and Pollution Research* 22 (15) (2015): 11360–11371; and Hannah Landecker, "Antibiotic resistance and the biology of history," *Body and Society* 22 (4) (2016): 19–52.

30. See, for example, Kelly S. Ramirez, Jonathan W. Leff, Albert Barberán, Scott Thomas Bates, Jason Betley, Thomas W. Crowther, Eugene F. Kelly, et al., "Biogeographic patterns in below-ground diversity in New York City's Central Park are similar to those observed globally," *Proceedings of the Royal Society B: Biological Sciences* 281 (1795) (2014): 20141988.

31. Esther Woolfson, *Field notes from a hidden city: an urban nature diary* (London: Granta, 2014).

32. Blake Morrison, *South of the river* (London: Chatto & Windus, 2007).

33. Jonathan Lethem, *Chronic city* (London: Faber and Faber, 2010). See also Susan Kollin, "Not yet another world: ecopolitics and urban natures in Jonathan Lethem's *Chronic city*," *Literature Interpretation Theory* 26 (4) (2015): 255–275.

34. See, for example, Loren Coleman, "Alligators-in-the-sewers: a journalistic origin," *Journal of American Folklore* 92 (365) (1979): 335–338; and Robert Daley, *The World Beneath the City* (Philadelphia: Lippincott, 1959).

35. Judith Kerr, *The tiger who came to tea* (London: HarperCollins, 2017 [1968]).

36. Chiyoko Nakatani, *The day Chiro was lost* (London: Bodley Head, 1969 [1965]).

37. Betty Smith, *A tree grows in Brooklyn* (Harmondsworth: Penguin, 1951 [1943]).

38. Rose Macaulay, *The world my wilderness* (London: Collins, 1950), 128–129.

39. W. G. Sebald, *Austerlitz*, trans. Anthea Bell (London: Penguin, 2002 [2001]).

40. On cemeteries and urban biodiversity see Matthew Gandy, "The fly that tried to save the world: saproxylic geographies and other-than-human ecologies," *Transactions of*

the Institute of British Geographers 44 (2) (2019): 392–406; and Ingo Kowarik, Sascha Buch-holz, Moritz von der Lippe, and Birgit Seitz, "Biodiversity functions of urban cemeteries: evidence from one of the largest Jewish cemeteries in Europe," *Urban Forestry and Urban Greening* 19 (2016): 68–78. In cities of the global South we also find significant examples of sacred spaces that have served as havens for urban biodiversity. See, for example, Amita Baviskar, "Urban nature and its publics: shades of green in the remaking of Delhi," in *Grounding urban natures: histories and futures of urban ecologies*, ed. Henrik Ernstson and Sverker Sörlin (Cambridge, MA: MIT Press, 2019), 233–246.

41. Bruce Braun, "A new urban dispositif? Governing life in an age of climate change," *Environment and Planning D: Society and Space* 32 (1) (2014): 49–64.

42. Alyssa Battistoni, "Bringing in the work of nature: from natural capital to hybrid labor," *Political Theory* 45 (1) (2017): 5–31. The role of nonhuman labor is deeply enmeshed with the circulatory metabolic dynamics of urban space, ranging from the role of plants and bacteria in wastewater treatment plants to the "scavenger ecologies" of pigs, goats, storks, and other animals associated with roadside accumulations of refuse or more distant landfill sites. See, for example, Jacob Doherty, "Filthy flourishing: para-sites, animal infrastruc-ture, and the waste frontier in Kampala," *Current Anthropology* 60 (S20) (2019): S321–S332; Colin Hoag, Filippo Bertoni, and Nils Bubandt, "Wasteland ecologies: undomestication and multispecies gains on an Anthropocene dumping ground," *Journal of Ethnobiology* 38 (1) (2018): 88–104; Kelsi Nagy and Phillip David Johnson II, eds., *Trash animals: how we live with nature's filthy, feral, invasive, and unwanted species* (Minneapolis: University of Minnesota Press, 2013); and Amy Zhang, "Circularity and enclosures: metabolizing waste with the black soldier fly," *Cultural Anthropology* 35 (1) (2020): 74–103.

43. Maan Barua, "Nonhuman labour, encounter value, spectacular accumulation: the geographies of a lively commodity," *Transactions of the Institute of British Geographers* 42 (2) (2017): 274–288.

44. See Ingo Kowarik, "Unkraut oder Urwald? Natur der Vierten Art auf dem Gleisdrei-eck," in *Dokumentation Gleisdreieck morgen. Sechs Ideen für einen Park*, ed. Bundesgartenschau (Berlin: Bezirksamt Kreuzberg, Berlin, 1995 [1991]), 45–55; and Kowarik, "Wild urban woodlands: towards a conceptual framework," in *Wild urban woodlands: new perspectives for urban forestry*, ed. Ingo Kowarik and Stefan Körner (Berlin: Springer, 2005), 1–32. Kowarik's "fourth nature" has become conceptually aligned with interest in "novel urban ecosystems." See, for example, Catarina Patoilo Teixeira and Cláudia Oliveira Fernandes, "Novel ecosystems: a review of the concept in non-urban and urban contexts," *Landscape Ecology* 35 (1) (2020): 23–39.

45. See Vera Vicenzotti, *Der "Zwischenstadt"-Diskurs: eine Analyse zwischen Wildnis, Kultur-landschaft und Stadt* (Bielefeld: transcript, 2014).

46. I take the term "spectral materialism" from Eric L. Santner, *On creaturely life: Rilke, Benjamin, Sebald* (Chicago: University of Chicago Press, 2006), xvi.

47. See Erle C. Ellis and Navin Ramankutty, "Putting people in the map: anthropogenic biomes of the world," *Frontiers in Ecology and the Environment* 6 (8) (2008): 439–447. On the delineation of "urban biomes" see also Stephanie Pincetl, "Cities as novel biomes: recognizing urban ecosystem services as anthropogenic," *Frontiers in Ecology and Evolution* 3 (2015): 140.

48. See Chris Otter, "Technosphere," in *Concepts of urban-environmental history*, ed. Sebastian Haumann, Martin Knoll, and Detlev Mares (Bielefeld: transcript, 2020), 21–32.

49. Ezequeil Martínez Estrada, *X-ray of the Pampa*, trans. Thomas F. McGann (Austin: University of Texas Press, (1971 [1933]), 270–271.

50. Matthew Gandy, "Where does the city end?," *Architectural Design* 82 (1) (2012): 128–132. In the case of London, interest in creating a greenbelt developed from the 1890s onwards, and is reflected in the London development plan of 1919, but full implementation only occurred after comprehensive advances in planning legislation in the 1930s and 1940s, as part of a wider reconfiguration of the public interest in land use policy.

51. Yimin Zhao, "Space as method: field sites and encounters in Beijing's green belts," *City* 21 (2) (2017): 190–206.

52. The landscapes north of Chennai present a vivid palimpsest of different cultural landscapes, with cement factories and other industrial facilities adjacent to creeks, ponds, and other remnants of the original landscape full of waterlilies and other flowers. For more on the urban-rural interface in India see, for example, Shubhra Gururani, "'When land becomes gold': changing political ecology of the commons in a rural-urban frontier," in *Land rights, biodiversity conservation and justice: rethinking parks and people*, ed. Sharlene Mollett and Thembela Kepe (London: Routledge, 2018), 107–125.

53. Yan Wang Preston, *Forest* (Ostfildern: Hatje Cantz, 2018).

54. Andrew Ross, "Bird on fire: lessons from the world's least sustainable city," *Places* (November 2011): 15. See also Andrew Ross, *Bird on fire: lessons from the world's least sustainable city* (Oxford: Oxford University Press, 2011).

55. See Andrew Needham, *Power lines: Pheonix and the making of the modern Southwest* (Princeton, NJ: Princeton University Press, 2014), 8.

56. See, for example, Marina Alberti, "Measuring urban sustainability," *Environmental Impact Assessment Review* 16 (4) (1996): 381–424; and Catalina Turcu, "Re-thinking sustainability indicators: local perspectives of urban sustainability," *Journal of Environmental Planning and Management* 56 (5) (2013): 695–719. The dispersed geographies of resource extraction, toxic waste, and the poisoning of local communities have been widely documented. See, for example, Martín Arboleda, *Planetary mine: territories of extraction under late capitalism* (London: Verso, 2020); Macarena Gómez-Barris, *The extractive zone: social ecologies and decolonial perspectives* (Durham, NC: Duke University Press, 2017); Josh Lepawski,

Reassembling rubbish: worlding electronic waste (Cambridge, MA: MIT Press, 2018); and Joan Martinez-Alier, "Urban 'unsustainability' and environmental conflict," in *The human sustainable city: challenges and perspectives from the Habitat Agenda*, ed. Luigi Fusco Girard, Bruno Forte, Maria Cerreta, Pasquale De Toro, and Fabiana Forte (London: Routledge, 2019 [2003]), 89–105.

57. See, for example, Ryo Kohsaka, Henrique M. Pereira, Thomas Elmqvist, Lena Chan, Raquel Moreno-Peñaranda, Yukihiro Morimoto, Takashi Inoue, et al., "Indicators for management of urban biodiversity and ecosystem services: city biodiversity index," in *Urbanization, biodiversity and ecosystem services: challenges and opportunities*, ed. Thomas Elmqvist et al. (Dordrecht: Springer, 2013), 699–718.

58. On urban experimentation see, for example, Harriet Bulkeley and Vanesa Castán Broto, "Government by experiment? Global cities and the governing of climate change," *Transactions of the Institute of British Geographers* 38 (3) (2013): 361–375; and James P. Evans, "Resilience, ecology and adaptation in the experimental city," *Transactions of the Institute of British Geographers* 36 (2) (2011): 223–237.

59. See, for example, Emanuel Gaziano, "Ecological metaphors as scientific boundary work: innovation and authority in interwar sociology and biology," *American Journal of Sociology* 101 (4) (1996): 874–907; and Ashwani Vasishth and David C. Sloane, "Returning to ecology: an ecosystem approach to understanding the city," in *From Chicago to LA: making sense of urban theory*, ed. Michael Dear (Thousand Oaks, CA: Sage, 2002), 343–366.

60. Ernest W. Burgess, "The growth of the city: an introduction to a research project," in *The city*, ed. Robert E. Park, Ernest W. Burgess, and R. D. McKenzie (Chicago: University of Chicago Press, 1925), 47–62. The Chicago school contribution should also be viewed in the wider context of "organicist" conceptions of urban space advanced by Frederick Law Olmsted Jr. and other influential figures within the emerging field of urban planning.

61. See Jennifer Wolch, Stephanie Pincetl, and Laura Pulido, "Urban nature and the nature of urbanism," in Dear, *From Chicago to LA*, 369–402.

62. Brian J. L. Berry and John D. Kasarda, *Contemporary urban ecology* (New York: Macmillan, 1977).

63. Koenraad Danneels, "Historicizing ecological urbanism: Paul Duvigneaud, the Brussels agglomeration and the influence of ecology on urbanism (1970–2016)," in *On reproduction: re-imagining the political ecology of urbanism*, U&U-9th International PhD Seminar in Urbanism and Urbanization, Ghent, Belgium, ed. Michiel Dehaene et al. (7–9 February 2018), 343–356; and Jens Lachmund, "The city as ecosystem: Paul Duvigneaud and the ecological study of Brussels," in *Spatializing the history of ecology: sites, journeys, mappings*, ed. Raf de Bont and Jens Lachmund (London: Routledge, 2017).

64. Paul Duvigneaud, "Étude écologique de l'écosystème urbain bruxellois: 1. L'écosystème 'urbs'," *Mémoires de la Société Royale de Botanique de Belgique* 6 (1974): 5–35. See also

the systems-based ecology developed by Pierre Dansereau, "Les dimensions écologiques de l'espace urbain," *Cahiers de Géographie du Québec* 31 (84) (1987): 333–395.

65. See Koenraad Danneels, "From sociobiology to urban metabolism: the interaction of urbanism, science and politics in Brussels (1900–1978)" (PhD thesis, University of Antwerp and KU Leuven, 2021).

66. Cited in Danneels, "From sociobiology to urban metabolism," 313.

67. Paul B. Sears, "Human ecology: a problem in synthesis," *Science* 120 (3128) (1954): 961.

68. Amos H. Hawley, "Ecology and human ecology," *Social Forces* 22 (4) (1944): 398–405.

69. Amos H. Hawley, *Human ecology: a theoretical essay* (Chicago: University of Chicago Press, 1986).

70. See, for example, Robert U. Ayres, "Industrial metabolism," *Technology and Environment* (February 1989): 23–49. A synthesis between elements of urban and industrial metabolism can be found in the "territorial ecology" concept developed by Sabine Barles. See, for example, Sabine Barles, "Écologie territoriale et métabolisme urbain: quelques enjeux de la transition socioécologique," *Revue d'Economie Régionale Urbaine* 5 (2017): 819–836.

71. The Vienna-based Institute for Social Ecology shares the same title with an earlier institute in the USA founded by Murray Bookchin, with links to "deep ecology" and associated forms of ecoactivism. The intellectual outlooks of the two institutes are markedly different: whilst the Vienna school is rooted in metabolic systems, the program of work led by Bookchin emanated from a radical critique of capitalist societies.

72. Marina Fischer-Kowalski and Helga Weisz, "The archipelago of social ecology and the island of the Vienna school," in *Social ecology: society-nature relations across time and space*, ed. Helmut Haberl, Marina Fischer-Kowalski, Fridolin Krausmann, and Verena Winiwarter (Basel: Springer, 2016), 4.

73. Marina Fischer-Kowalski and Walter Hüttler, "Society's metabolism: the intellectual history of materials flow analysis, part II, 1970–1998," *Journal of Industrial Ecology* 2 (4) (1998): 107–136. See also Fischer-Kowalski and Weisz, "The archipelago of social ecology and the island of the Vienna school."

74. Fischer-Kowalski and Weisz, "The archipelago of social ecology and the island of the Vienna school." On autopoiesis and systems theory see, for example, Francisco G. Varela, Humberto R. Maturana, and Ricardo Uribe, "Autopoiesis: the organization of living systems, its characterization and a model," *Biosystems* 5 (4) (1974): 187–196; and Humberto R. Maturana and Francisco G. Varela, *Autopoietic systems: a characterization of the living organization* (Urbana-Champaign: University of Illinois Press, 1975).

75. See Niklas Luhmann, *Social systems*, trans. John Bednarz and David Baecker (Stanford, CA: Stanford University Press, 1995 [1984]).

76. Luhmann, *Social systems*, 505.

77. Peter Baccini, "A city's metabolism: towards the sustainable development of urban systems," *Journal of Urban Technology* 4 (2) (1997): 27–39; and Ken Newcombe, Jetse D. Kalma, and Alan R. Aston, "The metabolism of a city: the case of Hong Kong," *Ambio* (1978): 3–15. See also Peter Baccini and Paul H. Brunner, *Metabolism of the anthroposphere: analysis, evaluation, design* (Cambridge, MA: MIT Press, 2012).

78. See Wilfried Endlicher, *Einführung in die Stadtökologie* (Stuttgart: Eugen Ulmer, 2012).

79. In terms of the origins of sustainable development discourse see, for example, Gro Harlem Brundtland, "World Commission on Environment and Development," *Environmental Policy and Law* 14 (1) (1985): 26–30. The Berlin-based environmental economist Udo Simonis, for example, elaborated on the need to integrate ecological principles into the heart of economic policymaking as part of a rationalist and homeostatic approach advanced through better-informed decision making. See, for example, Udo E. Simonis, "Ecological modernization of industrial society: three strategic elements," WZB Discussion Paper, No. FS II 88-401 (Berlin: Wissenschaftszentrum Berlin für Sozialforschung, 1988).

80. See, for example, Andrew Biro, ed., *Critical ecologies: The Frankfurt School and contemporary environmental crises* (Toronto: University of Toronto Press, 2011); Dana R. Fisher and William R. Freudenburg, "Ecological modernization and its critics: assessing the past and looking toward the future," *Society and Natural Resources* 14 (8) (2001): 701–709; and Christoph Görg, *Gesellschaftliche Naturverhältnisse* (Münster: Westfälisches Dampfboot, 1999).

81. Marina Alberti, John M. Marzluff, Eric Shulenberger, Gordon Bradley, Clare Ryan, and Craig Zumbrunnen, "Integrating humans into ecology: opportunities and challenges for studying urban ecosystems," *BioScience* 53 (12) (2003): 1169–1179.

82. Alberti et al., "Integrating humans into ecology," 1169.

83. Alberti et al., "Integrating humans into ecology," 1176.

84. See, for example, Marina Alberti, "Eco-evolutionary dynamics in an urbanizing planet," *Trends in Ecology and Evolution* 30 (2) (2015): 114–126; and Marina Alberti, John Marzluff, and Victoria M. Hunt, "Urban driven phenotypic changes: empirical observations and theoretical implications for eco-evolutionary feedback," *Philosophical Transactions of the Royal Society B: Biological Sciences* 372 (1712) (2017): 20160029. For an alternative perspective on the social dynamics of evolutionary change, see Marina Fischer-Kowalski and Jan Rotmans, "Conceptualizing, observing, and influencing social-ecological transitions," *Ecology and Society* 14 (2) (2009): 3.

85. James Evans, "Ecology in the urban century: power, place, and the abstraction of nature," in Ernstson and Sörlin, *Grounding urban natures*, 304. On the shifting terrains of ecology and urban ecology see also Robert A. Francis, Jamie Lorimer, and Mike Raco,

"Urban ecosystems as 'natural' homes for biogeographical boundary crossings," *Transactions of the Institute of British Geographers* 37 (2) (2012): 183–190; and Sharon E. Kingsland, *The evolution of American ecology 1890–2000* (Baltimore: Johns Hopkins University Press, 2005).

86. For an overview see Steward T. A. Pickett and Mary L. Cadenasso, "Advancing urban ecological studies: frameworks, concepts, and results from the Baltimore Ecosystem Study," *Austral Ecology* 31 (2) (2006): 114–125. See also similar approaches to "landscape mosaics" developed by Richard T. T. Forman, *Land mosaics: the ecology of landscapes and regions* (Cambridge: Cambridge University Press, 1995), and J. G. Wu and O. L. Loucks, "From balance of nature to hierarchical patch dynamics: a paradigm shift in ecology," *Quarterly Review of Biology* 70 (1995): 439–466. Other influences include Michael Hough, *City form and natural process: towards a new verancular* (Toronto: McGraw Hill, 1984); Anne Whiston Spirn, *The granite garden: urban nature and human design* (New York: Basic Books, 1984); and Ian L. McHarg, *Design with nature* (New York: American Museum of Natural History, 1969). On the significance of the aerial view for the emergence of a more systematic approach to "landscape ecology" see Chunglin Kwa, "The visual grasp of the fragmented landscape: plant geographers vs. plant sociologists," *Historical Studies in the Natural Sciences* 48 (2) (2018): 180–222.

87. J. Morgan Grove, Mary L. Cadenasso, Steward T. Pickett, Gary E. Machlis, and William R. Burch, *The Baltimore school of urban ecology: space, scale, and time for the study of cities* (New Haven, CT: Yale University Press, 2015). See also Steward T. Pickett, Mary L. Cadenasso, Matthew E. Baker, Lawrence E. Band, Christopher G. Boone, Geoffrey L. Buckley, Peter M. Groffman, et al., "Theoretical perspectives of the Baltimore Ecosystem Study: conceptual evolution in a social-ecological research project," *BioScience* 70 (4) (2020): 297–314.

88. Arthur G. Tansley, "The use and abuse of vegetational concepts and terms," *Ecology* 16 (3) (1935): 284–307. On the contrast between Tansley's interest in human-modified landscapes and Clements's emphasis on "wilderness" in a North Americam context see Laura Cameron, "Histories of disturbance," *Radical History Review* 74 (1999): 5–24.

89. Tansley, "The use and abuse of vegetational concepts and terms," 300.

90. Grove et al., *The Baltimore school of urban ecology*.

91. On the rise of nonequilibrium ecological theory see, for example, Wu and Loucks, "From balance of nature to hierarchical patch dynamics." On the wider implications see A. Dan Tarlock, "The nonequilibrium paradigm in ecology and the partial unraveling of environmental law," *Loyola of Los Angeles Law Review* 27 (1994): 1121–1144; and Mark J. McDonnell, "The history of urban ecology: an ecologist's perspective," in *Urban ecology: patterns, processes and applications*, ed. Jari Niemelä, Jürgen H. Breuste, Thomas Elmqvist, Glenn Guntenspergen, Philip James, and Nancy E. McIntyre (Oxford: Oxford University

Press, 2011), 5–13. On the difficulties in extending a "generalizable" ecological theory to human-dominated environments see also James P. Collins, Ann Kinzig, Nancy B. Grimm, William F. Fagan, Diane Hope, Jianguo Wu, and Elizabeth T. Borer, "A new urban ecology: modeling human communities as integral parts of ecosystems poses special problems for the development and testing of ecological theory," *American Scientist* 88 (5) (2000): 416–425; and Robert Mugerauer, "Toward a theory of integrated urban ecology: complementing Pickett et al.," *Ecology and Society* 15 (4) (2010): 31.

92. See Andrew Barry and Georgina Born, "Interdisciplinarity: reconfigurations of the social and natural sciences," in *Interdisciplinarity: reconfigurations of the social and natural sciences,* ed. Andrew Barry and Georgina Born (London: Routledge, 2013), 1–56.

93. Alberti et al., "Integrating humans into ecology."

94. An interesting elaboration of the contrast between systems-based approaches to the understanding of ecosystems and alternative perspectives emerging from nonequilibrium-derived modes of historical interpretation for specific landscapes can be found in William Balée, "The research program of historical ecology," *Annual Review of Anthropology* 35 (1) (2006): 75–98.

95. See, for example, O. Naegeli and A. Thellung, "Die Flora Kantons Zürich. I Teil: Die Ruderal- und Adventivflora des Kantons Zürich," *Vierteljahrsschrift der Naturforschenden Gesellschaft in Zürich* 50 (1905): 225–305; and Hans Höppner and Hans Preuss, *Flora des westfälisch-rheinischen Industriegebietes unter Einschluß der Rheinischen Bucht* (Dortmund: Friedrich Wilhelm Ruhfus, 1926). I would like to thank Susanne Hauser and Christoph Kueffer for bringing these interesting references to my attention.

96. Höppner and Preuss, *Flora des westfälisch-rheinischen Industriegebietes unter Einschluß der Rheinischen Bucht,* viii, ix. The significance of the early use of the term "adventive" is stressed by Susanne Hauser. See her interview in my documentary film *Natura urbana: the Brachen of Berlin* (2017).

97. Paul Jovet, "Evolution des groupements rudéraux 'parisiens,'" *Bulletin de la Société Botanique de France* 87 (1940): 305–312. See also Anne-Élisabeth Wolf, "L'herbier parisien de Paul Jovet: première analyse," and Jean-Marc Drouin, "Paul Jovet: les concepts de l'écologie végétale à l'épreuve de la ville," both in *Sauvages dans la ville. Actes du colloque, organisé pour le centenaire de la naissance de Paul Jovet,* ed. Bernadette Lizet, Anne-Élisabeth Wolf, and John Celecia (Paris: JATBA/Publications scientifiques du MNHN, 1997), 35–52 and 75–90.

98. See, for example, Herbert Sukopp, ed., *Stadtökologie. Das Beispiel Berlins* (Berlin: Dietrich Reimer, 1990); Herbert Sukopp, "On the early history of urban ecology in Europe," *Preslia, Praha* 74 (2002): 373–393; and Herbert Sukopp, Hans-Peter Blume, and Wolfram Kunick, "The soil, flora and vegetation of Berlin's waste lands," in *Nature in cities: the natural environment in the design and development of urban green space,* ed. Ian Laurie (Chichester:

Wiley, 1979), 115–134. The anthropologist Bettina Stoetzer sketches an interesting line of connection between a "ruderal analytic" and multispecies ethnographies. See Bettina Stoetzer, "Ruderal ecologies: rethinking nature, migration, and the urban landscape in Berlin," *Cultural Anthropology* 33 (2) (2018): 295–323.

99. See, for example, Leonie K. Fischer, Verena Rodorff, Moritz von der Lippe, and Ingo Kowarik, "Drivers of biodiversity patterns in parks of a growing South American megacity," *Urban Ecosystems* 19 (2016): 1–19; Divya Gopal, Moritz von der Lippe, and Ingo Kowarik, "Sacred sites, biodiversity and urbanization in an Indian megacity," *Urban Ecosystems* 22 (2019): 161–172; Divya Gopal, Harini Nagendra, and Michael Manthey, "Vegetation in Bangalore's slums: composition, species distribution, density, diversity, and history," *Environmental Management* 55 (2015): 1390–1401; Chi Yung Jim, "Old masonry walls as ruderal habitats for biodiversity conservation and enhancement in urban Hong Kong," in *Urban biodiversity and design*, ed. Norbert Müller, Peter Werner, and John G. Kelcey (Hoboken, NJ: Wiley-Blackwell, 2010), 323–347; Makoto Numata, ed., *Tokyo project: interdisciplinary studies of urban ecosystems in the metropolis of Tokyo* (Chiba: Chiba University, 1977); J. O. Rieley and S. E. Page, "Survey, mapping, and evaluation of green space in the Federal Territory of Kuala Lumpur, Malaysia," in *Urban ecology as the basis of urban planning*, ed. Herbert Sukopp, Makoto Numata, and A. Huber (The Hague: SPB Academic Publishing, 1995), 173–183; Muthulingam Udayakumar and Thangave Sekar, "Density, diversity and richness of woody plants in urban green spaces: a case study in Chennai metropolitan city," *Urban Forestry and Urban Greening* 11 (4) (2012): 450–459; Stefan Zerbe, Il-Ki Choi, and Ingo Kowarik, "Characteristics and habitats of non-native plant species in the city of Chonju, southern Korea," *Ecological Research* 19 (2004): 91–98; Shen Zhang, "The forest and urban greening in Shanghai," in *Urban ecology: plants and plant communities in urban environments*, ed. Herbert Sukopp, Slavomil Hejný, and Ingo Kowarik (The Hague: SPB Academic Publishing, 1990), 141–154.

100. Oliver Gilbert, "Wild figs by the river Don, Sheffield," *Watsonia* 18 (1990): 84–85. See also Oliver Gilbert, *The ecology of urban habitats* (London: Chapman and Hall, 1991 [1989]); and David Goode, *Nature in towns and cities* (London: Collins, 2014).

101. See the inclusion of previously published essays by Herbert Sukopp in John M. Marzluff, Eric Shulenberger, Wilfried Endlicher, Marina Alberti, Gordon Bradley, Clare Ryan, Craig ZumBrunnen, and Ute Simon, eds., *An international perspective on the interaction between humans and nature* (New York: Springer, 2008). Examples of the new wave of urban ecology literature since the late 1960s and early 1970s, spurred in particular by the increasing presence of birds and mammals in urban areas, include Don Gill and Penelope Bonnett, *Nature in the urban landscape: a study of city ecosystems* (Baltimore: York Press, 1973); and John Rublowsky, *Nature in the city* (New York: Basic Books, 1967). Recent examples of the global reach of the observational mode in urban ecology scholarship include Harini Nagendra, *Nature in the city: Bengaluru in the past, present, and future* (Delhi: Oxford University Press, 2016).

102. See, for example, Joanna Gliwicz, J. Goszczynski, and Maciej Luniak, "Character-istic features of animal populations under synurbization—the case of the Blackbird and of the Striped Field Mouse," *Memorabilia Zoologica* 49 (1994); Maciej Luniak, "The birds of the park habitats in Warsaw," *Acta ornithologica* 18 (6) (1981): 335–370; and Maciej Luniak, "Synurbization—adaptation of animal wildlife to urban development," in *Proceedings of the 4th International Urban Wildlife Symposium*, ed. William W. Shaw, Lisa K. Harris, and Larry Van Druff (Tucson: University of Arizona, 2004), 50–55.

103. Michael J. Watts, *Silent violence: food, famine, and peasantry in northern Nigeria* (Berkeley: University of California Press, 1983).

104. Watts, *Silent violence*, 21, 22.

105. See, for example, Joel A. Tarr, *The search for the ultimate sink: urban pollution in historical perspective* (Akron, OH: University of Akron Press, 1996); and Richard J. Evans, *Death in Hamburg: society and politics in the cholera years, 1830–1910* (Oxford: Clarendon Press, 1987).

106. See, in particular, Nik Heynen, Maria Kaïka, and Erik Swyngedouw, eds., *In the nature of cities: urban political ecology and the politics of urban metabolism* (London: Routledge, 2006); and Roger Keil, "Urban political ecology," *Urban Geography* 24 (8) (2003): 723–738.

107. See Peter A. Walker, "Political ecology: where is the ecology?," *Progress in Human Geography* 29 (1) (2005): 73–82. For early assessments of the field see also Bruce Braun, "Environmental issues: writing a more-than-human urban geography," *Progress in Human Geography* 29 (5) (2005): 635–650; and Noel Castree, "Environmental issues: relational ontologies and hybrid politics," *Progress in Human Geography* 27 (2) (2003): 203–211.

108. Examples of recent work that synthesizes elements of urban political ecology with new materialist approaches include Maan Barua and Anindya Sinha, "Animating the urban: an ethological and geographical conversation," *Social and Cultural Geography* 20 (8) (2019): 1160–1180; Cara Clancy, "Wild entanglements: exploring the visions and dilemmas of 'renaturing' urban Britain" (PhD dissertation, University of Plymouth, 2019); Marion Ernwein, *Les natures de la ville néolibérale. Une écologie politique du végétal urbain* (Grenoble: UGA Éditions, 2019); and Marion Ernwein, "Bringing urban parks to life: the more-than-human politics of urban ecological work," *Annals of the American Association of Geographers* 111 (2) (2020): 559–576.

109. See, for example, Kennan Ferguson, *William James: politics in the pluriverse* (Lanham, MD: Rowman and Littlefield, 2007); and Martin Savransky, "The pluralistic problematic: William James and the pragmatics of the pluriverse," *Theory, Culture and Society* 38 (2) (2021): 141–159. The term "pluriverse" begins to bubble up through the writings of John Law, Bruno Latour, Walter Mignolo, and others. A key moment in the reshaping of the concept was the workshop "Pluriverse and the Social Sciences" held at St John's, Newfoundland, in the autumn of 2010.

110. Arturo Escobar, *Designs for the pluriverse: radical interdependence, autonomy, and the making of worlds* (Durham, NC: Duke University Press, 2018). See also Mario Blaser and Marisol

de la Cadena, "Pluriverse: proposals for a world of many worlds," in *A world of many worlds*, ed. Marisol de la Cadena and Mario Blaser (Durham, NC: Duke University Press, 2018), 1–22; and Bernd Reiter, ed., *Constructing the pluriverse: the geopolitics of knowledge* (Durham, NC: Duke University Press, 2018).

111. See, for example, Steve Hinchliffe, "Reconstituting nature conservation: towards a careful political ecology," *Geoforum* 39 (1) (2008): 88–97. On heterogeneous subjectivities see also Catherine Johnston, "Beyond the clearing: towards a dwelt animal geography," *Progress in Human Geography* 32 (5) (2008): 633–649.

112. Philippe Descola, "The difficult art of composing worlds (and of replying to objections)," *HAU: Journal of Ethnographic Theory* 4 (3) (2014): 431–443. See also Philippe Descola, *Beyond nature and culture* (Chicago: University of Chicago Press, 2013).

113. Descola, *Beyond nature and culture*, 405.

114. See, for example, Eben Kirksey, *The multispecies salon* (Durham, NC: Duke University Press, 2014); and Cary Wolfe, *What is posthumanism?* (Minneapolis: University of Minnesota Press, 2010).

115. See in particular Patrick Bresnihan, "John Clare and the manifold commons," *Environmental Humanities* 3 (1) (2013): 71–91; and Katey Castellano, "Multispecies work in John Clare's 'Birds nesting' poems," in *Palgrave Advances in John Clare Studies*, ed. Simon Kövesi and Erin Lafford (Cham: Palgrave Macmillan/Springer Nature, 2020), 179–197.

116. Simon P. James, "Protecting nature for the sake of human beings," *Ratio* (2) (2016): 213–227.

Chapter 1

1. Wallace Stevens, "Thirteen ways of looking at a blackbird," in *Harmonium* (New York: Alfred A. Knopf, 1931 [1917]). Stevens's blackbird is most likely one of the icterid group of North American birds that includes the red-winged blackbird (*Agelaius phoeniceus*) as well as grackles, orioles, and cowbirds.

2. See, for example, Karl L. Evans, Kevin J. Gaston, Alain C. Frantz, Michelle Simeoni, Stewart P. Sharp, Andrew McGowan, Deborah A. Dawson, K. Walasz, et al., "Independent colonization of multiple urban centres by a formerly forest specialist bird species," *Proceedings of the Royal Society B* 276 (2009): 2403–2410; Karl L. Evans, Dan E. Chamberlain, Ben J. Hatchwell, Richard D. Gregory, and Kevin J. Gaston, "What makes an urban bird?," *Global Change Biology* 17 (2011): 32–44; Maciej Luniak, "Synurbization and adaptation of animal wildlife to urban development," in *Proceedings of the 4th International Urban Wildlife Symposium*, ed. William W. Shaw, Lisa K. Harris, and Larry Van Druff (Tucson: University of Arizona, 2004), 50–55; and Maciej Luniak, R. Mulsow, and K. Walasz,

"Urbanization of the European Blackbird—expansion and adaptations of urban population," in *Urban ecological studies in Central and Eastern Europe; international symposium Warsaw, Poland*, ed. Maciej Luniak (Warsaw: Polish Academy of Sciences, 1990), 187–198.

3. Jennifer Wolch, "Zoöpolis," *Capitalism, Nature, Socialism* 7 (2) (1996): 21–47.

4. Wolch, "Zoöpolis," 26.

5. See Jakob von Uexküll, *A foray into the worlds of animals and humans*, trans. Joseph D. O'Neil (Minneapolis: University of Minnesota Press, 2010 [1934]). The antimodern and organicist strands in von Uexküll's work are considered in Frederik Stjernfelt, "Simple animals and complex biology: von Uexküll's two-fold influence on Cassirer's philosophy," *Synthese* 179 (1) (2011): 169–186.

6. See, for example, Chandra Talpade Mohanty, "Under Western eyes: feminist scholarship and colonial discourses," *Feminist Review* 30 (1988): 61–88.

7. See Patricia Hill Collins, *Intersectionality* (Cambridge: Polity Press, 2016); and Rebecca Kukla, "Objectivity and perspective in empirical knowledge," *Episteme* 3 (1) (2006): 80–95.

8. See in particular Lorraine Daston and Peter Galison, *Objectivity*, 2nd ed. (New York: Zone Books, 2010 [2007).

9. Kukla, "Objectivity and perspective in empirical knowledge," 82. See also Sandra Harding, "'Strong objectivity': a response to the new objectivity question," *Synthese* 104 (3) (1995): 331–349.

10. Kukla, "Objectivity and perspective in empirical knowledge."

11. Tiffany Lethabo King, "Humans involved: lurking in the lines of posthumanist flight," *Critical Ethnic Studies* 3 (1) (2017): 162–185.

12. See Achille Mbembe, *Critique of black reason*, trans. Laurent Dubois (Durham, NC: Duke University Press, 2017). Questions of race, racism, and corporeal differentiation underlie Michel Foucault's conceptualization of the emergence of biopolitical strategies for social control. See, for example, Kim Su Rasmussen, "Foucault's genealogy of racism," *Theory, Culture and Society* 28 (5) (2011): 34–51; and Philipp Sarasin, "Zweierlei Rassismus? Die Selektion des Fremden als Problem in Michel Foucaults Verbindung von Biopolitik und Rassismus," in *Biopolitik und Rassismus*, ed. Martin Stingelin (Frankfurt: Suhrkamp, 2003), 55–79.

13. King, "Humans involved." On the problem of an "untenable elision" see Claire Jean Kim, *Dangerous crossings: race, species, nature in a multicultural age* (Cambridge: Cambridge University Press, 2015), 286. On the complexities of an "extended elision" between human and nonhuman forms of agency see, for example, Andrew Pickering, *The mangle of practice: time, agency, and science* (Chicago: University of Chicago Press, 1995).

14. See, for example, Jennifer Wolch, Kathleen West, and Thomas E. Gaines, "Transspecies urban theory," *Environment and Planning D: Society and Space* 13 (6) (1995): 735–760.

15. For a critique of the companion species thesis, and in particular Haraway's extension of a Levinasian ethical framework, see Zipporah Weisberg, "The broken promises of monsters: Haraway, animals and the humanist legacy," *Journal for Critical Animal Studies* 7 (2) (2009): 22–62.

16. Joanna Zylinska, *The end of man: a feminist counterapocalypse* (Minneapolis: University of Minnesota Press, 2018).

17. See, for example, Michael O'Neal Campbell, "An animal geography of avian ecology in Glasgow," *Applied Geography* 27 (2) (2007): 78–88.

18. See, for example, Maan Barua and Anindya Sinha, "Animating the urban: an ethological and geographical conversation," *Social and Cultural Geography* 20 (8) (2019): 1160–1180; Sindhu Radhakrishna, Michael A. Huffman, and Anindya Sinha, eds., *The macaque connection: cooperation and conflict between humans and macaques* (New York: Springer, 2012); and Anindya Sinha, "Not in their genes: phenotypic flexibility, behavioural traditions and cultural evolution in wild bonnet macaques," *Journal of Biosciences* 30 (1) (2005): 51–64.

19. See, for example, Jennifer Wolch, Alec Brownlow, and Unna Lassiter, "Constructing the animal worlds of inner-city Los Angeles," in *Animal spaces, beastly places: new geographies of human–animal relations*, ed. Chris Philo and Chris Wilbert (London: Routledge, 2000), 71–97. On the history of animals and cities see, for example, Phillip W. Bateman and Patricia A. Fleming, "Big city life: carnivores in urban environments," *Journal of Zoology* 287 (1) (2012): 1–23.

20. Ian Lovett, "Braying at the bard, appropriately and otherwise," *New York Times* (13 September 2011).

21. Mike Davis, *Ecology of fear: Los Angeles and the imagination of disaster* (New York: Metropolitan Books, 1998), 202.

22. Lila Higgins and Gregory B. Pauly, *Wild LA: explore the amazing nature in and around Los Angeles* (Portland, OR: Timber Press/Los Angeles County Natural History Museum, 2019), 14.

23. Readers react, "In praise of the intelligent, indestructible LA Coyote," *Los Angeles Times* (25 October 2018). See also Sharon Levy, "The new top dog," *Nature* 485 (2012): 296–297.

24. The *Times* editorial board, "Trapping and relocating coyotes? That's a really bad idea," *Los Angeles Times* (22 October 2018). See also Stanley D. Gehrt, Justin L. Brown, and Chris Anchor, "Is the urban coyote a misanthropic synanthrope? The case from Chicago," *Cities and the Environment (CATE)* 4 (1) (2011): 3.

25. Wolch, "Zoöpolis," 29.

26. Jane Bennett, "The force of things: steps toward an ecology of matter," *Political Theory* 32 (3) (2004): 347–372.

27. Neil Smith, "Nature at the millennium: production and re-enchantment," in *Remaking reality: nature at the millennium*, ed. Bruce Braun and Noel Castree (London: Routledge, 1998), 269–282.

28. Wolch, "Zoöpolis," 25.

29. Irus Braverman, "En-listing life: red is the color of threatened species lists," in *Critical animal geographies: politics, intersections, and hierarchies in a multispecies world*, ed. Kathryn Gillespie and Rosemary-Claire Collard (London: Routledge, 2015), 184–202. On the historiography of geography and animal studies see the pivotal contribution of Chris Philo, "Animals, geography and the city: notes on inclusions and exclusions," *Environment and Planning D: Society and Space* 13 (1995): 655–681.

30. See, for example, Diana Donald, "'Beastly Sights': the treatment of animals as a moral theme in representations of London c. 1820–1850," *Art History* 22 (4) (1999): 514–544.

31. Chris Otter, "The technosphere: a new concept for urban studies," *Urban History* 44 (1) (2017): 152.

32. See, for example, Nicola Davison, "The Anthropocene epoch: have we entered a new phase of planetary history," *Guardian* (30 May 2019); and Dinesh Wadiwel, "Chicken harvesting machine," *South Atlantic Quarterly* 117 (3) (2018): 527–549.

33. See Cary Wolfe's discussion of the genetic components of Estimated Breeding Values used for livestock in *Before the law: humans and other animals in a biopolitical frame* (Chicago: University of Chicago Press, 2013). On evidence of cruelty in food production and the transport of live animals see, for example, Nick Evershed and Calla Wahlquist, "Live exports: mass animal deaths going unpunished as holes in the system revealed," *Guardian* (10 April 2018). On the industrialized production of meat see Alex Blanchette, *Porkopolis: American animality, standardized life, and the factory farm* (Durham, NC: Duke University Press, 2020); William Boyd, "Making meat: science, technology, and American poultry production," *Technology and Culture* 42 (4) (2001): 631–664; and Timothy Pachirat, *Every twelve seconds: industrialized slaughter and the politics of sight* (New Haven, CT: Yale University Press, 2011).

34. Catherine McNeuer, *Taming Manhattan: environmental battles in the antebellum city* (Cambridge, MA: Harvard University Press, 2014), 8.

35. McNeuer, *Taming Manhattan*. See also Philip Howell, "Between wild and domestic, animal and human, life and death: the problem of the stray in the Victorian city," in *Animal history in the modern city: exploring liminality*, ed. Clemens Wischermann, Aline Steinbrecher, and Philip Howell (London: Bloomsbury, 2019), 145–160; and Harriet Ritvo, *The animal estate: the English and other creatures in the Victorian Age* (London: Penguin, 1987).

36. See Matthew Gandy, *Concrete and clay: reworking nature in New York City* (Cambridge, MA: MIT Press, 2002).

37. See Catherine Brinkley and Domenic Vitiello, "From farm to nuisance: animal agriculture and the rise of planning regulation," *Journal of Planning History* 13 (2) (2014): 113–135.

38. "Report of Colonial Surgeon 1877," *Hong Kong Government Gazette* (23 November 1878), 563, cited in Pui-yin Ho, *Making Hong Kong: a history of urban development* (Cheltenham: Edward Elgar, 2018), 60.

39. William Cronon, *Nature's metropolis: Chicago and the great west* (New York: W. W. Norton, 1991), 226.

40. Cronon, *Nature's metropolis*, 228, 229.

41. Cronon, *Nature's metropolis*, 256.

42. Anna Mazanik, "'Shiny shoes' for the city: the public abattoir and the reform of meat supply in imperial Moscow," *Urban History* 45 (2) (2018): 214–232.

43. James Oles, "In pursuit of Salamone: in the emptiness of the Argentine Pampas, a vision of modernity," *Cabinet* (Summer 2009) 34. I would like to thank Leandro Minuchin for further insights into Argentinian architecture.

44. Daniel Pick, *War machine: the rationalization of slaughter in the modern age* (New Haven, CT: Yale University Press, 1993), 180. See also Noëlie Vialles, *Animal to edible*, trans. J. A. Underwood (Cambridge: Cambridge University Press, 1994 [1987]).

45. George Blumer, "Some remarks on the early history of trichinosis (1822–1866)," *Yale Journal of Biology and Medicine* 11 (6) (1939): 581–588.

46. Upton Sinclair, *The Jungle* (London: Penguin, 1985 [1906]), 48.

47. Marco d'Eramo, *The pig and the skyscraper. Chicago: a history of our future*, trans. Graeme Thomson (London: Verso, 2002).

48. Sinclair, *The Jungle*, 44.

49. Dorothee Brantz, "Animal bodies, human health, and the reform of slaughterhouses in nineteenth-century Berlin," in *Meat, modernity and the rise of the slaughterhouse*, ed. Paula Young Lee (Durham: University of New Hampshire Press, 2008), 72.

50. Clay McShane and Joel A. Tarr, *The horse in the city: living machines in the nineteenth century* (Baltimore: Johns Hopkins University Press, 2007). On concerns with the treatment of horses and other animals in the nineteenth-century city see Paulus Ebner, "Nützen und Schützen. Städtischer Tierschutz im 19. Jahrhundert," in *Umwelt Stadt. Geschichte des Natur- und Lebensraumes Wien*, ed. Karl Brunner and Petra Schneider (Vienna: Böhlau, 2005), 433–437.

51. See Hilda Kean "Traces and representations: animal pasts in London's present," *London Journal* 36 (1) (2011): 54–71.

52. See Maan Barua, "Nonhuman labour, encounter value, spectacular accumulation: The geographies of a lively commodity," *Transactions of the Institute of British Geographers* 42 (2) (2017): 274–288; and Rosemary-Claire Collard, "Putting animals back together, taking commodities apart," *Annals of the Association of American Geographers* 104 (1) (2014): 151–165.

53. Eric Cardinale, Vincent Porphyre and Denis Bastianelli, "Methods to promote healthier animal production, examples in peri-urban poultry production around Dakar," paper presented at the conference Appropriate Methodologies for Urban Agriculture (Nairobi, Kenya, 2001), cited in Alice Hovorka, "Trans-species urban theory: chickens in an African city," *Cultural Geographies* 15 (1) (2008): 101.

54. James Sumberg, "Poultry production in and around Dar es Salaam, Tanzania: competition and complementarity," *Outlook on Agriculture* 27 (3) (1998): 177–185.

55. Hovorka, "Trans-species urban theory."

56. Zarin Ahmad, "Delhi's meatscapes: cultural politics of meat in a globalizing city," *IIM Kozhikode Society and Management Review* 3 (1) (2014): 21–31. On the intersections between animals, caste-based politics, and sociospatial modalities of abjection see also Yamini Narayanan, "Animating caste: visceral geographies of pigs, caste, and violent nationalisms in Chennai city," *Urban Geography* 42 (2021).

57. Ahmad, "Delhi's meatscapes." See also John Lever, "Halal meat and religious slaughter: from spatial concealment to social controversy—breaching the boundaries of the permissible?," *Environment and Planning C: Politics and Space* 37 (2019): 889–907.

58. Krithika Srinivasan, "Remaking more-than-human society: thought experiments on street dogs as 'nature,'" *Transactions of the Institute of British Geographers* 44 (2) (2019): 376–391.

59. See the documentary *Airborne* (dir.: Shaunak Sen, 2020). The Muslim symbolism of meat throwing for kites is also discussed by David Pinsault in "Raw meat skyward: pariah-kite rituals in Lahore," in *Notes from the fortune telling parrot* (London: Equinox, 2008), 108–121. See also Subhendu Mazumdar, Dipankar Ghose, and Goutam Kumar Saha, "Foraging strategies of Black Kites (*Milvus migrans govinda*) in urban garbage dumps," *Journal of Ethology* 34 (3) (2016): 243–247; and Nishant Kumar, Urvi Gupta, Harsha Malhotra, Yadvendradev V. Jhala, Qamar Qureshi, Andrew G. Gosler, and Fabrizio Sergio, "The population density of an urban raptor is inextricably tied to human cultural practices," *Proceedings of the Royal Society B* 286 (1900) (2019): 20182932.

60. See Jody Emel and Harvey Neo, *Political ecology of meat* (London: Routledge, 2015); and Pachirat, *Every twelve seconds*.

61. Burnett's focus on the experience of children in the urban landscape can be compared with the writer and poet Audre Lorde's reflections on memory, culture, and the meaning

of urban nature. See Kathleen R. Wallace, "'All natural things are strange': Audre Lorde, urban nature, and cultural place," in *The nature of cities: ecocriticism and urban environments*, ed. Michael Bennett and David W. Teague (Tucson: University of Arizona Press, 1999), 55–76. The place of the abattoir in urban African-American culture is touched on in Wolch, Brownlow, and Lassiter, "Constructing the animal worlds of inner-city Los Angeles." One of their interviewees "identified a chink in her father's hard emotional armour; discussing her father's experience at the slaughterhouse, she said: 'Daddy couldn't kill the sheep because of . . . the sound they would make when you kill 'em. The sheep broke his heart'" (92). The neglected urban alleys represented in Burnett's film are precisely the kind of marginal spaces that have become significant for the articulation of alternative black ecologies in Los Angeles, Detroit, and other North American cities. See, for example, Jennifer Wolch, Josh Newell, Mona Seymour, Hilary Bradbury Huang, Kim Reynolds, and Jennifer Mapes, "The forgotten and the future: reclaiming back alleys for a sustainable city," *Environment and Planning A* 42 (12) (2010): 2874–2896.

62. See, for example, S. Harris Ali and Roger Keil, "Global cities and the spread of infectious disease: the case of severe acute respiratory syndrome (SARS) in Toronto, Canada," *Urban Studies* 43 (3) (2006): 491–509; Creighton Connolly, Roger Keil, and S. Harris Ali, "Extended urbanisation and the spatialities of infectious disease: demographic change, infrastructure and governance," *Urban Studies* 58 (2) (2021): 245–263; Bryony A. Jones, Delia Grace, Richard Kock, Silvia Alonso, Jonathan Rushton, Mohammed Y. Said, Declan McKeever, et al., "Zoonosis emergence linked to agricultural intensification and environmental change," *Proceedings of the National Academy of Sciences* 110 (21) (2013): 8399–8404; and Meike Wolf, "Rethinking urban epidemiology: natures, networks and materialities," *International Journal of Urban and Regional Research* 40 (5) (2016): 958–982

63. See, for example, Mike Davis, *The monster at our door: the global threat of avian flu* (New York: New Press, 2005); Marius Gilbert, Scott H. Newman, John Y. Takekawa, Leo Loth, Chandrashekhar Biradar, Diann J. Prosser, Sivananinthaperumal Balachandran, et al. "Flying over an infected landscape: distribution of highly pathogenic avian influenza H5N1 risk in South Asia and satellite tracking of wild waterfowl," *EcoHealth* 7 (4) (2010): 448–458; Frédéric Keck, "Livestock revolution and ghostly apparitions: South China as a sentinel territory for influenza pandemics," *Current Anthropology* 60 (S20) (2019): S251–S259; and Robert G. Wallace, "Breeding influenza: the political virology of offshore farming," *Antipode* 41 (5) (2009): 916–951.

64. See, for example, Ivan V. Kuzmin, Brooke Bozick, Sarah A. Guagliardo, Rebekah Kunkel, Joshua R. Shak, Suxiang Tong, and Charles E. Rupprecht, "Bats, emerging infectious diseases, and the rabies paradigm revisited," *Emerging Health Threats Journal* 4 (1) (2011): 7159; Vineet D. Menachery, Boyd L. Yount, Amy C. Sims, Kari Debbink, Sudhakar S. Agnihothram, Lisa E. Gralinski, Rachel L. Graham, et al., "SARS-like WIV1-CoV poised for human emergence," *Proceedings of the National Academy of Sciences* 113 (11) (2016):

3048–3053; and Ye Qiu, Yuan-Bo Zhao, Qiong Wang, Jin-Yan Li, Zhi-Jian Zhou, Ce-Heng Liao, and Xing-Yi Ge, "Predicting the angiotensin converting enzyme 2 (ACE2) utilizing capability as the receptor of SARS-CoV-2," *Microbes and Infection* 22 (4–5) (2020): 221–225.

65. Christos Lynteris and Lyle Fearnley, "Why shutting down Chinese wet markets could be a terrible mistake," *The Conversation* (31 January 2020 [updated 3 March 2020]).

66. See Ying-kit Chan, "No room to swing a cat? Animal treatment and urban space in Singapore," *Southeast Asian Studies* 5 (2016): 305–329; Lucy Davis, "Zones of contagion: the Singapore body politic and the body of the street-cat," in *Considering animals: contemporary studies in human-animal relations*, ed. Carol Freeman, Elizabeth Leane, and Yvette Watt (London: Routledge, 2011), 183–198; and Cai Xuejiao, "Cats, dogs, caught in Coronavirus crossfire, worrying animal lovers," *Sixth Tone* (10 April 2020).

67. See Roger Frutos, Jordi Serra-Cobo, Tianmu Chen, and Christian A. Devaux, "COVID-19: time to exonerate the pangolin from the transmission of SARS-CoV-2 to humans," *Infection, Genetics and Evolution* 84 (2020): 104493.

68. Hannah Ellis-Petersen, "'No way to stop it': millions of pigs culled across Asia as swine fever spreads," *Guardian* (6 June 2019). See also Matthew Sparke and Dimitar Anguelov, "H1N1, globalization and the epidemiology of inequality," *Health and Place* 18 (4) (2012): 726–736.

69. Dennis Normile, "African swine fever keeps spreading in Asia, threatening food security," *Science* (14 May 2019).

70. Ann H. Kelly and Javier Lezaun, "Urban mosquitoes, situational publics, and the pursuit of interspecies separation in Dar es Salaam," *American Ethnologist* 41 (2) (2014): 371.

71. See Henry Buller, "Reconfiguring wild spaces: the porous boundaries of wild animal geographies," in *Routledge handbook of human-animal studies*, ed. Garry Marvin and Susan McHugh (London: Routledge, 2014), 233–245; Bernhard Kegel, *Tiere in der Stadt. Eine Naturgeschichte* (Cologne: DuMont, 2013); and Joëlle Zask, *Zoocities. Des animaux sauvages dans la ville* (Paris: Premier Parallèle, 2020).

72. See, for example, Philip W. Bateman and Patricia A. Fleming, "Big city life: carnivores in urban environments," *Journal of Zoology* 287 (1) (2012): 1–23; Seth P. D. Riley, Brian L. Cypher, and Stanley D. Gehrt, *Urban carnivores: ecology, conflict, and conservation* (Baltimore: Johns Hopkins University Press, 2010); and Martin Šálek, Lucie Drahníková, and Emil Tkadlec, "Changes in home range sizes and population densities of carnivore species along the natural to urban habitat gradient," *Mammal Review* 45 (1) (2015): 1–14.

73. Stanley Gehrt and Max McGraw, "Ecology of coyotes in urban landscapes," in *Wildlife Damage Management Conferences* 63 (2007): 303.

74. The coyotes of New York City are now the focus of the Gotham Coyote Project with an extensive camera trap network to study their movements and behavior along with DNA

analysis of their scat to determine aspects of diet and the degree of genetic interrelatedness between individuals. See, for example, Alexandra L. DeCandia, Carol S. Henger, Amelia Krause, Linda J. Gormezano, Mark Weckel, Christopher Nagy, Jason Munshi-South, and Bridgett M. von Holdt, "Genetics of urban colonization: neutral and adaptive variation in coyotes (*Canis latrans*) inhabiting the New York metropolitan area," *Journal of Urban Ecology* 5 (1) (2019); and Mark E. Weckel, Deborah Mack, Christopher Nagy, Roderick Christie, and Anastasia Wincorn, "Using citizen science to map human-coyote interaction in suburban New York, USA," *Journal of Wildlife Management* 74 (5) (2010): 1163–1171. On the shifting behavioral, ecological, and genetic interface between coyotes and wolves see Stephanie Rutherford, "The Anthropocene's animal? Coywolves as feral cotravelers," *Environment and Planning E: Nature and Space* 1 (1–2) (2018): 206–223.

75. On the changing place of the wolf in European environmental discourse see, for example, Henry Buller, "Safe from the wolf: biosecurity, biodiversity, and competing philosophies of nature," *Environment and Planning A* 40 (7) (2008): 1583–1597; and Martin Drenthen, "The return of the wild in the Anthropocene: wolf resurgence in the Netherlands," *Ethics, Policy and Environment* 18 (3) (2015): 318–337.

76. In the summer of 2018 the use of DNA analysis confirmed that a wolf had killed a sheep on the outskirts of Hamburg. See "Erster Wolfriss in Hamburger Stadtgebiet," *Der Spiegel* (8 August 2018).

77. See, for example, I. Beames, "The spread of the fox in the London area," *Ecologist* 2 (2) (1972): 25–26; and Stephen Harris and Jeremy M. V. Rayner, "Urban fox (*Vulpes vulpes*) population estimates and habitat requirements in several British cities," *Journal of Animal Ecology* 55 (2) (1986): 575–591.

78. W. G. Teagle, "The fox in the London suburbs," *London Naturalist* 46 (1967): 53.

79. See, for example, Angela Cassidy and Brett Mills, "'Fox tots attack shock': urban foxes, mass media and boundary-breaching," *Environmental Communication: A Journal of Nature and Culture* 6 (4) (2012): 494–511.

80. See Rob Kitchin, Justin Gleeson, Karen Keaveney, and Cian O'Callaghan, "A haunted landscape: housing and ghost estates in post-Celtic Tiger Ireland," *National Institute for Regional and Spatial Analysis Working Paper Series* 59 (1) (2010); and Rob Kitchin, Cian O'Callaghan, and Justin Gleeson, "The new ruins of Ireland? Unfinished estates in the post-Celtic Tiger era," *International Journal of Urban and Regional Research* 38 (3) (2014): 1069–1080.

81. See Peter S. Alagona, *After the grizzly: endangered species and the politics of place in California* (Berkeley: University of California Press, 2013), 3. In California a small number of human fatalities stemming from encounters with the cougar or mountain lion (*Puma concolor*) spurred the unsuccessful Proposition 197 that pitted a combination of hunting, agricultural, and landowner interests against a conservation emphasis on wilderness preservation.

See also Rosemary-Claire Collard, "Cougar-human entanglements and the biopolitical un/making of safe space," *Environment and Planning D: Society and Space* 30 (1) (2012): 23–42; and Davis, *Ecology of fear*, 245–247

82. See, for example, Marcus Wehner, "Der AfD will den Wolf jagen," *Frankfurter Allgemeine* (1 January 2020).

83. Ralph Acampora, "Nietzsche's feral philosophy: thinking through an animal imaginary," in *A Nietzschean bestiary: becoming animal beyond docile and brutal*, ed. Christa Davis Acampora and Ralph R. Acampora (Lanham, MD: Rowman and Littlefield, 2004), 1–16.

84. Ralph Acampora cited in Eduardo Mendieta, "The biotechnological *Scala Naturae* and interspecies cosmopolitanism: Patricia Piccinini, Jane Alexander, and Guillermo Gómez-Peña," in *Biopower: Foucault and beyond*, ed. Vernon W. Cisney and Nicolae Morar (Chicago: University of Chicago Press, 2015), 158–182.

85. Mendieta, "The biotechnological *Scala Naturae* and interspecies cosmopolitanism."

86. Mendieta, "The biotechnological *Scala Naturae* and interspecies cosmopolitanism," 162.

87. Margot Norris, *Beasts of the modern imagination: Darwin, Nietzsche, Kafka, Ernst, and Lawrence* (Baltimore: Johns Hopkins University Press, 1985), 231.

88. Richard Arnold, Ian Woodward, and Neil Smith, *Parrots in the London area: a London bird atlas supplement* (London: London Natural History Society, 2018).

89. O. J. N. Heald, C. Fraticelli, S. E. Cox, M. C. A. Stevens, S. C. Faulkner, T. M. Blackburn, and S. C. Le Comber, "Understanding the origins of the ring-necked parakeet in the UK," *Journal of Zoology* 312 (2020): 1–11.

90. Higgins and Pauly, *Wild LA*. The spread of the ring-necked parakeet in Europe has been the focus of ecological controversy. See, for example, Diederik Strubbe and Erik Matthysen, "Experimental evidence for nest-site competition between invasive ring-necked parakeets (*Psittacula krameri*) and native nuthatches (*Sitta europaea*)," *Biological Conservation* 142 (8) (2009): 1588–1594.

91. See, for example, Ian D. Rotherham, "Times they are a changin'—recombinant ecology as an emerging paradigm," *Urban Environments—History, Biodiversity and Culture* (2017): 1 –19.

92. Michael Soulé, "The onslaught of alien species, and other challenges in the coming decades," *Conservation Biology* 4 (3) (1990): 233–239.

93. The term "new natures" is derived from Martin Drenthen. See, for example, Martin Drenthen, Jozef Keulartz, and James Proctor, eds., *New visions of nature: complexity and authenticity* (Dordrecht: Springer, 2009). See also Steve Hinchliffe and Sarah Whatmore, "Living cities: towards a politics of conviviality," *Science as Culture* 15 (2) (2006): 123–138.

94. Emma Marris, *Rambunctious garden: saving nature in a post-wild world* (New York: Bloomsbury, 2011), 2.

95. William Attaway, *Blood on the forge* (New York: New York Review Book, 2005 [1941]). See also John Claborn, "From black Marxism to industrial ecosystem: racial and ecological crisis in William Attaway's *Blood on the forge*," *MFS Modern Fiction Studies* 55 (3) (2009): 566–595.

96. Attaway, *Blood on the forge*, 173.

97. Alan M. Beck, *The ecology of stray dogs: a study of free-ranging urban animals* (West Lafayette, IN: Purdue University Press, 1973).

98. Beck, *The ecology of stray dogs*, 11.

99. Rivke Jaffe, "Political animals: an interspecies approach to urban inequalities," paper presented to the Berkeley Black Geographies Symposium, University of California, Berkeley, 12–13 March 2020. On human-dog relations and subjectivities see also Bénédicte Boisseron, *Afro-dog: blackness and the animal question* (New York: Columbia University Press, 2018).

100. Clare Palmer, "Placing animals in urban environmental ethics," *Journal of Social Philosophy* 34 (1) (2003): 64–78.

101. Ger Duijzings, "Dictators, dogs, and survival in a post-totalitarian city," in *Urban constellations*, ed. Matthew Gandy (Berlin: jovis, 2011), 146–147.

102. Krithika Srinivasan, "The biopolitics of animal being and welfare: dog control and care in the UK and India," *Transactions of the Institute of British Geographers* 38 (1): 106–119. See also Eduardo Mendieta, "Interspecies cosmopolitanism: towards a discourse ethics grounding of animal rights," *Philosophy Today* 54 (2010): 208–216. For further conceptual insights also Yamini Narayanan, "Street dogs at the intersection of colonialism and informality: 'subaltern animism' as a posthuman critique of Indian cities," *Environment and Planning D: Society and Space* 35 (3) (2017): 475–494.

103. Beck, *The ecology of stray dogs*.

104. Gardiner Harris, "Where streets are thronged with strays bearing fangs," *New York Times* (6 August 2012). Since the early 2000s methods of dog control in Indian cities have moved from culling towards sterilization and other forms of population management.

105. Alexander R. Braczkowski, Christopher J. O'Bryan, Martin J. Stringer, James E. M. Watson, Hugh P. Possingham, and Hawthorne L. Beyer, "Leopards provide public health benefits in Mumbai, India," *Frontiers in Ecology and the Environment* 16 (3) (2018): 176–182. Other studies of keystone predators preying on stray dogs include, for example, Swapnil Kumbhojkar, Reuven Yosef, Jakub Z. Kosicki, Patrycja K. Kwiatkowska, and Piotr Tryjanowski, "Dependence of the leopard *Panthera pardus fusca* in Jaipur, India, on domestic

animals," *Oryx* (2020): 1–7; and James R. A. Butler, John D. C. Linnell, Damian Morrant, Vidya Athreya, Nicolas Lescureux, and Adam McKeown, "Dog eat dog: social-ecological dimensions of dog predation by wild carnivores," in *Free-ranging dogs and wildlife conservation*, ed. M. Gompper (Oxford: Oxford University Press, 2014), 117–143.

106. On the Chernobyl exclusion zone see, for example, Chris C. Park, *Chernobyl: the long shadow* (London: Routledge, 1989); Jonathon Turnbull, "Checkpoint dogs: photovoicing canine companionship in the Chernobyl exclusion zone," *Anthropology Today* 36 (6) (2020): 21–24; and Sarah C. Webster, Michael E. Byrne, Stacey L. Lance, Cara N. Love, Thomas G. Hinton, Dmitry Shamovich, and James C. Beasley, "Where the wild things are: influence of radiation on the distribution of four mammalian species within the Chernobyl exclusion zone," *Frontiers in Ecology and the Environment* 14 (4) (2016): 185–190. I would like to thank Jonny Turnbull for his many insights into the changing canine ecologies of the exclusion zone.

107. Synanthropic ecologies include a variety of human ectoparasites such as the bedbug (*Cimex lectularis*, replaced by *C. hemipterus* in warmer climates), the lice (*Pediculus humanus humanus*, *Pediculus humanus capitis*, and *Pthirus pubis*), the human flea (*Pulex irritans*), and many other species that have become closely associated with human settlements such as the house fly (*Musca domestica*), the stable fly (*Stomoxys calcitrans*), and the brown rat (*Rattus norvegicus*). The human body itself is also a distinctive biome supporting an array of mostly harmless bacteria and other organisms.

108. Eva Panagiotakopulu and Paul C. Buckland, "A thousand bites—insect introductions and late Holocene environments," *Quaternary Science Reviews* 156 (2017): 23–35. Some accounts highlight the emergence of pre-Neolithic forms of commensalism for crows, foxes, and other animals. See, for example, Terry O'Connor, *Animals as neighbors: the past and present of commensal species* (East Lansing: Michigan State University Press, 2009).

109. Stefan Ineichen and Max Ruckstuhl, *Stadtfauna Zürich* (Bern: Haupt Verlag, 2012). The presence of hair follicle mites on the human body is an example of commensalism since they do little or no harm, whilst dust mites such as the abundant *Dermatophagoides* that live on dead skin and other small particles of organic matter have been associated with respiratory illnesses such as asthma. A similarly comprehensive approach is taken by a recent field guide to the urban wildlife of North American cities that ranges from foxes and coyotes to bedbugs, lice, and ticks. See Julie Feinstein, *Field guide to urban wildlife* (Mechanicsburg, PA: Stackpole, 2011).

110. Bernhard Klausnitzer, *Ökologie der Großstadtfauna* (Stuttgart: Gustav Fischer, 1987).

111. George Ordish, *The living house* (London: Rupert Hart-Davis, 1960), 21.

112. Matthew A. Bertone, Misha Leong, Keith M. Bayless, Tara L. F. Malow, Robert R. Dunn, and Michelle D. Trautwein, "Arthropods of the great indoors: characterizing diversity inside urban and suburban homes," *PeerJ* 4 (2016): e1582.

113. Mosquitoes that occur naturally in water-filled tree holes that offer protection for their larvae from fish and other predators have gradually taken advantage of the aquatic micro-ecologies created by abandoned tires, saucers under flowerpots, or even the tiny habitats enabled by discarded bottle tops. See, for example, Alex M. Nading, *Mosquito trails: ecology, health, and the politics of entanglement* (Berkeley: University of California Press, 2014).

114. Shannon L. LaDeau, Paul T. Leisnham, Dawn Day Biehler, and Danielle Bodner, "Higher mosquito production in low-income neighborhoods of Baltimore and Washington, DC: understanding ecological drivers and mosquito-borne disease risk in temperate cities," *International Journal of Environmental Research and Public Health* 10 (4) (2013): 1505–1526.

115. Duane J. Gubler, "Dengue, urbanization and globalization: the unholy trinity of the 21st century," *Tropical Medicine and Health* 39 (4) (2011): 5–6.

116. Vinay Gidwani and Rajyashree N. Reddy. "The afterlives of 'waste': notes from India for a minor history of capitalist surplus," *Antipode* 43 (5) (2011): 1625–1658. See also Brent Z. Kaup, "Pathogenic metabolisms: a rift and the Zika Virus in Mato Grosso, Brazil," *Antipode* 53 (2) (2021): 567–586; and Nida Rehman, "Epidemiological landscapes: the spaces and politics of mosquito control in Lahore" (PhD thesis, University of Cambridge, 2020).

117. The bedbug is likely to have switched host from bats to humans when both species shared caves. See Klaus Reinhardt, *Bedbug* (London: Reaktion, 2018); and Clive Boase, "Bed bugs (Hemiptera: Cimicidae): an evidence-based analysis of the current situation," in *Proceedings of the Sixth International Conference on Urban Pests* (Veszprém, Hungary: OOK-Press, 2008), 7–14.

118. George Orwell, *Down and out in Paris and London* (Harmondsworth: Penguin, 1940 [1933]), 6.

119. UK Ministry of Health, *Report on the bed-bug*, Reports on Public Health and Medical Subjects, No. 72 (London: His Majesty's Stationery Office, 1934). See also Ben Campkin, "Terrors by night: bedbug infestations in London," in *Urban constellations*, ed. Matthew Gandy (London: jovis, 2011).

120. Michael F. Potter, "The history of bed bug management—with lessons from the past," *American Entomologist* 57 (1) (2011): 12–25. See also Dawn Day Biehler, "Permeable homes: a historical political ecology of insects and pesticides in US public housing," *Geoforum* 40 (6) (2009): 1014–1023.

121. Dawn Day Biehler, *Pests in the city: flies, bedbugs, cockroaches, and rats* (Washington: University of Washington Press, 2013).

122. William Beebe, *Unseen life of New York: as a naturalist sees it* (New York: Duell, Sloan and Pearce, 1953).

123. See, for example, Jennifer Northridge et al., "The role of housing type and housing quality in urban children with asthma," *Journal of Urban Health* 87 (2) (2010): 211–224; and Sampson B. Sarpong et al., "Socioeconomic status and race as risk factors for cockroach allergen exposure and sensitization in children with asthma," *Journal of Allergy and Clinical Immunology* 97 (6) (1996): 1393–1401.

124. See Oscar Zeta Acosta, *The revolt of the cockroach people* (San Francisco: Straight Arrow Press, 1973).

125. Pedro Pietri, "Suicide note from a cockroach in a low income housing project," in Pedro Pietri, *Puerto Rican obituary* (New York: Monthly Review Press, 1973).

126. Pierre Payment and Will Robertson, "The microbiology of piped distribution systems and public health," in *Safe piped water: managing microbial water quality in piped distribution systems*, ed. Richard G. Ainsworth (World Health Organization; London: IWA Publishing, 2004), 1–18.

127. See, for example, Jennifer Cope, Raoult C. Ratard, Vincent R. Hill, Theresa Sokol, Jonathan Jake Causey, Jonathan S. Yoder, Gayatri Mirani, et al., "The first association of a primary amebic meningoencephalitis death with culturable *Naegleria fowleri* in tap water from a US treated public drinking water system," *Clinical Infectious Diseases* 60 (8) (2015): e36–e42; and Mahwish Ali, Syed Babar Jamal, and Syeda Mehpara Farhat, "*Naegleria fowleri* in Pakistan," *Lancet Infectious Diseases* 20 (1) (2020): 27–28.

128. See Sammy Zahran, Shawn P. McElmurry, Paul E. Kilgore, David Mushinski, Jack Press, Nancy G. Love, Richard C. Sadler, and Michele S. Swanson, "Assessment of the Legionnaires' disease outbreak in Flint, Michigan," *Proceedings of the National Academy of Sciences* 115 (8) (2018): 1730–1739. On the history of Legionnaires' disease see Barry S. Fields, Robert F. Benson, and Richard E. Besser, "*Legionella* and Legionnaires' disease: 25 years of investigation," *Clinical Microbiology Reviews* 15 (3) (2002): 506–526.

129. Menno Schilthuizen, *Darwin comes to town: how the urban jungle drives evolution* (London: Quercus, 2018). See also Marina Alberti, Cristian Correa, John M. Marzluff, Andrew P. Hendry, Eric P. Palkovacs, Kiyoko M. Gotanda, Victoria M. Hunt, Travis M. Apgar, and Yuyu Zhou, "Global urban signatures of phenotypic change in animal and plant populations," *Proceedings of the National Academy of Sciences* 114 (34) (2017): 8951–8956; Marina Alberti, "Eco-evolutionary dynamics in an urbanizing planet," *Trends in Ecology and Evolution* 30 (2) (2015): 114–126; and Marina Alberti, John Marzluff, and Victoria M. Hunt, "Urban driven phenotypic changes: empirical observations and theoretical implications for eco-evolutionary feedback," *Philosophical Transactions of the Royal Society B: Biological Sciences*, 372 (1712) (2017): 20160029.

130. Katharine Byrne and Richard A. Nichols, "*Culex pipiens* in London Underground tunnels: differentiation between surface and subterranean populations," *Heredity* 82 (1)

(1999): 7–15; David N. Reznick, *The "Origin" then and now: an interpretive guide to the "Origin of species"* (Princeton, NJ: Princeton University Press, 2010).

131. Marco Di Luca, Luciano Toma, Daniela Boccolini, Francesco Severini, Giuseppe La Rosa, Giada Minelli, Gioia Bongiorno, et al., "Ecological distribution and CQ11 genetic structure of *Culex pipiens* complex (Diptera: Culicidae) in Italy," *PLoS One* 11 (1) (2016): e0146476.

132. See, for example, Rory J. Howlett and Michael E. N. Majerus, "The understanding of industrial melanism in the peppered moth (*Biston betularia*) (Lepidoptera: Geometridae)," *Biological Journal of the Linnaean Society* 30 (1) (1987): 31–44.

133. Laurence M. Cook, Bruce S. Grant, Ilik J. Saccheri, and Jim Mallet, "Selective bird predation on the peppered moth: the last experiment of Michael Majerus," *Biology Letters* 8 (4) (2012): 609–612.

134. Callum J. Macgregor, Darren M. Evans, Richard Fox, and Michael J. O. Pocock, "The dark side of street lighting: impacts on moths and evidence for the disruption of nocturnal pollen transport," *Global Change Biology* 23 (2) (2017): 697–707.

135. See, for example, Fabio Falchi, "The new world atlas of artificial night sky brightness," *Science Advances* 2 (2016): e1600377; Matthew Gandy, "Negative luminescence," *Annals of the American Association of Geographers* 107 (5) (2017): 1090–1107; and Dirk Sanders, Enric Frago, Rachel Kehoe, Christophe Patterson, and Kevin J. Gaston, "A meta-analysis of biological impacts of artificial light at night," *Nature Ecology and Evolution* 5 (1) (2021): 74–81.

136. See Jakob C. Müller, Jesko Partecke, Ben J. Hatchwell, Kevin J. Gaston, and Karl L. Evans, "Candidate gene polymorphisms for behavioural adaptations during urbanization in blackbirds," *Molecular Ecology* 22 (2013): 3629–3637. See also, for example, Hans Slabbekoorn and Margriet Peet, "Birds sing at higher pitch in urban noise," *Nature* 424 (2003): 267; and Hans Slabbekoorn and Ardie den Boer-Visser, "Cities change the songs of birds," *Current Biology* 16 (23) (2006): 2326–2331.

137. Mats Björklund, Iker Ruiz, and Juan Carlos Senar, "Genetic differentiation in the urban habitat: the great tits (*Parus major*) of the parks of Barcelona city," *Biological Journal of the Linnaean Society* 99 (1) (2010): 9–19.

138. See, for example, Eyal Shochat, Paige S. Warren, Stanley H. Faeth, Nancy E. McIntyre, and Diane Hope, "From patterns to emerging processes in mechanistic urban ecology," *Trends in Ecology and Evolution* 21 (4) (2006): 186–191; and Matthew Combs, Kaylee A. Byers, Bruno M. Ghersi, Michael J. Blum, Adalgisa Caccone, Federico Costa, Chelsea G. Himsworth, Jonathan L. Richardson, and Jason Munshi-South, "Urban rat races: spatial population genomics of brown rats (*Rattus norvegicus*) compared across multiple cities," *Proceedings of the Royal Society B: Biological Sciences* 285 (1880) (2018): 20180245.

139. See Pierre-Olivier Cheptou, O. Carrue, Soraya Rouifed, and Amélie M. Cantarel, "Rapid evolution of seed dispersal in an urban environment in the weed *Crepis sancta*," *Proceedings of the National Academy of Sciences* 105 (10) (2008): 3796–3799.

---

140. See, for example, Vanessa M. D'Costa, Emma Griffiths, and Gerard D. Wright, "Expanding the soil antibiotic resistome: exploring environmental diversity," *Current Opinion in Microbiology* 10 (5) (2007): 481–489; and Kevin J. Forsberg, Alejandro Reyes, Bin Wang, Elizabeth M. Selleck, Morten O. A. Sommer, and Gautam Dantas, "The shared antibiotic resistome of soil bacteria and human pathogens," *Science* 337 (6098) (2012): 1107–1111.

141. Bernice Sepers, Krista van den Heuvel, Melanie Lindner, Heidi Viitaniemi, Arild Husby, and Kees van Oers, "Avian ecological epigenetics: pitfalls and promises," *Journal of Ornithology* (2019): 1–21.

142. The geneticist Adrian Bird offers a modified definition of epigenetics as "the structural adaptation of chromosomal regions so as to register, signal or perpetuate altered activity states" without insisting on any heritable dimension. See Adrian Bird, "Perceptions of epigenetics," *Nature* 447 (7143) (2007): 396–398. See also Hannah Landecker and Aaron Panofsky, "From social structure to gene regulation, and back: a critical introduction to environmental epigenetics for sociology," *Annual Review of Sociology* 39 (2013): 333–357; and Jörg Niewöhner, "Epigenetics: embedded bodies and the molecularisation of biography and milieu," *BioSocieties* 6 (3) (2011): 279–298. Although Niewöhner's argument is concerned with the human environment, his reconceptualization of the body and its "context" raises significant issues for nonhuman occupants of the urban realm.

143. Michael J. Angilletta, S. Wilson Robbie Jr., Amanda C. Niehaus, Michael W. Sears, Carlos A. Navas, and Pedro L. Ribeiro, "Urban physiology: city ants possess high heat tolerance," *PLoS One* 2 (2) (2007): e258.

144. Kristin M. Winchell, R. Graham Reynolds, Sofia R. Prado-Irwin, Alberto R. Puente-Rolón, and Liam J. Revell, "Phenotypic shifts in urban areas in the tropical lizard *Anolis cristatellus*," *Evolution* 70 (5) (2016): 1009–1022.

145. See Bird, "Perceptions of epigenetics"; and Sabrina McNew, Daniel Beck, Ingrid Sadler-Riggleman, Sarah A. Knutie, Jennifer A. H. Koop, Dale H. Clayton, and Michael K. Skinner, "Epigenetic variation between urban and rural populations of Darwin's finches," *BMC Evolutionary Biology* 17 (1) (2017): 183.

146. See, for example, Srinivasan "The biopolitics of animal being and welfare."

147. See, for example, Katey Castellano, "Moles, molehills, and common right in John Clare's poetry," *Studies in Romanticism* 56 (2) (2017): 157–176. In the case of rats consider the Hindu temple of Karni Mata in northern Rajasthan where rats are revered and protected as sacred animals: these and other examples are indicative of arbitrary dimensions to the cultural categorization of animals. An interesting study of parakeets in the Paris region has explored how the population threshold at which biopolitical interventions might be culturally or politically feasible lies above the levels at which action remains a practical possibility. See Alizé Berthier, Philippe Clergeau, and Richard Raymond, "De la belle

exotique à la belle invasive: perceptions et appréciations de la Perruche à collier (*Psittacula kramerī*) dans la métropole parisienne," *Annales de géographie* 716 (2017): 408–434.

148. Sue Donaldson and Will Kymlicka, *Zoopolis: a political theory of animal rights* (Oxford: Oxford University Press, 2011).

149. José Lourenço, Maricelia Maia de Lima, Nuno Rodrigues Faria, Andrew Walker, Moritz U. G. Kraemer, Christian Julian Villabona-Arenas, Ben Lambert, et al., "Epidemiological and ecological determinants of Zika virus transmission in an urban setting," *Elife* 6 (2017): e29820.

150. See, for example, Ricardo Vieira Araujo, Marcos Roberto Albertini, André Luis Costa-da-Silva, Lincoln Suesdek, Nathália Cristina Soares Franceschi, Nancy Marçal Bastos, Gizelda Katz, et al., "São Paulo urban heat islands have a higher incidence of dengue than other urban areas," *Brazilian Journal of Infectious Diseases* 19 (2) (2015): 146–155; and Scott C. Weaver, "Urbanization and geographic expansion of zoonotic arboviral diseases: mechanisms and potential strategies for prevention," *Trends in Microbiology* 21 (8) (2013): 360–363. I would like to thank the biologist Joanna Coleman at the National University of Singapore for discussing some of these issues with me in greater detail (private communication with the author, 29 May 2020).

151. On the intersecting dynamics of insect-borne disease in urban areas see, for example, Enny S. Paixão, Maria Gloria Teixeira, and Laura C. Rodrigues, "Zika, chikungunya and dengue: the causes and threats of new and re-emerging arboviral diseases," *BMJ Global Health* 3 (1) (2018): e000530.

152. See, for example, Maria de Fatima P. Militão Albuquerque, Wayner V. de Souza, Antônio da Cruz G. Mendes, Tereza M. Lyra, Ricardo A. A. Ximenes, Thália V. B. Araújo, Cynthia Braga, Demócrito B. Miranda-Filho, Celina M. T. Martelli, and Laura C. Rodrigues, "Pyriproxyfen and the microcephaly epidemic in Brazil—an ecological approach to explore the hypothesis of their association," *Memórias do Instituto Oswaldo Cruz* 111 (12) (2016): 774–776; Raphael Parens, H. Frederik Nijhout, Alfredo Morales, Felipe Xavier Costa, and Yaneer Bar-Yam, "A possible link between pyriproxyfen and microcephaly," *PLoS Currents* 9 (2017). But compare with Thalia Velho Barreto de Araújo, Ricardo Arraes de Alencar Ximenes, Demócrito de Barros Miranda-Filho, Wayner Vieira Souza, Ulisses Ramos Montarroyos, Ana Paula Lopes de Melo, Sandra Valongueiro, et al., "Association between microcephaly, Zika virus infection, and other risk factors in Brazil: final report of a case-control study," *Lancet Infectious Diseases* 18 (3) (2018): 328–336.

153. See also Paige Marie Patchin, "Thresholds of empire: women, biosecurity, and the Zika chemical vector program in Puerto Rico," *Annals of the American Association of Geographers* 110 (4) (2020): 967–982.

154. Redardo Avila Vasquez, ed., *Report from physicians in the crop-sprayed villages regarding Dengue-Zika, microcephaly, and mass spraying with chemical poisons* (Córdoba: REDUAS, 2016), 4.

155. See Carlos E. Rodríguez-Díaz, Adriana Garriga-López, Souhail M. Malavé-Rivera, and Ricardo L. Vargas-Molina, "Zika virus epidemic in Puerto Rico: health justice too long delayed," *International Journal of Infectious Diseases* 65 (2017): 144–147; and Emily Kopp, "Fighting Zika: is aerial spraying effective?," *Miami Herald* (26 August 2016) (accessed 10 October 2020).

156. See, for example, Vanessa Agard-Jones, "Spray," *Somatosphere* (27 May 2014); and Michelle Murphy, "Alterlife and decolonial chemical relations," *Cultural Anthropology* 32 (4) (2017): 494–503.

157. See, for example, Uli Beisel, Ann H. Kelly, and Noémi Tousignant, "Knowing insects: hosts, vectors and companions of science," *Science as Culture* 22 (1) (2013): 1–15.

158. Roberto Esposito, *Bíos: politics and philosophy*, trans. T. Campbell (Minneapolis: University of Minnesota Press, 2008 [2004]).

159. Robert Mitchell, *Experimental life: vitalism in romantic science and literature* (Baltimore: Johns Hopkins University Press, 2013).

160. See Matthew Gandy, "The fly that tried to save the world: saproxylic geographies and other-than-human ecologies," *Transactions of the Institute of British Geographers* 44 (2) (2019): 392–406.

161. See, for example, Irus Braverman, *Wild life: the institution of nature* (Stanford, CA: Stanford University Press, 2015) and "Law's underdog: a call for more-than-human legalities," *Annual Review of Law and Social Science* 14 (2018): 127–144. On more-than-human ethics see also María Puig de la Bellacasa, *Matters of care: speculative ethics in more than human worlds* (Minneapolis: University of Minnesota Press, 2017). On strategies for avoiding without killing see, for example, Franklin Ginn, "Sticky lives: slugs, detachment and more-than-human ethics in the garden," *Transactions of the Institute of British Geographers* 39 (4) (2014): 532–544.

162. See Erin Luther, "Tales of cruelty and belonging: in search of an ethic for urban human-wildlife relations," *Animal Studies Journal* 2 (1) (2013): 35–54. On the history of cruelty towards urban animals see, for example, Michaela Laichmann, "Arbeitsvieh und Schoßtier. Hunde im mittelalterlichen und frühneuzeitlichen Wien," in *Umwelt Stadt. Geschichte des Natur-und Lebensraumes Wien*, ed. Karl Brunner and Petra Schneider (Vienna: Böhlau, 2005), 410–417.

163. Taimie L. Bryant, "Sacrificing the sacrifice of animals: legal personhood for animals, the status of animals as property, and the presumed primacy of humans," *Rutgers Law Journal* 39 (2007): 247–330.

164. See, for example, Palmer, "Placing animals in urban environmental ethics."

165. Etienne Benson, "The urbanization of the eastern gray squirrel in the United States," *Journal of American History* 100 (3) (2013): 691–710. On the squirrel as a symbolic marker for

"new ecologies" see Nicholas Holm, "Consider the squirrel: freaks, vermin, and value in the ruin(s) of nature," *Cultural Critique* 80 (2012): 56–95. On the ambivalent cultural status of "trash animals" see also Kelsi Nagy and Phillip David Johnson II, eds., *Trash animals: how we live with nature's filthy, feral, invasive, and unwanted species* (Minneapolis: University of Minnesota Press, 2013).

166. See Colin Jerolmack, "How pigeons became rats: the cultural-spatial logic of problem animals," *Social Problems* 55 (1) (2008): 72–94.

167. Patrick Barkham, "'It's very scary in the forest': should Finland's wolves be culled?," *Guardian* (25 February 2017). See also Buller, "Safe from the wolf"; and Ketil Skogen and Olve Krange, "A wolf at the gate: the anti-carnivore alliance and the symbolic construction of community," *Sociologia Ruralis* 43 (3) (2003): 309–325.

168. See, for example, Ralph Acampora, "Oikos and Domus: on constructive cohabitation with other creatures," *Philosophy and Geography* 7 (2) (2004): 223. See also Yi-Fu Tuan, *Dominance and affection* (New Haven, CT: Yale University Press, 2003).

169. See, for example, Dorothee Brantz and Sonja Dümpelmann, eds., *Greening the city: urban landscapes in the twentieth century* (Charlottesville: University of Virginia Press, 2011); Steve Hinchliffe, Matthew B. Kearns, Monica Degen, and Sarah Whatmore, "Urban wild things: a cosmopolitical experiment," *Environment and Planning D: Society and Space* 23 (2005): 643–658; Jamie Lorimer, "Nonhuman charisma," *Environment and Planning D: Society and Space* 25 (2007): 911–932; and Srinivasan, "The biopolitics of animal being and welfare."

170. Wolch, "Zoöpolis," 47. See also Kay Anderson, "Culture and nature at the Adelaide Zoo: at the frontiers of 'human' geography," *Transactions of the Institute of British Geographers* (1995): 275–294; and Michael Sorkin, "See you in Disneyland," in *Variations on a theme park: the new American city and the end of public space*, ed. Michael Sorkin (New York: Hill and Wang, 1992), 205–232.

CHAPTER 2

1. Richard Mabey, *The unofficial countryside* (London: Collins, 1973), 12.

2. The quote is taken from Pierre Carles's documentary about Pierre Bourdieu entitled *La sociologie est un sport de combat* (2001) (my translation).

3. Studies that emphasize the cultural richness of ostensibly "empty" spaces include Michèle Collin, "Nouvelles urbanités des friches," *Multitudes* 6 (2001): 148–155; Tim Edensor, *Industrial ruins: spaces, aesthetics and materiality* (London: Berg, 2005); Susanne Hauser, *Metamorphosen des Abfalls, Konzepte für alte Industrieareale* (Frankfurt: Campus, 2001); Andreas Huyssen, "The voids of Berlin," *Critical Inquiry* 24 (1) (1997): 57–81; Luc

Lévesque, "Montréal, l'informe urbanité des terrains vagues: pour une gestion créatrice du mobilier urbain," *Annales de la Recherche Urbaine* 85 (1999): 47–57; Bernadette Lizet, "Du terrain vague à la friche paysagée," *Ethnologie Française* 40 (4) (2010): 597–608; and Karen Till, "Interim use at a former death strip? Art, politics and urbanism at Skulpturen-park Berlin_Zentrum," in *After the Wall: Berlin in Germany and Europe*, ed. Marc Silberman (Basingstoke: Palgrave Macmillan, 2011), 99–122.

4. In English usage, from the twelfth century onwards, the noun *waste* is encountered as a description of empty, desolate, or inhospitable places. Alexander Pope, for instance, in his description of Windsor Forest, published in 1713, refers to "a waste for beasts," whilst just over a century later William Wordsworth recounts "the wastes of Rylstone Fell" in the *The white doe of Rylstone* (1815). See, for example, Vittoria Di Palma, *Wasteland: a history* (New Haven, CT: Yale University Press, 2014).

5. Blaise Pascal, *Expériences nouvelles touchant le vuide* (Paris: Pierre Margat, 1647).

6. Anthony Vidler, *Warped space: art, architecture, and anxiety in modern culture* (Cambridge, MA: MIT Press, 2000), 21.

7. Brownfield sites are often marked by abandoned buildings and fragments of infrastructure, along with toxic traces of cadmium, mercury, and other residues derived from former land uses. Far from being "dead zones," however, these heavily contaminated sites exhibit adaptive ecological formations that can be further stimulated by "phytoremediation" and the use of specific plants to capture and remove toxins from soil. See, for example, Hazrat Ali, Ezzat Khan, and Muhammad Anwar Sajad, "Phytoremediation of heavy metals— concepts and applications," *Chemosphere* 91 (7) (2013): 869–881; Karen E. Gerhardt, Xiao-Dong Huang, Bernard R. Glick, and Bruce M. Greenberg, "Phytoremediation and rhizoremediation of organic soil contaminants: potential and challenges," *Plant Science* 176 (1) (2009): 20–30; and A. G. Khan, C. Kuek, T. M. Chaudhry, C. S. Khoo, and W. J. Hayes, "Role of plants, mycorrhizae and phytochelators in heavy metal contaminated land remediation," *Chemosphere* 41 (1–2) (2000): 197–207. From a global perspective, of course, the putative "post-Fordist" transition is also a process of redistributing economic activities as well as creating postindustrial landscapes.

8. On the framing of void space as a pretext for sequestration see, for example, Jennifer Baka, "The political construction of wasteland: governmentality, land acquisition and social inequality in South India," *Development and Change* 44 (2) (2013): 409–428; and Debjani Bhattacharyya, *Empire and ecology in the Bengal Delta: the making of Calcutta* (Cambridge: Cambridge University Press, 2018).

9. See, for example, Mukul Kumar, K. Saravanan, and Nityanand Jayaraman, "Mapping the coastal commons: fisherfolk and the politics of coastal urbanisation in Chennai," *Economic and Political Weekly* 48 (2014): 46–53; and Bhavani Raman, "Sovereignty, property and land development: the East India Company in Madras," *Journal of the Economic and Social*

History of the Orient 61 (5–6) (2018): 976–1004. Niranjana Ramesh has pointed out to me that one possible origin for the word *poramboke* is a designation for land that lies "outside the book" in the context of colonial governmentality.

10. At a wasteland-themed event held at the Zentrum für Kunst und Urbanistik in Berlin in September 2012 there was a workshop devoted to "(Un)Common Language" hosted by Rebecca Beinhart and Wasteland Twinning Berlin, in which participants from diverse international locations discussed the meaning of different words associated with marginal urban spaces. On the diversity of terms used in relation to spontaneous urban ecologies see also Christoph D. D. Rupprecht and Jason A. Byrne, "Informal urban greenspace: a typology and trilingual systematic review of its role for urban residents and trends in the literature," *Urban Forestry and Urban Greening* 13 (4) (2014): 597–611.

11. The expression *terrain vague* in particular has recently gained conceptual currency after its elaboration by the Spanish architect Ignasi de Solà-Morales Rubió. See Ignasi de Solà-Morales Rubió, "Terrain vague," in *Anyplace*, ed. Cynthia Davidson (Cambridge, MA: MIT Press, 1993), 118–123. See also Herman de Vries, "Terrain vague," in *No art—no city! Stadtutopien in der zeitgenössischen kunst*, ed. Florian Matzner, Hans-Joachim Manske, and Rose Pfister (Ostfildern: Hatje Cantz, 2003) 151–153. For an overview of the field from a humanities perspective see also Jacqueline Maria Broich and Daniel Ritter, *Die Stadtbrache als "terrain vague": Geschichte und Theorie eines unbestimmten Zwischenraums in Literatur, Kino und Architektur* (Bielefeld: transcript, 2017).

12. The term "open mosaic habitat" is now being extensively used in the UK in preference to the term "brownfield" for a variety of scientific studies and also in some local government documentation such as Biodiversity Action Plans. For recent overviews of developments in urban ecology see Kevin J. Gaston, "Urban ecology," in *Urban ecology*, ed. Kevin J. Gaston (Cambridge: Cambridge University Press, 2010), 1–9; Ingo Kowarik, Leonie K. Fischer, Ina Säumel, Moritz von der Lippe, Frauke Weber, and Janneke R. Westermann, "Plants in urban settings: from patterns to mechanisms and ecosystem services," in *Perspectives in urban ecology: ecosystems and interactions between humans and nature in the metropolis of Berlin*, ed. Wilfried Endlicher (Heidelberg: Springer, 2011), 135–166; Jean-Pierre Savard, Philippe Clergeau, and Gwenaelle Mennechez, "Biodiversity concepts and urban ecosystems," *Landscape and Urban Planning* 48 (2000): 131–142; and Rudiger Wittig, "Biodiversity of urban-industrial areas and its evaluation: a critical review," in *Urban biodiversity and design*, ed. Norbert Müller, Peter Werner, and John G. Kelcey (Hoboken, NJ: Wiley-Blackwell, 2010), 37–55.

13. Joseph Pitton de Tournefort, *Histoire des plantes qui naissent aux environs de Paris, avec leur usage dans la medecine* (Paris: Imprimerie Royale, 1698).

14. See Richard S. R. Fitter, *London's natural history* (London: Collins, 1945); Herbert Sukopp, "On the early history of urban ecology in Europe," *Preslia, Praha* 74 (2002): 373–393.

15. John Eddington, "Early London botanists: '. . . in the fieldes aboute London, plentuously . . . ,'" *London Naturalist* 90 (2011): 21–45.

16. Thomas Johnson, *Botanical journeys in Kent and Hampstead: a facsimile reprint with introduction and translation of his Iter plantarum 1629 [and] Descriptio itineris plantarum 1632*, ed. John S. L. Gilmour (Pittsburgh: Hunt Botanical Library, 1972), 65.

17. Richard Mabey, *Weeds* (London: Profile, 2010), 219.

18. Richard Deakin, *Flora of the Colosseum of Rome; or illustrations and descriptions of four hundred and twenty plants growing spontaneously upon the ruins of the Colosseum of Rome* (London: Groombridge and Sons, 1855), vii.

19. Edmond Bonnet, *Petite flore parisienne* (Paris: Librairie F. Savy, 1883), v.

20. Richard Jefferies, *Nature near London* (London: John Clare Books, 1980 [1893]), v.

21. J. C. Shenstone, "The flora of London building-sites," *Journal of Botany* 50 (1912): 117–124.

22. See Paul Jovet, "Evolution des groupements rudéraux 'parisiens,'" *Bulletin de la Société Botanique de France* 87 (1940): 305–312; Bernadette Lizet, Anne-Élisabeth Wolf, and John Celecia, eds., *Sauvages dans la ville. Actes du colloque, organisé pour le centenaire de la naissance de Paul Jovet* (Paris: JATBA/Publications scientifiques du MNHN, 1997).

23. See the interview with Herbert Sukopp in the documentary *Natura urbana: the Brachen of Berlin* (dir.: Matthew Gandy, 2017).

24. See Jens Lachmund, *Greening Berlin: the co-production of science, politics, and urban nature* (Cambridge, MA: MIT Press, 2013); Herbert Sukopp, ed., *Stadtökologie. Das Beispiel Berlins* (Berlin: Dietrich Reimer, 1990).

25. Paul Duvigneaud, "Étude écologique de l'écosystème urbain bruxellois: 1. L'écosystème 'urbs'," *Mémoires de la Société Royale de Botanique de Belgique* 6 (1974): 5–35.

26. Ludwig Trepl, "City and ecology," *Capitalism, Nature, Socialism* 7 (2) (1996): 85–94.

27. Adorno raises a similar point: "Even as bourgeois consciousness naïvely condemns the ugliness of a torn-up industrial landscape, a relation is established that reveals a glimpse of the domination of nature, where nature shows humans its facade of having yet to be mastered." Theodor W. Adorno, *Aesthetic theory*, trans. Robert Hullot-Kentor (London: Continuum, 1997 [1970]), 61.

28. Richard Fitter's *London's natural history*, first published in 1945, already lists 126 species of plants found on "extensive areas of open waste ground." Fitter, *London's natural history*, 230. See also Seth Denizen, "The flora of bombed areas (an allegorical key)," in *The botanical city*, ed. Matthew Gandy and Sandra Jasper (Berlin: jovis, 2020), 38–45; Leo Mellor, "Words from the bombsites: debris, modernism and literary salvage," *Critical Quarterly* 46

(4) (2004): 77–90; and Edward Salisbury, "The flora of bombed areas," *Nature* 1943 (151) (3834): 462–466.

29. W. G. Sebald, *The natural history of destruction*, trans. Anthea Bell (New York: Modern Library, 2004 [1999]), 39–40.

30. Lachmund, *Greening Berlin*.

31. See Gerhard Hard, *Spuren und Spurenleser. Zur Theorie und Ästhetik des Spurenlesens in der Vegetation und anderswo* (Osnabrück: Universitätsverlag Rasch, 1995); Gerhard Hard, *Ruderalvegetation. Ökologie und Ethnoökologie, Ästhetik und "Schutz,"* Notizbuch 49 der Kasseler Schule (Kassel: Arbeitsgemeinschaft Freiraum und Vegetation, 1998).

32. Sébastien Filoche, Gérard Arnal, and Jacques Moret, *La biodiversité du département de la Seine-Saint-Denis. Atlas de la flore sauvage* (Mèze: Biotope; Paris: Muséum National d'Histoire Naturelle, 2006).

33. Hua-Feng Wang, Jordi López-Pujol, Laura A. Meyerson, Jiang-Xiao Qiu, Xiao-Ke Wang, and Zhi-Yun Ouyang, "Biological invasions in rapidly urbanizing areas: a case study of Beijing, China," *Biodiversity and Conservation* 20 (11) (2011): 2483–2509.

34. See Ingo Kowarik, Leonie K. Fischer, Ina Säumel, Moritz von der Lippe, Frauke Weber, and Janneke R. Westermann, "Plants in urban settings: from patterns to mechanisms and ecosystem services," in *Perspectives in urban ecology: ecosystems and interactions between humans and nature in the metropolis of Berlin*, ed. Wilfried Endlicher (Heidelberg: Springer, 2011), 135–166; Dieter Rink, "Wilderness: the nature of urban shrinkage?," *Nature and Culture* 4 (2009): 275–292; and Savard et al. "Biodiversity concepts and urban ecosystems."

35. Audrey Muratet, Nathalie Machon, Frédéric Jiguet, Jacques Moret, and Emmanuelle Porcher, "The role of urban structures in the distribution of wasteland flora in the Greater Paris Area, France," *Ecosystems* 10 (2007): 661–671.

36. Mabey, *Weeds*; Diane Saint-Laurent, "Approches biogéographiques de la nature en ville: parcs, espaces verts et friches," *Cahiers de Géographie du Québec* 44 (2000): 147–166. Examples of infrastructure systems converted into parks include parts of the Rec Comtal irrigation network in Barcelona and the flood retention basins in Bilancourt Park, Paris. In all these cases we find a complicated intersection between different ecological design discourses, some geared towards aesthetic ambience and others framed by technical concerns.

37. David Takacs, *The idea of biodiversity: philosophies of paradise* (Baltimore: Johns Hopkins University Press, 1996).

38. See the interviews with Ingo Kowarik and Jens Lachmund in my documentary *Natura urbana: the Brachen of Berlin*.

39. See Danielle Dagenais, "The garden of movement: ecological rhetoric in support of gardening practice," *Studies in the History of Gardens and Designed Landscapes* 24 (4) (2004):

313–340; Matthew Gandy, "Entropy by design: Gilles Clément, Parc Henri Matisse and the limits to avant-garde urbanism," *International Journal of Urban and Regional Research* 37 (1) (2013): 259–278; Allan Ruff, "Holland and the ecological landscape," *Garden History* 30 (2) (2002): 239–251; and Jan Woudstra, "The eco-cathedral: Louis Le Roy's expression of a free landscape architecture," *Die Gartenkunst* 20 (1) (2008): 185–202.

40. The Irchelpark, under construction between 1979 and 1986, was designed by the architect Eduard Neuenschwander, working in partnership with Stern studio, who developed a distinctive synthesis of modernist aesthetics and ecological ideas.

41. Those species most closely linked to, or even dependent on, human settlements for food, shelter, and other needs, termed synanthropic species, include a variety of pests as well as the benign ubiquity of many birds and insects such as the house sparrow, *Passer domesticus*, or the honeybee, *Apis mellifera*.

42. Jamie Lorimer, "Living roofs and brownfield wildlife: towards a fluid biogeography of UK nature conservation," *Environment and Planning A* 40 (2008): 2042–2060.

43. See Kowarik et al., "Plants in urban settings"; Norbert Kühn, "Intentions for the unintentional: spontaneous vegetation as the basis for innovative planting design in urban areas," *Journal of Landscape Architecture* (Autumn 2006): 46–53.

44. I would like to thank Moritz von der Lippe at the Technical University, Berlin, for showing me the remnant flora of the Charlottenburg Palace Garden in the spring of 2013.

45. See Hansjörg Küster, *Die Entdeckung der Landschaft: Einführung in eine neue Wissenschaft* (Munich: C. H. Beck, 2012).

46. Paul H. Gobster, "Visions of restoration: conflict and compatibility in urban park restoration," *Landscape and Urban Planning* 56 (2001): 35–51.

47. See, for example, Ingo Kowarik, "Wild urban woodlands: towards a conceptual framework," in *Wild urban woodlands: new perspectives for urban forestry*, ed. Ingo Kowarik and Stefan Körner (Berlin: Springer, 2005), 1–32.

48. Patricia Pellegrini and Sandrine Baudry, "Streets as new places to bring together both humans and plants: examples from Paris and Montpellier (France)," *Social and Cultural Geography* 15 (8) (2014): 871–900.

49. See Susanne Frank, "Rückkehr der Natur. Die Neuerfindung von Natur und Landschaft in der Emscherzone," in *EMSCHERplayer* (October 2010); and Susanne Frank, "Auf der Suche nach der 'optimalen Landschaft'. Regionalentwicklung durch Landschaftswandel: Das Beispiel Niederlausitz," mimeo (2010).

50. I would like to thank Tom Baker for drawing my attention to the neopastoral dimension of the High Line. See also Natalie Gulsrud and Henriette Steiner, "When urban greening becomes an accumulation strategy: exploring the ecological, social and economic calculus of the High Line," *Journal of Landscape Architecture* 14 (3) (2019): 82–87.

51. Allen Carlson, "Nature, aesthetic appreciation, and knowledge," *Journal of Aesthetics and Art Criticism* 53 (4) (1995): 393. See also Allen Carlson, "Appreciating art and appreciating nature," in *Landscape, natural beauty and the arts*, ed. Salim Kemal and Ivan Gaskell (Cambridge: Cambridge University Press, 1993), 199–227.

52. Pierre Bourdieu, *On Television*, trans. Priscilla Parkhurst Ferguson (New York: New Press, 1998 [1996]).

53. Important exceptions include Berlin's Langer Tag der StadtNatur (Long day of urban nature), an annual event under way since 2006, which is a science-led program of public activities organized by the Berlin Conservation Foundation.

54. Studies of wastelands have often revealed high levels of species diversity for aculeate hymenoptera and other warmth-loving insects adapted to sand dunes or coastal environments. See, for example, M. Eyre, M. Luff, and J. Woodward, "Beetles (Coleoptera) on brownfield sites in England: an important conservation resource?," *Journal of Insect Conservation* 7 (4) (2003): 223–231; C. Gibson, *Brownfield: red data. The values artificial habitats have for uncommon invertebrates* (London: English Nature, 1998); and Gyongyver Kadas, "Rare invertebrates colonising green roofs in London," *Urban Habitats* 4 (1) (2006): 66–86. In some cases, depending on soil conditions, there can be unique combinations of habitat mimicry, so that species associated with disparate ecotopes such as heathlands or chalk downland may occur together on the same site. Colin Plant, consultant entomologist, discussion with the author (12 February 2012). For detailed studies of urban habitat diversity see, for example, Sébastien Filoche, Gérard Arnal, and Jacques Moret, *La biodiversité du département de la Seine-Saint-Denis. Atlas de la flore sauvage* (Mèze: Biotope; Paris: Muséum National d'Histoire Naturelle, 2006); Muratet et al., "The role of urban structures in the distribution of wasteland flora in the Greater Paris Area"; Franz Rebele, "Urban ecology and special features of urban ecosystems," *Global Ecology and Biogeography Letters* 4 (6) (1994): 173–187; Ute Schadek, Barbara Strauss, Robert Biedermann, and Michael Kleyer, "Plant species richness, vegetation structure and soil resources of urban brownfield sites linked to successional age," *Urban Ecosystems* 12 (2009): 115–126; and Herbert Sukopp, ed., *Stadtökologie. Das Beispiel Berlins* (Berlin: Dietrich Reimer, 1990).

55. Natalie Blanc, *Vers uns esthétique environmentale* (Paris: Éditions Quæ, 2008).

56. Terry C. Daniel, "Whither scenic beauty? Visual landscape quality in the 21st century," *Landscape and Urban Planning* 54 (2001): 267–281; Klaus C. Ewald, "The neglect of aesthetics in landscape planning in Switzerland," *Landscape and Urban Planning* 54 (2001): 255–266. The attempt to articulate an "ecological aesthetic" marks a corollary of the wider "ecologization" of public policy through the enhanced role of scientific discourse in decision making. See, for example, James Evans, "Resilience, ecology and adaptation in the experimental city," *Transactions of the Institute of British Geographers* 36 (2011): 223–237.

57. Cheryl Foster, "The narrative and the ambient in environmental aesthetics," *Journal of Aesthetics and Art Criticism* 56 (2) (1998): 127–137.

58. Arnold Berleant, *The aesthetics of environment* (Philadelphia: Temple University Press, 1992).

59. Mark Frost, "Entering the 'circles of vitality': beauty, sympathy, and fellowship," in *Vital beauty: reclaiming aesthetics in the tangle of technology and nature*, ed. Joke Brouwer; Arjen Mulder; and Lars Spuybroek (Rotterdam: V2_Publishing, 2012), 137.

60. Frost, "Entering the 'circles of vitality,'" 151.

61. Henri Bergson, *Creative evolution*, trans. Arthur Mitchell, ed. Keith Ansell-Pearson, Michael Kolkman, and Michael Vaughan (Basingstoke: Palgrave-Macmillan, 2007 [1907]).

62. See Gilles Deleuze, "Bergson's conception of difference," in *The new Bergson*, ed. John Mullarkey (Manchester: Manchester University Press, 1999 [1956]), 42–65; Elizabeth Grosz, "Bergson, Deleuze, and the becoming of unbecoming," *Parallax* 11 (2) (2005): 4–13.

63. In Merleau-Ponty's exegesis of Bergson's conception of nature, for example, he emphasizes the poetic dimension. "He [Bergson] stands both against Berkeley's idealism, for which everything is a representation, and against a realism that admits that the thing has an aseity, but which posits that this is other than what appears." See Maurice Merleau-Ponty, *Nature: course notes from the Collège de France*, compiled by D. Ségland, trans. R. Vallier (Evanston, IL: Northwestern University Press 2003 [1956–1960]), 53. Elsewhere Merleau-Ponty seeks to distinguish his "phenomenal psychology" from what he terms Bergson's "introspective psychology." See Maurice Merleau-Ponty, *Phenomenology of perception*, trans. C. Smith (London: Routledge and Kegan Paul, 1962 [1945]), 59.

64. See J. Baird Callicott, "The land aesthetic," *Renewable Resources Journal* 10 (1992): 12–17.

65. Ronald Rees, "The taste for mountain scenery," *History Today* 25 (5) (1975): 305–312.

66. Richard Rorty, *Philosophy and the mirror of nature* (Princeton, NJ: Princeton University Press, 1979).

67. See, for example, Thomas McCarthy, "Private irony and public decency: Richard Rorty's new pragmatism," *Critical Inquiry* 16 (2) (1990): 355–370.

68. See Paul H. Gobster, Joan I. Nassauer, Terry C. Daniel, and Gary Fry, "The shared landscape: what does aesthetics have to do with ecology?," *Landscape Ecology* 22 (2007): 959–972.

69. Nick Bertrand, urban botanist, discussion with the author, Creekside Centre, London (13 November 2011).

70. Yuriko Saito, "The aesthetics of unscenic nature," *Journal of Aesthetics and Art Criticism* 56 (2) (1998): 101–111.

71. See Matthew Gandy, "The fly that tried to save the world: saproxylic geographies and other-than-human ecologies," *Transactions of the Institute of British Geographers* 44 (2) (2019): 392–406.

72. In much of London, for example, specialist teams of municipal workers devoted to looking after urban trees have been partially or completely laid off since the 1980s, leading to a loss of arboriculture skills, premature death or damage to urban trees, and longer-term implications for the character of urban green space. Russell Miller, arboriculturalist and chair of Sustainable Hackey, private communication with the author (16 December 2011).

73. See Robert B. Feagan and Michael Ripmeester, "Contesting natural(ized) lawns: a geography of private green space in the Niagara region," *Urban Geography* 20 (7) (1999): 617–634.

74. Mariana Valverde, "The ethic of diversity: local law and the negotiation of urban norms," *Law and Social Inquiry* 33 (4) (2008): 895–923.

75. Karen A. Franck and Quentin Stevens, eds., *Loose space: possibility and diversity in urban life* (London: Routledge, 2007). On the landscape design possibilities offered by wild urban nature see also Jill Desimini, "Notions of nature and a model for managed urban wilds," in *Terrain vague: interstices at the edge of the pale*, ed. Patrick Barron and Manuela Mariani (London: Routledge, 2014), 173–186.

76. See Matthew Gandy, "At a tangent: delineating a new ecological imaginary in Berlin's Park am Gleisdreieck," *Architectural Design* 90 (1): 106–113. Other examples of urban park design inspired by the ecological diversity of abandoned spaces include Frank Bruggemann and Hans Engelbrecht's project The New Garden at Het Nieuwe Institute, Rotterdam, completed in 2015.

77. Werner Nohl, "Sustainable landscape use and aesthetic perception—preliminary reflections on future landscape aesthetics," *Landscape and Urban Planning* 54 (2001): 224.

78. See Denis Cosgrove, "Prospect, perspective and the evolution of the landscape idea," *Transactions of the Institute of British Geographers* 10 (1) (1985): 45–62; Gerhard Hard, "Vegetationsgeographie und sozialökologie einer Stadt," *Geographische Zeitung* 75 (1985): 125–144.

79. The term "neoromanticism" has been most extensively deployed within art history and literary criticism. See, for example, David Mellor, ed., *A paradise lost: the neo-romantic imagination in Britain 1935–1955* (London: Lund Humphries and Barbican Art Gallery, 1987).

80. For a reflection on alternative modes of aesthetic judgment see Sianne Ngai, *Our aesthetic categories: zany, cute, interesting* (Cambridge, MA: Harvard University Press, 2012).

81. See Zachary J. S. Falck, *Weeds: an environmental history of metropolitan America* (Pittsburgh: University of Pittsburgh Press, 2010). For an alternative perspective on the wastelands of Detroit see Paul Draus and Juliette Roddy, "Weeds, pheasants and wild dogs:

resituating the ecological paradigm in postindustrial Detroit," *International Journal of Urban and Regional Research* 42 (5) 2018: 807–827; Rebecca J. Kinney, *Beautiful wasteland: the rise of Detroit as America's postindustrial frontier* (Minneapolis: University of Minnesota Press, 2016); Nate Millington, "Post-industrial imaginaries: nature, representation and ruin in Detroit, Michigan," *International Journal of Urban and Regional Research* 37 (1) (2013): 279–296; Alesia Montgomery, *Greening the Black urban regime: the culture and commerce of sustainability in Detroit* (Detroit: Wayne State University Press, 2020); and Sara Safransky, "Greening the urban frontier: race, property, and resettlement in Detroit," *Geoforum* 56 (2014): 237–248.

82. See Elisabeth Wilson, "The invisible flâneur," *New Left Review* 191 (1992): 90–110.

83. See Oliver Gilbert, *The flowering of the cities: the natural flora of "urban commons"* (Peterborogh: English Nature, 1992); Anna Jorgensen and Marian Tylecote, "Ambivalent landscapes—wilderness in the urban interstices," *Landscape Research* 32 (4) (2007): 443–462.

84. See Martin Franz, Orhan Güles, and Gisela Prey, "Place-making and 'green' reuses of brownfields in the Ruhr," *Tijdschrift voor Economische en Sociale Geografie* 99 (3) (2008): 316–328; Andreas Keil, "Use and perception of postindustrial urban landscapes in the Ruhr," in Kowarik and Körner, *Wild urban woodlands*, 117–130; and Juliane Mathey and Dieter Rink, "Urban wastelands—a chance for biodiversity in cities? Ecological aspects, social perceptions and acceptance of wilderness by residents," in *Urban biodiversity and design*, ed. Norbert Müller, Peter Werner, and John G. Kelcey (Hoboken, NJ: Wiley-Blackwell, 2010), 406–424.

85. Lara Almarcegui, lecture given to the conference Art and the Environment, Tate Britain, London (30 June 2010).

86. See Susanne Schroeder, ed., *Skulpturenpark Berlin_Zentrum* (Berlin: Walther König, 2010).

87. Walter Benjamin, for example, describes the "style of the modern flâneur as one who goes botanizing on the asphalt," drawing connections between natural history and the "poetic imagination." See Walter Benjamin, "The Paris of the Second Empire in Baudelaire," in Benjamin, *The writer of modern life: essays on Charles Baudelaire* (Cambridge, MA: Belknap Press, 2006 [1938]), 68. His childhood recollections of urban nature in Berlin are to be found in Walter Benjamin, *Beroliniana* (Munich: Koehler and Amelang, 2001 [1932–1938]). See also Nigel Clark, "'Botanizing on the asphalt'? The complex life of cosmopolitan bodies," *Body and Society* 6 (3–4) (2000): 12–33; and Peter Del Tredici, "Spontaneous urban vegetation: reflections of change in a globalizing world," *Nature and Culture* 5 (3) (2010): 299–315.

88. Karl Marx, "Feuerbach. Opposition of the materialist and idealist outlook," in Karl Marx and Friedrich Engels, *The German ideology*, part one, trans. W. Lough (London: Lawrence & Wishart, 1965 [1844]), 39–95.

89. See in particular Michael L. McKinney, "Urbanisation as a major cause of biotic homogenisation," *Biological Conservation* 127 (2006): 247–260.

90. Nigel Clark, "The demon-seed: bioinvasion as the unsettling of environmental cosmopolitanism," *Theory, Culture and Society* 19 (1–2) (2002): 101–125.

91. Ingo Kowarik, *Biologische Invasionen. Neophyten und Neozoen in Mitteleuropa* (Stuttgart: Ulmer, 2003).

92. Ute Eser, *Der Naturschutz and das Fremde: ökologische und normative Grundlagen der Umweltethik* (Frankfurt: Campus, 1999). For an overview of the implications of "invasion ecology" for the urban realm see also Joëlle Salomon Cavin and Christian A. Kull, "Invasion ecology goes to town: from disdain to sympathy," *Biological Invasions* 19 (12) (2017): 3471–3487.

93. Karl S. Zimmerer, "Human geography and the 'new ecology': the prospect and promise of integration," *Annals of the Association of American Geographers* 84 (1) (1994): 111. See also Karl S. Zimmerer, "The reworking of conservation geographies: non-equilibrium landscapes and nature-society hybrids," *Annals of the Association of American Geographers* 90 (2) (2000): 356–369.

94. Marina Alberti, John M. Marzluff, Eric Shulenberger, Gordon Bradley, Clare Ryan, and Craig Zumbrunnen, "Integrating humans into ecology: opportunities and challenges for studying urban ecosystems," *BioScience* 53 (12) (2003): 1169–1179.

95. Evans, "Resilience, ecology and adaptation in the experimental city."

96. See Kevin J. Gaston, "Urban ecology," in *Urban ecology*, ed. Kevin J. Gaston (Cambridge: Cambridge University Press, 2010), 1–9; and Robert E. Kohler, *Landscapes and labscapes: exploring the lab-field border in biology* (Chicago: University of Chicago Press, 2002).

CHAPTER 3

1. The exhibition was held at the former gallery of the Deutsche Akademischer Austauschdienst (DAAD) in the Kurfürstenstraße, West Berlin.

2. During Gette's visit to Berlin in October 1971, for example, he mapped the distribution of the aromatic weed *Artemesia vulgaris* growing on waste ground in the Lützow Platz, one of the key ruderal sites for the development of urban ecology in the city, which would become the focus of intense land use conflict between proponents of an urban nature park and speculative real estate interests. Perhaps the key example of Gette's studies of *terrain vague* during this period is his 1975 study of the Beaubourg site in Paris designated for the new Pompidou Center (a work which is now displayed within the museum itself). Gette's explorations of urban space belong to a lineage of various types of "physiologies" undertaken in the nineteenth century, yet the political dimensions to ambulatory and

enumerative approaches to the urban realm remain a point of contestation. Walter Benjamin, for example, regarded these largely descriptive approaches as a "suspect genre." See Tom McDonough, "The crimes of the flâneur," *October* 102 (Autumn 2002): 104.

3. Gette's earlier project *Jardins botaniques* (1974) had emphasized the arbitrary distinction between those plants that were highlighted for didactic display and those whose presence was either ignored or had been the focus of meticulous eradication. Günter Metken, for example, had also referred to *Galinsoga parviflora* as a "cosmopolitan species" by virtue of its global reach. Günter Metken, "Gettes parallele Wissenschaft," in Paul-Armand Gette, *Arbeiten 1959–1979* (Munich: Städtische Galerie im Lenbachhaus, 1979), 22.

4. To most gallery visitors the species of plants depicted, and especially the use of scientific nomenclature, would have been obscure, but Gette's use of scientific rather than vernacular names connects with his interest in "scientific poetry" derived from the curious cadence and etymology of Latin names. See Metken, "Gettes parallele Wissenschaft." In a 1975 performance at the University of Nanterre, Paris X, entitled *La nomenclature binaire: homage à Carl von Linné*, Gette spent six hours simply reading out a list of scientific names with their original describers.

5. See, for example, Ann Laura Stoler, "Rethinking colonial categories: European communities and the boundaries of rule," *Comparative Studies in Society and History* 31 (1) (1989): 135–161; and Alan Bewell, *Natures in translation: romanticism and colonial natural history* (Baltimore: Johns Hopkins University Press, 2017).

6. The characteristics of urban vegetation remain a critical indicator for the class-based topography of the industrial city. In this context weeds or wastelands can be read as signs of neglect as well as sites of cultural and scientific appropriation. See, for example, Jürgen von Reuß, "Freiflächenpolitik als Sozialpolitik," in *Martin Wagner 1885–1975. Wohnungsbau und Weltstadtplanung. Die Rationalisierung des Glücks*, ed. Klaus Homann, Martin Kieren, and Ludovica Scarpa (Berlin: Akademie der Künste, 1987), 49–65.

7. For further details on the ecology of *Ailanthus* in Berlin see, for example, T. Bachman, "*Ailanthus altissima* (Mill.) Swingle und *Acer negundo* L. als typische Neophyten urbaner Pflanzengemeinschaften" (PhD thesis, Technical University, Berlin, 2005); and Ingo Kowarik and Ina Säumel, "Biological flora of Central Europe: *Ailanthus altissima* (Mill.) Swingle," *Perspectives in Plant Ecology, Evolution and Systematics* 8 (2007): 207–237. See also Günter Metken, "*Ailanthus altissima* in Wedding," in Paul-Armand Gette, *Exotik als Banalität / De l'exotisme en tant que banalité* (Berlin: DAAD, 1980), i–iii. In North America *Ailanthus* is widely known as the "ghetto palm," denoting its presence as a common street tree where none have been planted.

8. An arboricultural overview of Berlin street trees published in 1961 still lists *Ailanthus* as a suggested choice for planting because of its tolerance for dry conditions. See Rudolf Kühn, *Die Strassenbaüme* (Hannover: B. Patzer, 1961).

9. Jens Lachmund, *Greening Berlin: the co-production of science, politics, and urban nature* (Cambridge, MA: MIT Press, 2013), 20. On the history of urban nature in Berlin see also Dorothee Brantz, "The urban politics of nature: two centuries of green spaces in Berlin, 1800–2014," in *Green Landscapes in the European City, 1750 to 2010*, ed. Peter Clark, Marjaana Niemi, and Catharina Nolin (London: Routledge, 2017), 141–159.

10. Block, for example, played a key role in the DAAD's Berlin program. Katarzyna Bittner, Berliner Künstlerprogramme, DAAD, interview with the author (6 February 2013). See also René Block, "Planquadrat SoHo. Europa in Soho," in *New York–Downtown Manhattan: SoHo* (Berlin: Akademie der Künste/Berliner Festwochen, 1976), 6–90; Stefanie Endlich and Rainer Höynick, eds., *Blickwechsel: 25 Jahre Berliner Künstler Programm* (Berlin: Argon, 1988).

11. Letter from Paul-Armand Gette to Helga Retzer dated 8 August 1979, DAAD archive, Berlin. Gette's interest in the spontaneous ecological dynamics of urban space also finds resonance with early examples of "ecological art" such as Hans Haacke's *Bowery seeds* (1970), where the spontaneous dynamics of bare earth were observed in an urban setting over several weeks.

12. Günter Metken, "Gettes parallele Wissenschaft," in Gette, *Arbeiten 1959–1979*, 22. See also Günter Metken, *Spurensicherung* (Cologne: DuMont, 1977).

13. There is an interesting parallel here with Gordon Matta-Clark's studies of interstitial spaces in *Fake Estates* (1973–1974).

14. See Gette, *Exotik als Banalität*.

15. Lucius Burckhardt, "Wissenschaft ohne Fragen," in *Transect and some other attitudes toward landscape* (Ipswich: European Visual Arts Centre, c. 1988), 22–25. See also Lucius Burckhardt, "documenta urbana: sichtbar machen," in *Landschafts-theoretische Aquarelle und Spaziergangs-wissenschaft*, ed. Noah Regenass, Markus Ritter, and Martin Schmitz (Berlin: Martin Schmitz, 2017 [1982]), 225–254.

16. In the early nineteenth century Humboldt shifted the emphasis of taxonomy to a more systematic interest in geographical differences in vegetation patterns. His work was later elaborated by Joachim Frederik Schouw (1789–1852) and August Grisebach (1814–1879), although the term "ecology" was first used in 1867 by Ernst Haeckel (1871–1919) and the idea of the "ecosystem" was popularized from the 1930s onwards through the writings of Arthur Tansley. More recently, some ecologists, such as Robert Whittaker, have questioned the existence of "plant communities" that are greater than their individual elements, so that the field of "plant sociology" remains a contentious epistemological zone. On the cultural resonance of Humboldt in nineteenth-century Germany see also Susanne Zantop, *Colonial fantasies: conquest, family, and nation in precolonial Germany, 1770–1870* (Durham, NC: Duke University Press, 1997).

17. The German landscape is of course not "natural" at all but formed under precise sets of historical circumstances, as Marx reminds us in his critique of the philosophical idealism advanced by Ludwig Feuerbach. On the geopolitical notion of "natural" borders for Mitteleuropa see, for example, Hans-Dietrich Schultz, "Deutschlands 'natürliche' Grenzen. 'Mittellage' und 'Mitteleuropa' in der Diskussion der Geographen seit dem Beginn des 19. Jahrhunderts," *Geschichte und Gesellschaft* 15 (2) (1989): 248–281.

18. See, for example, the work of the geographer Josef Schmithüsen, who would later take a key role in the UNESCO "Man and Biosphere" research program initiated in the late 1960s. See also Klaus Fehn, "'Germanisch-deutsche Kulturlandschaft'—Historische Geographie und NS-Forschung," *Petermanns geographische Mitteilungen* 146 (6) (2002): 64–69.

19. Reinhold Tüxen, "Pflanzengesellschaften als Gestaltungsstoff," *Gartenkunst* 52 (1939): 209–216, cited in Gert Gröning and Joachim Wolschke-Bulmahn, "Politics, planning and the protection of nature: political abuse of early ecological ideas in Germany, 1933–45," *Planning Perspectives* 2 (2) (1987): 127–148.

20. Cited in Gröning and Wolschke-Bulmann, "Politics, planning and the protection of nature," 143. For a contrasting, if somewhat unconvincing, perspective on the historiography of landscape ideals in Nazi Germany see Frank Uekötter, "Native plants: a Nazi obsession?," *Landscape Research* 32 (3) (2007): 379–383.

21. See, for example, Jacob Borut, "Struggles for spaces: where could Jews spend free time in Nazi Germany?," *Leo Baeck Institute Year Book* 56 (1) (2011): 307–350; Saul Friedländer, *Nazi Germany and the Jews: the years of persecution, 1933–1939* (New York: Harper Collins, 1997); Matthew Gandy, "Borrowed light: a journey through Weimar Berlin," in *The fabric of space: water, modernity, and the urban imagination* (Cambridge, MA: MIT Press, 2014), 55–79; Marion A. Kaplan, *Between dignity and despair: Jewish life in Nazi Germany* (Oxford: Oxford University Press, 1998); and Jürgen Matthäus, "Antisemitic symbolism in early Nazi Germany 1933–1935," *Leo Baeck Institute Yearbook* 45 (1) (2000): 183–203.

22. On the background to environmental thought in the Nazi era see, for example, Joachim Radkau and Frank Uekötter, eds., *Naturschutz und Nationalsozialismus* (Frankfurt: Campus Verlag, 2003).

23. Some measure of Ellenberg's enthusiasm for the Nazi project is indicated by his 1941 essay on the agricultural landscapes of northern Germany, published in the *Geographische Zeitschrift*, which concludes with the exclamatory declaration "Blut und Boden!" Heinz Ellenberg, "Deutsche Bauernhaus-Landschaften als Ausdruck von Natur, Wirtschaft und Volkstum," *Geographische Zeitschrift* 47 (2) (1941): 72–87.

24. After 1945 Tüxen remained in post as director of the ZVR, now renamed as the Bundesanstalt für Vegetationskartierung, a precursor of the current Bundesamt für Naturschutz, with responsibility for nature conservation, whilst Ellenberg rivaled the

Swiss botanist Josias Braun-Blanquet, with whom he had worked in the 1930s, as the most influential plant sociologist of the postwar era.

25. Kurt Hueck, "Vorschläge für die Wiederbepflanzung der Grünanlagen auf pflanzensoziologischer Grundlage im Landschaftsraum von Gross Berlin" (no date). I am grateful to Jens Lachmund for sharing a copy of this document with me.

26. See Kurt Hueck, *Waldbäume und Waldtypen aus NW-Argentinien* (Berlin: Fritz Haller Verlag, 1954) and *Urlandschaft, Taublandschaft und Kulturlandschaft in der Provinz Tucumán im nordwestlichen Argentinien* (Bonn: Geographisches Institut, 1953). In 1960 Hueck returned to Germany as an emeritus professor of botany at the University of Munich.

27. See Wolfgang Müller-Stoll, *Der Pflanzenwelt Brandenburgs* (Berlin-Kleinmachnow: Gartenverlag, 1955).

28. See Jennifer Evans, "Life among the ruins: sex, space, and subculture in zero hour Berlin," in *Berlin: divided city, 1945–1989*, ed. Philip Broadbent and Sabine Hake (New York: Berghahn, 2010), 11–22; and Wolfgang Schivelbusch, *In a cold crater: cultural and intellectual life in Berlin, 1945–1948*, trans. Kelly Barry (Berkeley: University of California Press, 1998). On spaces of forgetting see also W. G. Sebald, "Zwischen Geschichte und Naturgeschichte. Über die literarische Beschreibung totaler Zerstörung," in *Campo santo* (Frankfurt am Main: Fischer, 2006), 69–100.

29. Hildemar Scholz, "Die Ruderalvegetation Berlins" (PhD dissertation, Free University Berlin, 1956). See also Hildemar Scholz, "Die Veränderungen in der Ruderalflora Berlins. Ein Beitrag zur jüngsten Florengeschichte," *Willdenowia* 2 (3) (1960): 379–397.

30. See Herbert Sukopp, "Beitrage zur Ökologie von *Chenopodium botrys* L. I. Verbreitung und Vergesellschaftung," in *Verhandlungen des botanischen Vereins der Provinz Brandenburg* 108 (1971): 2–25, and other essays in the collection.

31. In the case of Hans Scharoun's modernist design for the Berlin Philharmonie, completed in 1963, it was suggested that the vegetation in the surrounding wastelands provided an additional acoustic buffer for the state-of-the-art concert hall. See Sandra Jasper, "Acoustic ecologies: architecture, nature, and modernist experimentation in West Berlin," *Annals of the American Association of Geographers* 110 (4) (2020): 1114–1133.

32. In the 1920s, for example, the botanist Johannes Mattfield tried to produce a complete survey of all the plant species of Germany down to a scale of 250 meters, but the project was abandoned. As Humboldt, Ritter, and other figures in the history of geobotany found, the task of encompassing everything within a unified empirical schema is impossible as well as philosophically overwrought.

33. Burckhardt, "Wissenschaft ohne Fragen."

34. The Institute's narrow epistemological frame—namely the exclusion of animals and most crucially people—is also an outcome of the institutional landscape and disciplinary

subdivisions within the Berlin university system, in addition to any conscious analytical strategy on the part of the TU-based ecologists.

35. Lachmund, *Greening Berlin*.

36. Lachmund, *Greening Berlin*, 81.

37. See Ludwig Trepl, "City and ecology," *Capitalism, Nature, Socialism* 7 (2) (1996): 85–94. The systems-based approach to urban ecology at this time can be illustrated by the work of influential Belgian botanist Paul Duvigneaud. See, for example, Paul Duvigneaud, "Étude écologique de l'écosystème urbain bruxellois: 1. L'écosystème 'urbs'," *Mémoires de la Société Royale de Botanique de Belgique* 6 (1974): 5–35.

38. See, for example, Gerhard Hard, "Vegetationsgeographie und sozialökologie einer Stadt," *Geographische Zeitung* 75 (1985): 125–144. We should be careful, however, to differentiate between the methodological marginalization of the "cultural landscape" as an object of analysis and the emergence of critical perspectives on the ideological role of landscape in modern German history. See, for example, Winfried Schenk, "'Landschaft' und 'Kulturlandschaft'—'getönte' Leitbegriffe für aktuelle Konzepte geographischer Forschung und räumlicher Planung," *Petermanns geographische Mitteilungen* 146 (6) (2002): 6–13.

39. "Der Pflanzenprofessor. Mit Herbert Sukopp durch Berlin," Rundfunk Berlin-Brandenburg (20 October 2008).

40. See Steven Vertovec, "Berlin multikulti: Germany, 'foreigners' and world openness," *New Community* 22 (3) 1996: 381–399.

41. The implicit movement from the postwar presence of "guest workers," inhabiting a parallel but largely hidden social realm, to an emerging engagement with "cultural guests" forms part of a wider discourse of "multicultural capitalism" in which the politics of race and ethnicity has been effectively recast as an adjunct to new forms of urbanity.

42. Vertovec, "Berlin multikulti."

43. Günter Grass, "In Kruezberg fehlt ein Minarett," in *Berlin, ach Berlin*, ed. Hans Werner Richter (Berlin: Severin und Siedler, 1981 [1971]), 140–141. See also Peter Hielscher, "Anmerkungen zu Seiltänzern: Wolfgang Krowlows Bilder aus Kreuzberg," in *Grossstadt Dschungel: Neuer Realismus aus Berlin*, ed. Wolfgang Jean Stock (Munich: Kunstverein, 1983), 79.

44. See Brenda S. A. Yeoh, "Cosmopolitanism and its exclusions in Singapore," *Urban Studies* 41 (12) (2004): 2431–2445.

45. Jacoby provided the captions to his own photographs in Richter, *Berlin, ach Berlin*, 124–125.

46. Krolow's photograph was used to promote *Schulprojekt SO 36* in the Skalitzer Straße, Kreuzberg.

47. See, for example, Andreas Huyssen, "Nation, race, and immigration: German identities after unification," in *Twilight memories: marking time in a culture of amnesia* (London: Routledge, 2012), 67–84; Ruth Mandel, *Cosmopolitan anxieties: Turkish challenges to citizenship and belonging in Germany* (Durham, NC: Duke University Press, 2008); and Nora Räthzel, "Germany: one race, one nation," *Race and Class* 32 (3) (1990): 31–48.

48. See Sebastian Conrad, *Deutsche Kolonialgeschichte* (Munich: C. H. Beck, 2008).

49. Vertovec, "Berlin multikulti." See also Eva Kolinsky, "Non-German minorities in contemporary German society," in *Turkish culture in German society today*, ed. David Horrocks and Eva Kolinsky (Providence, RI: Berghahn, 1996), 71–111.

50. In Rainer Werner Fassbinder's film *Katzelmacher* (1969), for example, we encounter a mix of curiosity and hostility displayed towards a Greek guestworker, played by the director himself, who moves into a working-class housing estate in Munich.

51. Achille Mbembe, "Provincializing France," *Public Culture* 23 (1) (2011): 86.

52. See Werner Hülsberg, *The German Greens: a social and political profile*, trans. Gus Fagan (London: Verso, 1988).

53. In its first political contest in March 1979 the Alternative Liste exceeded the 5 percent electoral threshold in four city districts, and from 1981 it was represented in the citywide Berlin Senate. After a further breakthrough in the elections of January 1989, with nearly 12 percent of the vote, it entered into a governing coalition with the Sozialdemokratische Partei Deutschlands (SPD), and an AL councillor, Michaele Schreyer, was named as environmental commissioner for the city. In the event, however, this "red-green" coalition proved short-lived, and shattered in November 1990 over an SPD-sanctioned police raid against squatters in Mainzerstraße in Friedrichshain. See, for example, Gudrun Heinrich, "Rot-Grün in Berlin 1989–1990," in *Die Grünen. Wie sie wurden, was sie sind*, ed. Joachim Raschke (Cologne: Köln, 1993), 809–822.

54. See Alternative Liste, *Zum Thema: Stadtentwicklung. Die "grüne Mitte." Das Konzept der Alternativen Liste zum zentralen Bereich* (Berlin: Alternative Liste, 1984). See also Jouni Häkli, "Culture and politics of nature in the city: the case of Berlin's 'green wedge'," *Capitalism Nature Socialism* 7 (2) (1996): 125–138.

55. Ingo Kowarik, director of the TU Institute of Ecology, interview with the author (26 April 2011).

56. Bezirksamt Mitte, "Park am Nordbahnhof eröffnet," Pressemitteilung Nr. 211/2009 (13 May 2009).

57. See, for example, T. J. Demos, "The politics of sustainability: art and ecology," in *Radical nature: art and architecture for a changing planet, 1969–2009*, ed. Francesco Manacorda and Ariella Yedgar (London: Koenig/Barbican, 2009), 17–30; and Susanne Witzgall,

Kunst nach der Wissenschaft. Zeitgenössische Kunst im Diskurs mit den Naturwissenschaften (Nuremburg: Verlag für moderne Kunst, 2003).

58. kathrynamiller.com (accessed 15 May 2020). For a critique of nativist environmentalism in a North American context see, for example, Lisa Sun-Hee Park and David N. Pellow, *The slums of Aspen: immigrants vs. the environment in America's Eden* (New York: New York University Press, 2011). On exclusionary dimensions to "ecological urbanism" see also Andrew Newman, *Landscape of discontent: urban sustainability in immigrant Paris* (Minneapolis: University of Minnesota Press, 2015).

59. A cartographic approach is also deployed by Hans Haacke in his detailed study of Manhattan slum tenements in *Schapolsky et al. Manhattan Real Estate Holdings, a Real-Time Social System as of 1. May 1971* (1971). See Rosalyn Deutsche, "Property values: Hans Haacke, real estate, and the museum," in *Evictions: art and spatial politics* (Cambridge, MA: MIT Press, 1996), 159–192.

60. See, for example, Susanne Schroeder, ed., *Skulpturenpark Berlin_Zentrum* (Cologne: Walther König, 2010).

61. On walking and creativity see, for example, David Evans, ed., *The art of walking* (London: Black Dog, 2012); and Frédéric Gros, *A philosophy of walking*, trans. John Howe (London: Verso, 2014 [2009]).

62. Having first appeared as a verb in the mid-seventeenth century, the English word *transect* begins to acquire usage as a noun in the early twentieth century, and becomes closely associated with the emergence of ecological science, the elucidation of vegetation patterns, and the application of modern botanical methods. *Oxford English Dictionary* (www.oed.com) (accessed 30 July 2019).

63. Frederic Edward Clements, *Research methods in ecology* (Lincoln, NE: University Publishing Company, 1905), 176.

64. Robert H. Whittaker, "Gradient analysis of vegetation," *Biological Reviews* 42 (2) (1967): 207–264.

65. Dorothy Brown, *Methods of surveying and measuring vegetation* (Farnham: Commonwealth Agricultural Bureaux, 1954).

66. Tim Ingold, *Lines* (London: Routledge, 2007), 85, 87.

67. Sara Ahmed, *Queer phenomenology: orientation, objects, others* (Durham, NC: Duke University Press, 2006), 121.

68. Cited in Blake W. Burleson, *Jung in Africa* (New York: Continuum, 2005), 114.

69. Ingold, *Lines*, 85, 87.

70. See Kevin Austin, "Botanical processes in urban derelict spaces" (PhD thesis, University of Birmingham, 2002).

71. See Herbert Sukopp, ed., *Stadtökologie. Das Beispiel Berlins* (Berlin: Dietrich Reimer, 1990).

72. See, for example, Andrés Duany and Emily Talen, "Transect planning," *Journal of the American Planning Association* 68 (3) (2002): 245–266; and Nicolas Tixier, "Le transect urbain. Pour une écriture corrélée des ambiances et de l'environnement," in *Écologies urbaines. Sur le terrain*, ed. Sabine Barles and Nathalie Blanc (Paris: Economica-Anthropos, 2016), 130–148.

73. In the hands of geographers such as Gerhard Hard, botanical methods not only enable interpretation of the unusual characteristics of urban vegetation but also become a means to interrogate the history of landscape itself in all its ideological complexity. See, for example, Gerhard Hard, "Die Natur, die Stadt und die Ökologie. Reflexionen über 'Stadtnatur' und 'Stadtökologie,'" in *Dimensionen geographischen Denkens* (Osnabrück: V & R unipress, 2003 [1994]), 341–370.

74. Gette first used the transect method derived from "sociobotany" or *Pflanzensoziologie* in 1974. In the spring of 1975, for example, he presented a 16 mm black-and-white film entitled *Le transect* in the exhibition "Ögon-Blickar/New media 1" held at the Malmö Konsthall in southern Sweden. The film, and its associated diagrams, sketches, and photographs, were derived from a study of the Ribersborgstranden beach near Malmö based on the use of a transect, or study line, passing from the lower shoreline up to the windswept dune system overlooking the sea. Working slowly along the transect, Gette recorded all the species of plants he could find along with observations on other biophysical parameters such as elevation, humidity, and temperature. Gette also used a transect for his contribution to documenta 6 in 1977, based on a 35 km walk between Chalon and the industrial town of Le Creusot. See Franz W. Kaiser, "Von einigen Haltungen gegenüber Landschaft sowie über den Künstler Paul-Armand Gette," in *Transect and some other attitudes toward landscape* (Ipswich: European Visual Arts Centre, c. 1988), 5–18; Didier Paschal-Lejeune, "Premiers éléments pour un catalogue chronologique des travaux de Paul-Armand Gette repris sous le titre général de Contribution à l'étude des lieux restreints," in Paul-Armand Gette, *de quelques lisières: prolégomènes à un essai de définition de la notion d'écotone* (Paris: cheval d'attaque, 1977), 63–73.

75. See, for example, Richard Phillips, "Georges Perec's experimental fieldwork; Perecquian fieldwork," *Social and Cultural Geography* 19 (2) (2018): 171–191.

76. See François Bon, *Paysage fer* (Lagrasse: Verdier, 2000); and François Maspero, *Roissy Express: a journey through the Paris suburbs*, trans. Paul Jones (London: Verso, 1994 [1990]).

77. Mustafa Dikeç, *Badlands of the republic: space, politics, and urban policy* (Oxford: Blackwell, 2007).

78. Isabelle Stengers, *Another science is possible: a manifesto for slow science*, trans. Stephen Muecke (Cambridge: Polity, 2017 [2013]).

79. On aspects of the *dérive* as methodology see also Philip Conway's review of Stengers, *Another science is possible*, in *Society and Space* (2018).

80. Lauren Elkin, *Flâneuse: women walk the city in Paris, New York, Tokyo, Venice, and London* (London: Chatto and Windus, 2016), 288.

81. Morag Rose, "Confessions of an anarcho-flâneuse, or psychogeography the Mancunian way," in *Walking inside out: contemporary British psychogeography*, ed. Tina Richardson (London: Rowman & Littlefield, 2015), 158, 159, 160. See also Stephanie Springgay and Sarah E. Truman, *Walking methodologies in a more-than-human world: WalkingLab* (London: Routledge, 2018).

82. See, for example, Claire Colebrook, "On the very possibility of queer theory," in *Deleuze and queer theory*, ed. Chrysanthi Nigianni and Merl Storr (Edinburgh: Edinburgh University Press, 2009), 11–23.

83. Ahmed, *Queer phenomenology*.

84. Gayatri Gopinath, *Unruly visions: the aesthetic practices of queer diaspora* (Durham, NC: Duke University Press, 2018).

85. Susan Stryker, Paisley Currah, and Lisa Jean Moore, "Introduction: trans-, trans, or transgender?," *Women's Studies Quarterly* 36 (3/4) (2008): 12, 13.

86. See, for example, Julie Livingston and Jasbir K. Puar, "Interspecies," *Social Text* 106 (29) (1) (2011): 1–14; and Camille Nurka, "Animal techne: transing posthumanism," *TSQ: Transgender Studies Quarterly* 2 (2) (2015): 209–226.

87. Londa Schiebinger, "Gender and natural history," in *Cultures of natural history*, ed. Nicholas Jardine, James A. Secord, and Emma C. Spary (Cambridge: Cambridge University Press, 1996), 163–177. On plants, politics, and empire building see also Londa Schiebinger, *Plants and empire: colonial bioprospecting in the Atlantic world* (Cambridge, MA: Harvard University Press, 2004).

88. Raymond F. Dasmann cited in Stuart Davey, "Role of wildlife in an urban environment," in *Transactions of the Thirty-Second North American Wildlife and Natural Resources Conference*, ed. J. B. Trefethen (Washington, DC: Wildlife Management Institute, 1967), 50–59. See also Anne I. Dagg, "Wildlife in an urban area," *Le Naturaliste Canadien* 97 (1970): 201–212; and Raymond F. Dasmann, *A different kind of country* (New York: Macmillan, 1968).

89. For an example of the problematic handling of race in urban design discourse see Ian L. McHarg, *Design with nature* (New York: American Museum of Natural History, 1969).

90. Nathan Hare, "Black ecology," *Black Scholar* 1 (6) (1970): 2–8.

91. See also Andrew Hurley, *Environmental inequalities: class, race, and industrial pollution in Gary, Indiana, 1945–1980* (Chapel Hill: University of North Carolina Press, 1995); Kevin

DeLuca and Anne Demo, "Imagining nature and erasing class and race: Carleton Watkins, John Muir, and the construction of wilderness," *Environmental History* 6 (4) (2001): 541–560; Robert Gioielli, *Environmental activism and the urban crisis: Baltimore, St. Louis, Chicago* (Philadelphia: Temple University Press, 2014); and Paul Mohai and Bunyan Bryant, "Is there a 'race' effect on concern for environmental quality?," *Public Opinion Quarterly* (December 1998): 475–505.

92. Julian Agyeman, "Black people in a white landscape: social and environmental justice," *Built Environment* (1990) 16 (3): 232–236.

93. Dorceta E. Taylor, "Minority environmental activism in Britain: from Brixton to the Lake District," *Qualitative Sociology* 16 (3) (1993): 263–295.

94. James Fenton, "Native or alien," *Ecos* 7 (2) (1986): 20–23.

95. Julian Agyemen, "Entering cosmopolis: crossing over, hybridity, conciliation and the intercultural city ecosystem," *Minding Nature* 7 (1) (2014): 22.

96. Giovanni Aloi, "Sorely visible: plants, roots, and national identity," *Plants, People, Planet* 1 (3) (2019): 204–211.

97. Discussion with Gert Gröning held at the Universität der Künste, Berlin, 20 May 2014.

98. On the dangers facing black birdwatchers see J. Drew Lanham, "9 rules for the black birdwatcher," *Orion* (October 2013); Adam Schatz, "America explodes," *London Review of Books* 42 (12) (18 June 2020); Jacqueline L. Scott, "What you should know about Black birders," *The Conversation* (2 June 2020); and Andrea Thompson, "Black birders call out racism, say nature should be for everyone," *Scientific American* (5 June 2020).

99. Carolyn Finney, *Black faces, white spaces: reimagining the relationship of African Americans to the great outdoors* (Chapel Hill: University of North Carolina Press, 2014).

100. See, for example, Alesia Montgomery, *Greening the Black urban regime: the culture and commerce of sustainability in Detroit* (Detroit: Wayne State University Press, 2020).

101. See Jason Byrne, Jennifer Wolch, and Jin Zhang, "Planning for environmental justice in an urban national park," *Journal of Environmental Planning and Management* 52 (3) (2009): 365–392; Jason Byrne, "When green is White: The cultural politics of race, nature and social exclusion in a Los Angeles urban national park," *Geoforum* 43 (3) (2012): 595–611; and Patrick C. West, "Urban region parks and black minorities: subculture, marginality, and interracial relations in park use in the Detroit metropolitan area," *Leisure Sciences* 11 (1) (1989): 11–28. On race, public space, and park management in North American cities see also Alec Brownlow, "An archaeology of fear and environmental change in Philadelphia," *Geoforum* 37 (2) (2006): 227–245; and Kevin Loughran, "Race and the construction of city and nature," *Environment and Planning A* 49 (9) (2017): 1948–1967.

102. See, for example, Bernd Berlin, *Ethnobiological classification: principles of categorization of plants and animals in traditional societies* (Princeton, NJ: Princeton University Press, 1992).

103. See Andrea Pieroni and Ina Vanderbroek, *Traveling cultures and plants: the ethnobiology and ethnopharmacy of human migration* (New York: Berghahn, 2007).

104. The term "pluriverse," first coined by the philosopher William James, has gradually diffused from metaphysics into anthropology, economics, and a variety of other fields. See Arturo Escobar, *Designs for the pluriverse: radical interdependence, autonomy, and the making of worlds* (Durham, NC: Duke University Press, 2018). See also Julio Alberto Hurrell and María Lelia Pochettino, "Urban ethnobotany: theoretical and methodological contributions," in *Methods and techniques in ethnobiology and ethnoecology*, ed. Ulysses Paulino Albuquerque et al. (NewYork: Springer / Humana Press, 2014), 293–309.

105. See, for example, Edward Voss and Anton A. Reznicek, *Field manual of Michigan flora* (Ann Arbor: University of Michigan Press, 2012).

106. Divya P. Tolia-Kelly, "Landscape, race and memory: biographical mapping of the routes of British Asian landscape values," *Landscape Research* 29 (3) (2004): 290.

107. See Zachary J. S. Falck, *Weeds: an environmental history of metropolitan America* (Pittsburgh: University of Pittsburgh Press, 2010).

108. Stuart Hall, "When was 'the post-colonial'? Thinking at the limit," in *The postcolonial question*, ed. Iain Chambers and Lidia Curti (London: Routledge, 2002), 248–266.

CHAPTER 4

1. Eyal Weizman, "Introduction: forensis," in *Forensis: the architecture of public truth* (Berlin: Sternberg Press, Berlin, 2014), 27.

2. Tim Choy, *Ecologies of comparison: an ethnography of endangerment in Hong Kong* (Durham, NC: Duke University Press, 2011), 28.

3. Yevgenia Chirikova, "The battle for Khimki forest," *Open Democracy* (23 August 2010).

4. *Red Book Database* (in Russian), available at http://www.sevin.ru/redbooksevin/ (accessed 10 December 2019).

5. Although Beketof's life was initially saved, he lost a leg, three fingers, and suffered permanent brain damage. He eventually died from his injuries in 2013.

6. CEE Bankwatch Network Response to VINCI's claims regarding the Moscow-St. Petersburg motorway (2011).

7. Chirikova, "The battle for Khimki forest." See also Alfred B. Evans, "Protests and civil society in Russia: the struggle for the Khimki forest," *Communist and Post-Communist Studies* 45 (3–4) (2012): 233–242; and Joshua P. Newell and Laura A. Henry "The state of environmental protection in the Russian Federation: a review of the post-Soviet era," *Eurasian Geography and Economics* 57 (6) (2016): 779–801.

8. Michael Marder, "The life of plants and the limits of empathy," *Dialogue: Canadian Philosophical Review / Revue Canadienne de Philosophie* 5 (2) (2012): 260.

9. Marder, "The life of plants and the limits of empathy," 261.

10. Marder, "The life of plants and the limits of empathy," 260.

11. See, for example, Irus Braverman, "Anticipating endangerment: the biopolitics of threatened species lists," *BioSocieties* 12 (2017): 132–157.

12. The term "forensic ecology" has also been deployed in relation to war-damaged landscapes. See Joseph Pugliese, *Biopolitics of the more-than-human: forensic ecologies of violence* (Durham, NC: Duke University Press, 2020).

13. See, for example, Mark Benecke, "A brief history of forensic entomology," *Forensic Science International* 120 (1–2) (2001): 2–14.

14. Jean-Pierre Mégnin, *Le faune des cadavres: application de l'entomologie à la médecine légale* (Langes: Klincksieck, 2015 [1894]).

15. Patricia Yancey Martin, John R. Reynolds, and Shelley Keith, "Gender bias and feminist consciousness among judges and attorneys: a standpoint theory analysis," *Signs: Journal of Women in Culture and Society* 27 (3) (2002): 665–701.

16. Alex Biedermann, "The role of the subjectivist position in the probabilization of forensic science," *Journal of Forensic Science and Medicine* 1 (2) (2015): 145. On the history of probability see in particular Ian Hacking, *The emergence of probability: a philosophical study of early ideas about probability, induction and statistical inference* (Cambridge: Cambridge University Press, 2006 [1975]).

17. Paolo Garbolino, "The scientification of forensic practice," in *New challenges to philosophy of science*, ed. Hanne Andersen, Dennis Dieks, Wenceslao J. Gonzalez, Thomas Uebel, and Gregory Wheeler (Dordrecht: Springer, 2013), 287–297.

18. The analysis of beetle remains found in dated core samples, for example, has proved an extremely helpful guide to changes in climate and vegetation over the last 110,000 years. See, for example, T. C. Atkinson, K. R. Briffa, G. R. Coope, M. J. Joachim, and D. W. Perry, "Climatic calibration of Coleopteran data," in *Handbook of Holocene Palaeoecology and Palaeohydrology*, ed. B. E. Berglund (Chichester: Wiley, 1986), 851–858; Russell G. Coope and Scott A. Elias, "The environment of Upper Palaeolithic (Magdalenian and Azilian) hunters at Hauterive-Champréveyres, Neuchâtel, Switzerland, interpreted from Coleopteran remains," *Journal of Quaternary Science* 15 (2000): 157–175; and Michael L. Rosenzweig, *Species diversity in space and time* (Cambridge: Cambridge University Press, 1995).

19. See Caspar A. Hallmann, Martin Sorg, Eelke Jongejans, Henk Siepel, Nick Hofland, Heinz Schwan, Werner Stenmans, et al., "More than 75 percent decline over 27 years in total flying insect biomass in protected areas," *PLoS One* 12 (10) (2007): e0185809.

20. Rob Nixon, *Slow violence and the environmentalism of the poor* (Cambridge, MA: Harvard University Press, 2011).

21. Eyal Weizman, *Forensic architecture: violence at the threshold of detectability* (New York: Zone Books, 2017), 118.

22. Weizman, *Forensic architecture*, 65. See also Susan Schuppli, *Material witness: media, forensics, evidence* (Cambridge, MA: MIT Press, 2020).

23. Weizman, *Forensic architecture*.

24. Yve-Alain Bois, Michel Feher, Hal Foster, and Eyal Weizman, "On forensic architecture: a conversation with Eyal Weizman," *October* 156 (2016): 116–140.

25. See also Paolo Tavares, "Nonhuman rights," in *Forensis: the architecture of public truth*, ed. Eyal Weizman (Berlin: Sternberg Press, 2014), 553–572.

26. Bois et al., "On forensic architecture."

27. See, for example, Kate Nesbitt, *Theorizing a new agenda for architecture: an anthology of architectural theory 1965–1995* (Princeton, NJ: Princeton Architectural Press, 1996).

28. Bois et al., "On forensic architecture."

29. Gastón R. Gordillo, *Rubble: the afterlife of destruction* (Durham, NC: Duke University Press, 2014).

30. See also Schuppli, *Material witness*. Schuppli, who has worked with Weizman as part of the forensic architecture project, develops the theme of nonhuman material traces through the concept of the "material witness" that comprises "non-human entities and machinic ecologies that archive their complex interactions with the world." These material traces form part of an archival process "that can be forensically decoded and reassembled back into history." Following Timothy Morton's conceptualization of the "hyperobject," Schuppli also explores the aesthetic contours of aesthetic or damaged environments and the tensions between localized manifestions of environmental phenomena and the vastness of underlying totalities that cannot be grasped. See Susan Schuppli, "Dirty pictures," in *Living earth: field notes from the Dark Ecology Project 2014–2016*, ed. Mirna Belina (Amsterdam: Sonic Acts Press, 2016), 196.

31. Weizman, *Forensic architecture*, 119.

32. Weizman, *Forensic architecture*, 123.

33. Robert E. Witcher, "On Rome's ecological contribution to British flora and fauna: landscape, legacy and identity," *Landscape History* 34 (2) (2013): 5–26. The study of urban soils can also take ecology into the field of forensic archaeology, enabling the reconstruction of urban stratigraphies. See Yannick Devos, Luc Vrydaghs, Ann Degraeve, and Sylvianne Modrie, "Unravelling urban stratigraphy. The study of Brussels'(Belgium) Dark Earth. An archaeopedological perspective," *Medieval and Modern Matters* 2 (2011): 51–76.

34. See, for example, David L. Hawksworth and Francis Rose, *Lichens as pollution monitors* (London: Edward Arnold, 1976); J. R. Laundon, "London's lichens," *London Naturalist* 49 (1970): 20–68; and T. H. Nash and C. Gries, "Lichens as indicators of air pollution," in *Air pollution* (Berlin: Springer, 1991), 1–29. Laundon, for example, describes a steady disappearance of many lichen species during London's rapid nineteenth-century expansion.

35. William Nylander, "Les lichens du Jardin du Luxembourg," *Bulletin de la Société Botanique de France* 13 (7) (1866), 365 (my translation).

36. Jennifer Gabrys, "Sensing lichens: from ecological microcosms to environmental subjects," *Third Text* 32 (2–3) (2018): 352.

37. Gabrys, "Sensing lichens," 354.

38. Eyal Weizman, "Matters of calculation: the evidence of the Anthropocene," Eyal Weizman in conversation with Heather Davis and Etienne Turpin, in *Architecture in the Anthropocene: encounters among design, deep time, science and philosophy*, ed. Etienne Turpin (Ann Arbor, MI: Open Humanities Press, 2013), 69.

39. Frédéric Keck, "Sentinels for the environment: birdwatchers in Taiwan and Hong Kong," *China Perspectives* 2 (2015): 43–52.

40. Frédéric Keck, *Avian reservoirs: virus hunters and birdwatchers in Chinese sentinel posts* (Durham, NC: Duke University Press, 2020).

41. Keck, "Sentinels for the environment," 51.

42. Keck, *Avian reservoirs*.

43. In the use of Red Lists to highlight degrees of endangerment, there is a distinction between what is deemed "listable" (predominately vertebrates) and the bulk of nature that remains "listless." If we shift our focus towards the web of life as a whole, however, there is less emphasis on individuals, populations, or even the flagship species that have underpinned the parameters of conservation biology. See, for example, Irus Braverman, "Enlisting life: red is the color of threatened species lists," in *Critical animal geographies: politics, intersections, and hierarchies in a multispecies world*, ed. Kathryn Gillespie and Rosemary-Claire Collard (London: Routledge, 2015), 184–202.

44. Andrew Lakoff, "The indicator species: tracking ecosystem collapse in arid California," *Public Culture* 28 (2016): 237–259. The question of measuring attributes of nature, such as ecological assemblages, also holds implications for attempts to integrate different policy domains. See, for example, Morgan M. Robertson, "The nature that capital can see: science, state, and market in the commodification of ecosystem services," *Environment and Planning D: Society and Space* 24 (3) (2006): 367–387.

45. See, for example, Roderick P. Neumann, "Life zones: the rise and decline of a theory of the geographic distribution of species," in *Spatializing the history of ecology: sites, journeys,*

mappings, ed. Raf de Bont and Jens Lachmund (London: Routledge, 2017), 37–55. The vulnerability of a particular species is partly a matter of context specific since many species of plants and animals are much rarer towards the edge of their range. The question of extinction is also complicated by the disappearance of subspecies, the loss of genetic biodiversity, and the possibilities of sustaining some life forms in artificial environments such as zoos or laboratories. In practice, however, the understanding of what a species represents extends to factors such as hybridization or the capacity of individual species to use different means of reproduction. In the case of organisms that rely on phylogenetic reproduction such as fungi, the use of sequence variants from DNA analysis has revealed much higher levels of biological diversity than previously thought. The use of DNA barcoding has unsettled the science of taxonomy by shifting the emphasis of classification away from the morphological features of whole organisms towards a more narrow focus on nucleotide sequences. The revelation that many organisms hitherto regarded as a single species are actually parts of a "species complex" complicates the relationship between conservation biology and public policy. See, for example, Georgina M. Mace, "The role of taxonomy in species conservation," *Philosophical Transactions: Biological Sciences* 359 (1444) (2004): 711–719; James Malet, "Subspecies, semispecies, superspecies," in *Encyclopedia of biodiversity*, ed. Simon A. Levin (London: Academic Press, 2007); E. Smith and David J. Read, *Mycorrhizal symbiosis* (London: Academic Press, 2010); and Claire Waterton, *Barcoding nature: shifting cultures of taxonomy in an age of biodiversity loss* (London: Routledge, 2017).

46. Choy, *Ecologies of comparison*, 71.

47. Fernando Vidal and Nélia Dias, "Introduction: the endangerment sensibility," in *Endangerment, biodiversity, and culture*, ed. Fernando Vidal and Nélia Dias (New York: Routledge, 2016), 1.

48. Kathryn Yusoff, "Aesthetics of loss: biodiversity, banal violence and biotic subjects," *Transactions of the Institute of British Geographers* 37 (4) (2012): 578–592.

49. Kristie Dotson and Kyle Whyte, "Environmental justice, unknowability and unqualified affectability," *Ethics and the Environment* 18 (2) (2013): 55. See also María Puig de la Bellacasa, *Matters of care: speculative ethics in more than human worlds* (Minneapolis: University of Minnesota Press, 2017).

50. See, for example, Ulrich Heink and Ingo Kowarik, "What are indicators? On the definition of indicators in ecology and environmental planning," *Ecological Indicators* 10 (3) (2010): 584–593; and Ahmed Siddig, Aaron Ellison, Alison Ochs, Claudia Villar-Leeman, and Matthew K. Lau, "How do ecologists select and use indicator species to monitor ecological change? Insights from 14 years of publication in *Ecological Indicators*," *Ecological Indicators* 60 (2016): 223–230.

51. See, for example, Ash E. Samuelson, Richard J. Gill, Mark J. F. Brown, and Ellouise Leadbeater, "Lower bumblebee colony reproductive success in agricultural compared with

urban environments," *Proceedings of the Royal Society B: Biological Sciences* 285 (1881) (2018): 20180807.

52. Choy, *Ecologies of comparison*, 7.

53. Miranda Fricker, *Epistemic injustice: power and the ethics of knowing* (New York: Oxford University Press, 2007).

54. Kristie Dotson, "Tracking epistemic violence, tracking practices of silencing," *Hypatia* 26 (2) (2011): 242.

55. Zygmunt J. B. Plater, "From the beginning, a fundamental shift of paradigms: a theory and short history of environmental law," *Loyola of Los Angeles Law Review* 27 (3) (1994): 981–1008.

56. Plater, "From the beginning, a fundamental shift of paradigms," 983.

57. Plater, "From the beginning, a fundamental shift of paradigms," 991.

58. See, for example, Steve Lerner, *Diamond: a struggle for environmental justice in Louisiana's chemical corridor* (Cambridge, MA: MIT Press, 2006); and Richard Misrach and Kate Orff, *Petrochemical America* (New York: Aperture, 2012).

59. See, for example, Thom Davies, "Toxic space and time: slow violence, necropolitics, and petrochemical pollution," *Annals of the American Association of Geographers* 108 (6) (2018): 1537–1553; and Alice Mah and Xinhong Wang, "Accumulated injuries of environmental injustice: living and working with petrochemical pollution in Nanjing, China," *Annals of the American Association of Geographers* 109 (6) (2019): 1961–1977.

60. See in particular Thomas F. Gieryn, "Boundary-work and the demarcation of science from non-science: strains and interests in professional ideologies of scientists," *American Sociological Review* 48 (6) (1983): 781–795.

61. See, for example, Gin Armstrong and Derek Seidman, "Fossil fuel industry pollutes black and brown communities while propping up racist policing," *LittleSis* (27 July 2020).

62. See, for example, Lavanya Rajamani, "Public interest environmental litigation in India: exploring issues of access, participation, equity, effectiveness and sustainability," *Journal of Environmental Law* 19 (3) (2007): 293–321; and Balakrishnan Rajagopal, "Limits of law in counter-hegemonic globalization: the Indian Supreme Court and the Narmada Valley struggle," in *Law and globalization from below: toward a cosmopolitan legality*, ed. Boaventura de Sousa Santos and César A. Rodríguez-Garavito (Cambridge: Cambridge University Press, 2005), 183–217.

63. See, for example, Anuj Bhuwania, *Courting the people: public interest litigation in post-emergency India* (Cambridge: Cambridge University Press, 2017); D. Asher Ghertner, "Analysis of new legal discourse behind Delhi's slum demolitions," *Economic and Political Weekly* 43 (20) (2008): 57–66; and Marie-Hélène Zérah, "Conflict between green space

preservation and housing needs: the case of the Sanjay Gandhi National Park in Mumbai," *Cities* 24 (2) (2007): 122–132.

64. See Sudipta Kaviraj, "Filth and the public sphere: concepts and practices about waste in Calcutta," *Public Culture* 10 (1) (1997): 83–113.

65. See, for example, Amita Baviskar, "Urban nature and its publics: shades of green in the remaking of Delhi," in *Grounding urban natures: histories and futures of urban ecologies*, ed. Henrik Ernstson and Sverker Sörlin (Cambridge, MA: MIT Press, 2019), 233–246; Emma Mawdsley, "India's middle classes and the environment," *Development and Change* 35 (1) (2004): 79–103; and Anne M. Rademacher and Kalyanakrishnan Sivaramakrishnan, eds., *Ecologies of urbanism in India: metropolitan civility and sustainability* (Hong Kong: Hong Kong University Press, 2013). Calin Cotoi also offers an interesting example from Bucharest where a process of "ecologization" through the transformation of a marginal site into a legally recognized nature reserve reflects a wider class-based recomposition of environmental discourse. See Calin Cotoi, "We should have asked what year we were in! Wastelands and wilderness in the Văcărești Park," *Antipode* 53 (2021).

66. See Zérah, "Conflict between green space preservation and housing needs."

67. Examples include opposition to the health-threatening effects of a copper smelting facility in the city Thoothukudi, Tamil Nadu, where police killed 11 protestors at a demonstration in 2018. See Michael Safi, "Indian copper plant shut down days after deadly protests," *Guardian* (28 May 2018). See also Saumya Uma and Arvind Narrain, "Human rights and its future: some reflections," *Jindal Global Law Review* 9 (2) (2018): 287–297; and Reva Yunus, "Unpacking the histories, contours and multiplicity of India's women's movement(s): an interview with Uma Chakravarti," *Journal of Law, Social Justice and Global Development* 23 (2019): 55–73.

68. See, for example, Nick Middeldorp and Philippe Le Billon, "Deadly environmental governance: authoritarianism, eco-populism, and the repression of environmental and land defenders," *Annals of the American Association of Geographers* 109 (2) (2019): 324–337.

69. See, for example, Anna Lora-Wainwright, *Resigned activism: living with pollution in rural China* (Cambridge, MA: MIT Press, 2017).

70. Nicholas K. Blomley, *Law, space, and the geographies of power* (New York: Guilford Press, 1994), xiii.

71. Boaventura de Sousa Santos, "Law: a map of misreading. Toward a postmodern conception of law," *Journal of Law and Society* 14 (3) (1987): 297–302.

72. See Pierre Bourdieu, "The force of law: toward a sociology of the juridical field," *Hastings Law Journal* 38 (5) (1987): 814–853.

73. Lynn K. Nyhart, "Natural history and the 'new' biology," in *Cultures of natural history*, ed. Nicholas Jardine, James A. Secord, and Emma C. Spary (Cambridge: Cambridge University Press, 1996), 426–443.

74. See, for example, Elizabeth B. Keeney, *The botanizers: amateur scientists in nineteenth-century America* (Chapel Hill: University of North Carolina Press, 1992); Robert E. Kohler, *All creatures: naturalists, collectors, and biodiversity, 1850–1950* (Princeton, NJ: Princeton University Press, 2006); and Willis Conner Sorensen, *Brethren of the net: American entomology, 1840–1880* (Tuscaloosa: University of Alabama Press, 1995).

75. James A. Secord, "The crisis of nature," in Jardine et al., *Cultures of natural history*, 447–459.

76. On the global reach of urban ornithology see, for example, John M. Marzluff, "A decadal review of urban ornithology and a prospectus for the future," *Ibis* 159 (2016): 1–13.

77. I am grateful to Niranjana Ramesh for assistance with the interpretation of some of the contemporary Tamil-language ornithological sources to illuminate the emerging environmental agenda for biodiversity protection in the Chennai metropolitan region.

78. See Jean-Marc Drouin and Bernadette Bensaude-Vincent, "Nature for the people," in Jardine et al., *Cultures of natural history*, 408–425; and Anne Secord, "Science in the pub: artisan botanists in early 19th century Lancashire," *History of Science* 32 (3) (1994): 269–315.

79. Frank Eugene Lutz, *A lot of insects: entomology in a suburban garden* (New York: Putnam and Sons, 1941).

80. Jennifer Owen, *The ecology of a garden: the first fifteen years* (Cambridge: Cambridge University Press, 1991); and Jennifer Owen, *Wildlife of a garden: a thirty-year study* (London: Royal Horticultural Society, 2010).

81. Richard Lewington, *Guide to garden wildlife*, 2nd ed. (London: Bloomsbury, 2019).

82. Owen, *Wildlife of a garden*.

83. See, for example, Kathleen S. Murphy, "Collecting slave traders: James Petiver, natural history, and the British slave trade," *William and Mary Quarterly* 70 (4) (2013): 637–670.

84. Francesco Careri, *Walkscapes: walking as aesthetic practice* (Ames, IA: Culicidae Architectural Press, 2017).

85. See, for example, David Frisby, "The metropolis as text: Otto Wagner and Vienna's 'second renaissance,'" *Culture, Theory and Critique* 40 (1) (1997): 1–16.

86. Siegfried Kracauer, "Aus dem Fenster gesehen," in *Straßen in Berlin und anderswo* (Berlin: Das Arsenal, 1987 [1931]), 50–52.

87. On Benjamin's approach to the interpretation of urban modernity see Graeme Gilloch, *Myth and metropolis: Walter Benjamin and the city* (Cambridge: Polity, 1996). On history, memory, and visual culture see, for example, Giuliana Bruno, *Streetwalking on a ruined map: cultural theory and the city films of Elvira Notari* (Princeton, NJ: Princeton University Press, 1993).

88. See, for example, Rebecca Ellis and Claire Waterton, "Caught between the cartographic and the ethnographic imagination: the whereabouts of amateurs, professionals, and nature in knowing biodiversity," *Environment and Planning D: Society and Space* 23 (5) (2005): 673–693.

89. Jean-Claude Abadie, Camila Andrade, Nathalie Machon, and Emmanuelle Porcher, "On the use of parataxonomy in biodiversity monitoring: a case study on wild flora," *Biodiversity and Conservation* 17 (14) (2008): 3485–3500.

90. See Rick Bonney, Caren B. Cooper, Janis L. Dickinson, Steve Kelling, Tina Phillips, Kenneth V. Rosenberg, and Jennifer Shirk, "Citizen science: a developing tool for expanding science knowledge and scientific literacy," *BioScience* 59 (11) (2009): 977–984; and Janis L. Dickinson, Benjamin Zuckerberg, and David N. Bonter, "Citizen science as an ecological research tool: challenges and benefits," *Annual Review of Ecology, Evolution, and Systematics* 41 (2010): 149–172.

91. See Ellis and Waterton, "Caught between the cartographic and the ethnographic imagination"; and Claire Waterton, "From field to fantasy: clarifying nature, constructing Europe," *Social Studies of Science* 32 (2002): 177–204.

92. See Rebecca Lave, "The future of environmental expertise," *Annals of the Association of American Geographers* 105 (2) (2015): 244–252.

93. Jean-François Lyotard, *The postmodern condition: a report on knowledge*, trans. Geoffrey Bennington and Brian Massumi (Minneapolis: University of Minnesota Press, 1984 [1979]).

94. Thomas F. Gieryn, *Cultural boundaries of science: credibility on the line* (Chicago: University of Chicago Press, 1999), x.

95. See Cathy C. Conrad and Krista G. Hilchey, "A review of citizen science and community-based environmental monitoring: issues and opportunities," *Environmental Monitoring and Assessment* 176 (1–4) (2011): 273–291.

96. See, for example, Maarten Hajer, "Policy without polity? Policy analysis and the institutional void," *Policy Sciences* 36 (2) (2003): 175–195.

97. See in particular Robert D. Bullard, *Dumping in Dixie: race, class, and environmental quality* (Boulder, CO: Westview Press, 2000).

98. Site visits and discussion with local activist, Denton, Texas, undertaken in April 2018. On citizen science and regulatory voids see also Abby J. Kinchy and Simona L. Perry, "Can volunteers pick up the slack? Efforts to remedy knowledge gaps about the watershed impacts of Marcellus Shale gas development," *Duke Environmental Law and Policy Forum* 22 (2) (2012): 303–339.

99. Of the 7 to 10 million species estimated to currently exist on earth, most will probably never be known. See Camilo Mora, Derek P. Tittensor, Sina Adl, Alastair G. B. Simpson,

and Boris Worm, "How many species are there on Earth and in the ocean?," *PLoS Biology* 9 (8) (2011): e1001127.

100. Andrew Feenberg, *Technosystem: the social life of reason* (Cambridge, MA: Harvard University Press, 2017).

101. On risks to natural history museums see, for example, Hanna Pennock, "Natural history museum security," in *The future of natural history museums*, ed. Eric Dorfman (London: Routledge, 2018), 49–64.

102. Jens Lachmund, *Greening Berlin: the co-production of science, politics, and urban nature* (Cambridge, MA: MIT Press, 2013), 5.

103. Lachmund, *Greening Berlin*. Lachmund's conceptualization of a "biotope protection regime" draws on the anthropologist Arturo Escobar's characterization of the "nature regime" as a historically and geographically specific mode of cultural articulation with nature. Escobar identifies three types of regimes—the organic, the capitalist, and the technonature—which though framed by an antiessentialist account of nature also acknowledge the role of "biophysical laws" in their configuration. The organic regime, most strongly represented in Indigenous cultures, rests on a unified ontological world view, and is widely present in anthropological investigations of local knowledge. The capitalist regime of nature encompasses processes of commodification in combination with new modes of governmentality, whilst the technonature regime resides in what Escobar terms "a domain of artificiality" that connects with emerging interest in the posthuman. What Escobar initiates is not an overarching theory but rather a heuristic typology that traces connections between anthropology and political ecology amid the constructivist debates of the late 1990s. See Arturo Escobar, "After nature: steps to an antiessentialist political ecology," *Current Anthropology* 40 (1) (1999): 1–30.

104. Lachmund, *Greening Berlin*. The Alternative Liste had gained sufficient support by the late 1980s to briefly enter into a "red-green" urban coalition (see also note 53 on p. 306).

105. The Grüntangente (Green tangent) campaign, for example, sought to block a major road scheme with an alternative plan based on an interconnected series of parks running through the city.

106. Lachmund, *Greening Berlin*.

107. Marianne E. Krasny, Cecilia Lundholm, Soul Shava, Eunju Lee, and Hiromi Kobori, "Urban landscapes as learning arenas for biodiversity and ecosystem services management," in *Urbanization, biodiversity and ecosystem services: challenges and opportunities*, ed. Thomas Elmqvist et al. (Dordrecht: Springer, 2013), 629–664.

108. Richard Primack, Hiromi Kobori, and Seiwa Mori, "Dragonfly pond restoration promotes conservation awareness in Japan," *Conservation Biology* 14 (5) (2000): 1553–1554. See also Kumiko Kiuchi, "Tokyo ecology: the Akabane Nature Observation Park," in *The botanical city*, ed. Matthew Gandy and Sandra Jasper (Berlin: jovis, 2020), 253–265.

109. See Begüm Özkaynak, Cem İskender Aydın, Pınar Ertör-Akyazı, and Irmak Ertör, "The Gezi Park resistance from an environmental justice and social metabolism perspective," *Capitalism Nature Socialism* 26 (1) (2015): 99–114; and Çağan H. Şekercioğlu, Sean Anderson, Erol Akçay, Raşit Bilgin, Özgün Emre Can, Gürkan Semiz, Çağatay Tavşanoğlu, et al., "Turkey's globally important biodiversity in crisis," *Biological Conservation* 144 (12) (2011): 2752–2769. On the destruction of urban neighbourhoods in Istanbul see also the documentary *Rantİstanbul* (dir.: Devrim CK, Alper Şen, 2021).

110. See, for example, Henrik Ernstson, Sverker Sörlin, and Thomas Elmqvist, "Social movements and ecosystem services—the role of social network structure in protecting and managing urban green areas in Stockholm," *Ecology and Society* 13 (2) (2008).

111. Alan Irwin, *Citizen science: a study of people, expertise and sustainable development* (London: Routledge, 1995).

112. Andrew Feenberg, *Transforming technology: a critical theory revisited* (Oxford: Oxford University Press, 2002).

113. Jacques Ellul, *The technological society* (New York: Alfred A. Knopf, 1964), 4.

114. Feenberg, *Transforming technology*.

115. The waging of "ecological warfare" involved the "destruction of vegetation, the transformation of the physical characteristics of the soil, the deliberate precipitation of new erosional processes," and also "the rupture of hydrological systems in order to change the level of the water table (so as to dry up wells and rice paddies)." Yves Lacoste, "An illustration of geographical warfare: bombing of the dikes on the Red River, North Vietnam," in *Radical geography: alternative viewpoints on contemporary social issues*, ed. Richard Peet (London: Methuen, 1978 [1972]), 246. See also Yves Lacoste, *La géographie ça sert d'abord à faire la guerre* (Paris: François Maspero, 1976).

116. Erik Harms, "Knowing into oblivion: clearing wastelands and imagining emptiness in Vietnamese New Urban Zones," *Singapore Journal of Tropical Geography* 35 (3) (2014): 312–327.

117. Jorges Luis Borges, *Collected fictions*, trans. Andrew Hurley (London: Penguin, 2000).

118. Lachmund, *Greening Berlin*. Another city that has pursued long-term biotope mapping programs is Frankfurt am Main. See, for example, Thomas Gregor, Dirk Bönsel, Indra Starke-Ottich, and Georg Zizka, "Drivers of floristic change in large cities—A case study of Frankfurt/Main (Germany)," *Landscape and Urban Planning* 104 (2) (2012): 230–237; Indra Ottich, Dirk Bönsel, Thomas Gregor, Andreas Malten, Georg Zizka, Uwe Barth, Kurt Baumann, and Klaus Hoppe, *Natur vor der Haustür—Stadtnatur in Frankfurt am Main* (Stuttgart: E. Schweizerbart'sche Verlagsbuchhandlung, 2009).

119. See Sabine Barles, *La ville délétère: médecins et ingénieurs dans l'espace urbain XVIIIe–XIXe siècle* (Seyssel: Champ Vallon, 1999); Michael Hebbert and Vladimir Jankovic, "Cities and

climate change: the precedents and why they matter," *Urban Studies* 50 (7) (2013): 1332–1347; and Albert Kratzer, "Das Klima der Städte," *Geographische Zeitschrift* 41 (9) (1935): 321–339.

120. See, for example, L. A. Conrads, "Observations of meteorological urban effects. The heat island of Utrecht" (PhD thesis, Rijksuniversiteit Utrecht, 1975); and A. Peppler, "Die Temperaturverhältnisse von Karlsruhe an heissen Sommertagen," *Deutsches meteorologisches Jahrbuch Baden* 61 (1929): 59–60.

121. Frank Zacharias, "Blühphaseneintritt an Straßenbäumen (insbesondere *Tilia x euchlora* KOCH) und Temperaturverteilung in Westberlin" (PhD dissertation, Technical University, Berlin, 1972).

122. Paul Duvigneaud, "Étude écologique de l'écosystème urbain bruxellois: 1. L'écosystème 'urbs'," *Mémoires de la Société Royale de Botanique de Belgique* 6 (1974): 5–35.

123. Eric W. Sanderson, *Mannahatta: a natural history of New York City* (New York: Abrams, 2009).

124. Sanderson, *Mannahatta*, 68.

125. Elena M. Bennett, Martin Solan, Reinette Biggs, Timon McPhearson, Albert V. Norström, Per Olsson, Laura Pereira, et al., "Bright spots: seeds of a good Anthropocene," *Frontiers in Ecology and the Environment* 14 (8) (2016): 441–448.

126. See also Matthias Kohler, "Aerial architecture," *Log* 25 (2012): 23–30.

127. Hans-Helmut Poppendieck, Gisela Bertram, and Barbara Engelschall, eds., *Der botanische Wanderführer für Hamburg und Umbegung* (Hamburg and Munich: Dörling and Galitz, 2016), 13.

128. Philippe Vasset, *Un livre blanc* (Paris: Fayard, 2007).

129. See Eduardo Brito-Henriques, "Arruinamento e regeneração do espaço edificado na metrópole do século XXI: o caso de Lisboa," *EURE* 43 (128): 251–272; and Eduardo Brito-Henriques, Daniel Paiva, and Pablo Costa, "Cyborg urbanization beyond the human: the construction and ruination of the Matinha gasworks site," *Urban Geography* 40 (10) (2019): 1596–1615. On multisensory and multispecies immersion in marginal spaces see also Lucilla Barchetta, *La rivolta del verde. Nature e rovine a Torino* (Milan: Agenzia X, 2021); and Mathilda Rosengren, "There's life in dead wood: tracing a more-than-human urbanity in the spontaneous nature of Gothenburg," in *The botanical city*, ed. Matthew Gandy and Sandra Jasper (Berlin: jovis, 2020), 229–236. For an ironic extension of taxonomic imaginaries to material traces of modernity see the documentary *Calx ruderalis* subsp. *Istanbulensis* (dir.: Elif Kendir-Beraha, Aslıhan Demirtaş, and Ali Mahmut Demirel, 2021).

130. Daniel Paiva and Eduardo Brito-Henriques, "A podcast on urban ruins, or the aural weaving of theory and field," *Cultural Geographies* 26 (4) (2019): 535–540.

131. The reliability or otherwise of "nonscientific" data remains a focus of contention. See, for example, J. P. Cohn, "Citizen science: can volunteers do real research?," *BioScience* 58 (3) (2008): 192–197.

132. Thordis Arrhenius, "Preservation and protest: counterculture and protest in 1970s Sweden," *Future Anterior* (Winter 2010): 106–123.

133. Stephen Castle, "Toxic tea and other tales from an English tree war," *New York Times* (14 March 2018). See also Matthew Flinders and Matthew Wood, "Ethnographic insights into competing forms of co-production: a case study of the politics of street trees in a northern English city," *Social Policy and Administration* 53 (2) (2019): 279–294; Ian D. Rotheram, "'The city that hates trees': standing up to the Sheffield street tree slaughter," *Ecos* 39 (3) (2018) (accessed 10 August 2020); and Philip B. Whyman, "Street trees, the private finance initiative and participatory regeneration: policy innovation or incompatible perspectives," *Political Quarterly* 91 (1) (2020): 156–164. My reflections on the Sheffield street tree conflict also benefited from site visits and interviews with activists undertaken on 5 September 2017.

134. See Nicolas Soulier, *Reconquérir les rues. Exemples à travers le monde et pistes d'actions* (Paris: Ulmer, 2012).

135. Michel Serres, *The natural contract*, trans. Elizabeth MacArther and William Paulson (Ann Arbor: University of Michigan Press, 1995).

136. Michel Serres, "Revisiting *The natural contract*," lecture given to the Institute of Humanities, trans. Anne-Marie Feenberg-Dibon, Simon Fraser University (4 May 2006).

137. See Erin Fitz-Henry, "The natural contract: from Lévi-Strauss to the Ecuadorian constitutional court," *Oceania* 82 (3) (2012): 264–277.

138. See David R. Boyd, "Recognizing the rights of nature: lofty rhetoric or legal revolution," *Natural Resources and Environment* 32 (4) (2018): 13–17; and T. J. Demos, "Rights of nature: the art and politics of earth jurisprudence," gallery notes for exhibition held at the Nottingham Contemporary, 2015.

139. Brian Favre, "Is there a need for a new, an ecological, understanding of legal animal rights?," *Journal of Human Rights and the Environment* 11 (2) (2020): 297–319.

140. Gunther Teubner, "Rights of non-humans? Electronic agents and animals as new actors in politics and law," *Journal of Law and Society* 33 (4) (2006): 497–521. It is interesting to note that Teubner's use of the term "ecologization" is radically different from that of Deleuzian scholars such as Erich Hörl. See Erich Hörl, "Introduction to general ecology: the ecologization of thinking," in *General ecology: the new ecological paradigm*, ed. Erich Hörl and James Edward Burton (London: Bloomsbury, 2017), 1–74.

141. Teubner, "Rights of non-humans?"

142. See Robyn Eckersley, "Geopolitan democracy in the Anthropocene," *Political Studies* 65 (4) (2017): 983–999.

CHAPTER 5

1. J. G. Ballard, *Memories of the space age* (Sauk City, WI: Arkham House, 1988 [1982]).

2. Amitav Ghosh, *The great derangement: climate change and the unthinkable* (Chicago: University of Chicago Press, 2016), 115.

3. Globally there are more than ten major migratory routes now recognized, with particular conservation emphasis on birds that are dependent on water bodies for their survival en route between destinations. There is evidence that some bird species are opting to partially remain in situ and no longer undertake seasonal migrations: increasing numbers of birds have been observed overwintering in European cities, for example, to take advantage of both warmer temperatures and safety from predators.

4. The Research Centre for Climate Change and Adaptation Research at Chennai's Anna University predicts widespread regional disruption in Tamil Nadu from sea level rise, increasing temperatures, and reduced or erratic rainfall.

5. A. J. Urfi, Monalisa Sen, A. Kalam, and T. Meganathan, "Counting birds in India: methodologies and trends," *Current Science* (December 2005): 1997–2003. See also Richard Grimmett and Tim Inskipp, *Birds of southern India* (London: Christopher Helm, 2005).

6. See Pushpa Arabindoo, "Unprecedented natures? An anatomy of the Chennai floods," *City* 20 (6) (2016): 800–821; and V. Gajendran, "Chennai's peri-urban: accumulation of capital and environmental exploitation," *Environment and Urbanization Asia* 7 (1) (2016): 1–19.

7. Dipesh Chakrabarty, "The climate of history: four theses," *Critical Inquiry* 35 (2) (2009): 197–222.

8. Roy Bhaskar, *A realist theory of science* (London: Verso, 1997 [1975]), 256.

9. See, for example, Cornelius Castoriadis, "Radical imagination and the social instituting imaginary," in *The Castoriadis reader*, ed. David A. Curtis (Oxford: Blackwell, 1997).

10. On Ballard's "topographic" imagination see, for example, W. Warren Wagar, "J. G. Ballard and the transvaluation of utopia," *Science Fiction Studies* (1991): 53–70.

11. Mark Williams, Jan Zalasiewicz, Peter K. Haff, Christian Schwägerl, Anthony D. Barnosky, and Erle C. Ellis, "The Anthropocene biosphere," *Anthropocene Review* 2 (3) (2015): 212.

12. Steffen Richter and Andreas Rötzer, eds., *Dritte Natur: Technik Kapital Umwelt* (Berlin: Matthes & Seitz, 2018).

13. See Anna Tsing, *The mushroom at the end of the world: on the possibility of life in capitalist ruins* (Princeton, NJ: Princeton University Press, 2015).

14. See Dennis Dean, *James Hutton and the history of geology* (Ithaca, NY: Cornell University Press, 2019).

15. See, for example, Henry Gee, *In search of deep time: beyond the fossil record to a new history of life* (Ithaca, NY: Cornell University Press, 2000).

16. Georges Cuvier, *Essay on the theory of the earth*, 2nd ed., trans. Robert Kerr (Cambridge: Cambridge University Press, 2009 [1813]), 3, 4.

17. David Knight, "Introduction," in *The evolution debate 1813–1870*, ed. David Knight, vol. 1: *Georges Cuvier, Essay on the theory of the earth* (London: Routledge, 2003), vii–xiv.

18. Cuvier, *Essay on the theory of the earth*, 6.

19. Julia Pongratz, C. H. Reick, T. Raddatz, and Martin Claussen, "Effects of anthropogenic land cover change on the carbon cycle of the last millennium," *Global Biogeochemical Cycles* 23 (4) (2009).

20. Simon L. Lewis and Mark A. Maslin, "Defining the Anthropocene," *Nature* 519 (7542) (2015): 171–180.

21. See, for example, Arun Saldanha, "A date with destiny: racial capitalism and the beginnings of the Anthropocene," *Environment and Planning D: Society and Space* 38 (1) (2020): 12–34; and Kathryn Yusoff, *A billion black Anthropocenes or none* (Minneapolis: University of Minnesota Press, 2018).

22. Andreas Malm, *Fossil capital: the rise of steam power and the roots of global warming* (London: Verso, 2016), 25.

23. Benjamin Kunkel, "The capitalocene," *London Review of Books* (2 March 2017).

24. Timothy Mitchell, *Carbon democracy: political power in the age of oil* (London: Verso, 2011).

25. J. R. McNeill and Peter Engelke, *The great acceleration: an environmental history of the Anthropocene since 1945* (Cambridge, MA: Harvard University Press, 2014).

26. Jeremy Davies, *The birth of the Anthropocene* (Berkeley: University of California Press, 2015).

27. McNeill and Engelke, *The great acceleration*, 128.

28. Yuval Noah Harari, *Sapiens: a brief history of humankind* (New York: Random House, 2014).

29. Sebastian Conrad, *What is global history?* (Princeton, NJ: Princeton University Press, 2016).

30. Kathryn Yusoff, "Politics of the Anthropocene: formation of the commons as a geologic process," *Antipode* 50 (1) (2018): 255–276. See also Nigel Clark and Yasmin

Gunaratnam, "Earthing the Anthropos? From 'socializing the Anthropocene' to geologizing the social," *European Journal of Social Theory* 20 (1) (2017): 146–163; and Kathryn Yusoff, "Anthropogenesis: origins and endings in the Anthropocene," *Theory, Culture and Society* 33 (2) (2016): 3–28.

31. See, for example, "Manifiesto Antropoceno en Chile: hacia en nuevo Pacto de Convivencia" (Las Cruces, 2017), www.antropoceno.co (accessed 26 October 2020). See also Manuel Tironi, "Lithic abstractions: geophysical operations against the Anthropocene," *Distinktion: Journal of Social Theory* 20 (3) (2019): 284–300.

32. Chris Thomas, *Inheritors of the earth: how nature is thriving in an era of extinction* (London: Allen Lane, 2017), 250.

33. Thomas, *Inheritors of the earth*, 243, 247.

34. Thomas, *Inheritors of the earth*, 107. See also Benjamin Baiser, Julian D. Olden, Sydne Record, Julie L. Lockwood, and Michael L. McKinney, "Pattern and process of biotic homogenization in the New Pangaea," *Proceedings of the Royal Society B: Biological Sciences* 279 (1748) (2012): 4772–4777.

35. Thomas, *Inheritors of the earth*, 103, 242.

36. Emma Marris, *Rambunctious garden: saving nature in a post-wild world* (London: Bloomsbury, 2013), 1.

37. Fredric Jameson, "The brick and the balloon: architecture, idealism and land speculation," *New Left Review* 228 (March 1998): 43.

38. Wendy Wolford, "The Plantationocene: a lusotropical contribution to the theory," *Annals of the American Association of Geographers* (2021).

39. Davies, *The birth of the Anthropocene*, 6.

40. Davies, *The birth of the Anthropocene*, 6.

41. See Joe R. Feagin, "The secondary circuit of capital: office construction in Houston, Texas," *International Journal of Urban and Regional Research* 11 (2) (1987): 172–192; and David Harvey, *Consciousness and the urban experience: studies in the history and theory of capitalist urbanization*, vol. 1 (Baltimore: Johns Hopkins University Press, 1985).

42. David Harvey, *The limits to capital* (Oxford: Blackwell, 1982), 395, cited in Feagin, "The secondary circuit of capital," 188. See also Kevin Loughran, "Urban parks and urban problems: an historical perspective on green space development as a cultural fix," *Urban Studies* 57 (11) (2020): 2321–2338.

43. See Brett Christophers, "Revisiting the urbanization of capital," *Annals of the Association of American Geographers* 101 (6) (2011): 1347–1364.

44. Ernest Mandel, *Late capitalism*, trans. Joris De Bres (London: Verso, 1978 [1972]), 504.

45. Mandel, *Late capitalism*, 504.

46. James R. O'Connor, *Natural causes: essays in ecological Marxism* (London: Guilford, 1998).

47. See, for example, David Wallace-Wells, *The uninhabitable earth: a story of the future* (London: Penguin, 2019).

48. See Erik Swyngedouw, "Into the sea: desalination as hydro-social fix in Spain," *Annals of the Association of American Geographers* 103 (2) (2013): 261–270. See also Erik Swyngedouw and Henrik Ernston, "Interrupting the anthropo-obScene: immuno-biopolitics and depoliticizing ontologies in the Anthropocene," *Theory, Culture and Society* 35 (6) (2018): 3–30.

49. See Melinda E. Cooper, *Life as surplus: biotechnology and capitalism in the neoliberal era* (Seattle: University of Washington Press, 2011).

50. John Asafu-Adjaye, Linus Blomquist, Stewart Brand, Barry W. Brook, Ruth DeFries, Erle Ellis, Christopher Foreman, et al., *An ecomodernist manifesto* (Oakland, CA: Breakthrough Institute, 2015).

51. Erle C. Ellis, Jed O. Kaplan, Dorian Q. Fuller, Steve Vavrus, Kees Klein Goldewijk, and Peter H. Verburg, "Used planet: a global history," *Proceedings of the National Academy of Sciences* 110 (20) (2013): 7978–7985.

52. Swyngedouw and Ernston, "Interrupting the Anthropo-obScene."

53. See Fredric Jameson, *The seeds of time* (New York: Columbia University Press, 1993).

54. Benjamin H. Bratton, *The stack: on software and sovereignty* (Cambridge, MA: MIT Press, 2015), 75–76.

55. Jussi Parikka, *The Anthrobscene* (Minneapolis: University of Minnesota Press, 2014), 6. See also T. J. Demos, *Against the Anthropocene: visual culture and environment today* (Berlin: Sternberg, 2017); and Timothy W. Luke, "Reconstructing social theory and the Anthropocene," *European Journal of Social Theory* 20 (1) (2017): 80–94.

56. Jussi Parikka, "Deep times of planetary trouble," *Cultural Politics* 12 (3) (2016): 279–292.

57. See, for example, Henrik Ernston and Erik Swyngedouw, eds., *Urban political ecology in the anthro-obscene: interruptions and possibilities* (London: Routledge, 2018).

58. Frédéric Neyrat, *The unconstructable earth: an ecology of separation*, trans. Drew S. Burk (New York: Fordham University Press, 2019 [2016]).

59. Jason W. Moore, "Anthropocene or Capitalocene? Nature, history, and the crisis of capitalism," in *Anthropocene or Capitalocene? Nature, history, and the crisis of capitalism*, ed. Jason W. Moore (Oakland, CA: PM Press, 2016), 1–13.

60. See in particular John Bellamy Foster, "Marx's theory of metabolic rift: classical foundations for environmental sociology," *American Journal of Sociology* 105 (2) (1999): 366–405.

61. Justus von Liebig, *Letters on modern agriculture*, ed. John Blyth (London: Walton and Maberly, 1859), 83–84, 132–133.

62. See Baron Haussmann, *Mémoire sur les eaux de Paris, présenté à la commission municipale par m. le préfet de la Seine* (4 August 1854) (Paris: Vinchon) and Baron Haussmann, *Second mémoire sur les eaux de Paris présenté par Le Préfet de la Seine au Conseil Municipal* (16 July 1858) (Paris: Typographie de Charles de Mourgues Fréres).

63. Karl Marx cited in Foster, "Marx's theory of metabolic rift," 379.

64. See Karl Marx, *Capital*, vol. 1 (London: Lawrence & Wishart, 1977 [1887]), 474–475.

65. Jason W. Moore, *Capitalism in the web of life: ecology and the accumulation of capital* (London: Verso, 2015).

66. Erik Swyngedouw, *Social power and the urbanization of water: flows of power* (Oxford: Oxford University Press, 2004), 175.

67. See also Brian M. Napoletano, John Bellamy Foster, Brett Clark, Pedro S. Urquijo, Michael K. McCall, and Jaime Paneque-Gálvez, "Making space in critical environmental geography for the metabolic rift," *Annals of the American Association of Geographers* 109 (6) (2019): 1811–1828.

68. Michael Ekers and Scott Prudham "The socioecological fix: fixed capital, metabolism, and hegemony," *Annals of the American Association of Geographers* 108 (1) (2017): 17–34.

69. See also Sianne Ngai, *Our aesthetic categories: zany, cute, interesting* (Cambridge, MA: Harvard University Press, 2012).

70. Jameson, "The brick and the balloon," 26.

71. Jameson, "The brick and the balloon," 43.

72. See, for example, Michael Hudson, "From Marx to Goldman Sachs: the fictions of fictitious capital, and the financialization of industry," *Critique* 38 (3) (2010): 419–444.

73. Jameson, "The brick and the balloon," 43.

74. Jeff Cox, "Government debt hits record $66 trillion, 80% of global GDP, Fitch says," CNBC (23 January 2019) (accessed 27 April 2020).

75. Christophe Bonneuil and Jean-Baptiste Fressoz, *The shock of the Anthropocene* (London: Verso, 2016), 211.

76. See, for example, David I. Stern, "The environmental Kuznets curve," in *Encyclopedia of energy*, vol. 2, ed. Cutler Cleveland (Amsterdam: Elsevier, 2004), 517–525.

77. See, for example, Michael L. McKinney, "Urbanization as a major cause of biotic homogenization," *Biological Conservation* 127 (2006): 247–260; and Julian D. Olden, "Biotic homogenization: a new research agenda for conservation biogeography," *Journal of Biogeography* 33 (2006): 2027–2039.

78. Robert H. Whittaker, "Evolution and measurement of species diversity," *Taxon* 21 (2–3) (1972): 213–251. More recent contributions to the measurement-of-biodiversity debate include Lou Jost, "Partitioning diversity into independent alpha and beta components," *Ecology* 88 (10) (2007): 2427–2439.

79. Petr Pyšek, "Alien and native species in Central European urban floras: a quantitative comparison," *Journal of Biogeography* 25 (1998): 155–163.

80. Sergei L. Mosyakin and Oksana G. Yavorska, "The nonnative flora of the Kiev (Kyiv) urban area, Ukraine: a checklist and brief analysis," *Urban Habitats* 1 (1) (2002): 45–65.

81. Stefan Zerbe, Ute Maurer, Solveig Schmitz, and Herbert Sukopp, "Biodiversity in Berlin and its potential for nature conservation," *Landscape and Urban Planning* 62 (3) (2003): 139–148.

82. Alfred W. Crosby, *Ecological imperialism: the biological expansion of Europe, 900–1900* (Cambridge: Cambridge University Press, 2004).

83. Marris, *Rambunctious garden.*

84. Henrik Ernstson, "Urban plants and colonial durabilities," in *The botanical city*, ed. Matthew Gandy and Sandra Jasper (Berlin: jovis, 2020), 71–81.

85. Amita Baviskar, "Urban nature and its publics: shades of green in the remaking of Delhi," in *Grounding urban natures: histories and futures of urban ecologies*, ed. Henrik Ernstson and Sverker Sörlin (Cambridge, MA: MIT Press, 2019), 233–246.

86. Mirijam Gaertner, Brendon M. H. Larson, Ulrike M. Irlich, Patricia M. Holmes, Louise Stafford, Brian W. van Wilgen, and David M. Richardson, "Managing invasive species in cities: a framework from Cape Town, South Africa," *Landscape and Urban Planning* 151 (2016): 1–9.

87. Compare Nicholas Holm, "Consider the possum: foes, anti-animals, and colonists in paradise," *Animal Studies Journal* 4 (1) 2015: 32–56, with Peta Carey, "Making possums pay," *New Zealand Geographic* 63 (May–June 2003).

88. See Clare Palmer, *Animal ethics in context* (New York: Columbia University Press, 2010).

89. See Marris, *Rambunctious garden.*

90. Clive A. Stace and Michael J. Crawley, *Alien plants* (London: HarperCollins, 2015), 426.

91. I am grateful to Henriette Steiner for pointing out these connections at the Urban Salon event held at the Bartlett School of Planning, University College London, 18 February 2020. On the cultural and historical complexities of urban forests see also John F. Dwyer, Herbert W. Schroeder, and Paul H. Gobster, "The deep significance of urban trees and forests," in *The ecological city: preserving and restoring urban biodiversity*, ed. Rutherford H. Platt, Rowan A. Rowntree, and Pamela C. Muick (Amherst: University of Massachusetts Press, 1994), 137–150.

92. Gunnar Keppel, Kimberly P. Van Niel, Grant W. Wardell-Johnson, Colin J. Yates, Margaret Byrne, Ladislav Mucina, Antonius G. T. Schut, Stephen D. Hopper, and Steven E. Franklin, "Refugia: identifying and understanding safe havens for biodiversity under climate change," *Global Ecology and Biogeography* 21 (4) (2012): 393–404.

93. Gilles Clément, *Manifeste du tiers paysage* (Paris: Éditions Sujet/Objet, 2004).

94. Matthew Gandy, "Entropy by design: Gilles Clément, Parc Henri Matisse and the limits to avant-garde urbanism," *International Journal of Urban and Regional Research* 37 (1) (2013): 259–278.

95. Maria Thereza Alves, *The long road to Xico. El largo camino a Xico, 1991–2015* (Berlin: Sternberg Press, 2017). See also Mara Polgovsky Ezcurra, "The flight of seeds," in *The botanical city*, ed. Matthew Gandy and Sandra Jasper (Berlin: jovis, 2020), 122–130.

96. See John Hogg, "On the ballast-flora of the coasts of Durham and Northumberland," *Annals and magazine of natural history: zoology, botany, and geology* 19 (3) (1867): 38–43; and Chas. E. Perkins, "Ballast plants in Boston and vicinity," *Botanical Gazette* 8 (3) (1883): 188–190.

97. Alves, *The long road to Xico*. On "spectral materialism" see Eric L. Santner, *On creaturely life: Rilke, Benjamin, Sebald* (Chicago: University of Chicago Press, 2006).

98. On the emerging concept of multispecies justice see also Danielle Celermajer, Sria Chatterjee, Alasdair Cochrane, Stefanie Fishel, Astrida Neimanis, Anne O'Brien, Susan Reid, Krithika Srinivasan, David Schlosberg, and Anik Waldow, "Justice through a multispecies lens," *Contemporary Political Theory* (March 2020): 1–38.

99. Ursula K. Heise, "The case for 'sanctuary cities' for endangered species," citylab (16 June 2018), KCET and the Laboratory for Environmental Strategies (LENS) at the University of California, Los Angeles. See also the documentary *Urban ark Los Angeles* (dir.: Stefan Wanigatunga, 2017).

100. Maike van Stiphout, *First guide to nature inclusive design* (nextcity.nl, 2019).

101. Kate Orff, *Toward an urban ecology* (New York: Monacelli Press, 2016).

102. Matt Hickman, "Plant-covered residential towers in Chengdu, attract mosquitoes, repel tenants," *Architect's Newspaper* (17 September 2020).

103. Jamie Lorimer and Clemens Driessen, "Wild experiments at the Oostvaardersplassen: rethinking environmentalism in the Anthropocene," *Transactions of the Institute of British Geographers* 39 (2) (2014): 169–181.

104. Frans W. M. Vera, *Grazing ecology and forest history* (Wallingford: CABI, 2000). For a reflection on Vera's experimental ecology see, for example, Jonathan Prior and Kim J. Ward, "Rethinking rewilding: a response to Jørgensen," *Geoforum* 69 (2016): 132–135.

105. Herbert Sukopp, Hans-Peter Blume, and Wolfram Kunick, "The soil, flora, and vegetation of Berlin's waste lands," in *Nature in cities: the natural environment in the design and development of urban green space*, ed. Ian Laurie (London: Wiley, 1979), 130.

106. Jelle Reumer, *Wildlife in Rotterdam: nature in the city*, trans. Anthony Runia (Rotterdam: Natuurhistorisch Museum Rotterdam, 2014), 13. A similar argument is made by Steven D. Garber, *The urban naturalist* (New York: John Wiley, 1987).

107. Michael J. Angilletta, S. Wilson Robbie Jr., Amanda C. Niehaus, Michael W. Sears, Carlos A. Navas, and Pedro L. Ribeiro, "Urban physiology: city ants possess high heat tolerance," *PLoS One* 2 (2) (2007): e258.

108. Ingo Kowarik and Ina Säumel, "Biological flora of central Europe: *Ailanthus altissima* (Mill.) Swingle," *Perspectives in Plant Ecology, Evolution and Systematics* 8 (4) (2007): 207–237.

109. See, for example, Bettina Stoetzer, "*Ailanthus altissima*, or the botanical afterlives of European power," in *The botanical city*, ed. Matthew Gandy and Sandra Jasper (Berlin: jovis, 2020), 82–91.

110. Darren Patrick, "Queering the urban forest: invasions, mutualisms, and eco-political creativity with the tree of heaven (*Ailanthus altissima*)," in *Urban forests, trees, and greenspace*, ed. L. Anders Sandberg, Adrina Bardekjian, and Sadia Butt (London: Routledge, 2015), 202.

111. Anne Matthews, *Wild nights: the nature of New York City* (New York: North Point Press, 2001), 7.

112. Matthews, *Wild nights*, 89.

113. Scott Gottlieb, "West Nile virus detected in mosquitoes in Central Park," *Bulletin of the World Health Organization* 78 (9) (2000): 1168.

114. Marris, *Rambunctious garden*.

115. See, for example, Richard D. G. Irvine and Mina Gorji, "John Clare in the Anthropocene," *Cambridge Journal of Anthropology* 31 (1) (2013): 119–132; and Jason Sperb, "The end of Detropia: Fordist nostalgia and the ambivalence of poetic ruins in visions of Detroit," *Journal of American Culture* 39 (2) (2016): 212–227. On recent developments in ecocriticism see, for example, Timothy Clark, *Ecocriticism on the edge: the Anthropocene as a threshold concept*

(London: Bloomsbury, 2015); and David Farrier, *Anthropocene poetics: deep time, sacrifice zones, and extinction* (Minneapolis: University of Minnesota Press, 2019).

116. Franz Hohler, *Die Rückeroberung* (Munich: btb, 2012 [1982]).

117. Alan Weisman, *The world without us* (London: Virgin, 2007).

118. Costas M. Constantinou and Evi Eftychiou, "The Cyprus buffer zone as a socio-ecological landscape," *Satoyama Initiative* 2 (2014).

119. Clifford D. Simak, *City* (Garden City, NY: Doubleday, 1952).

120. Simak, cited in Michael Page, "Evolution and apocalypse in the Golden Age," in *Green planets: ecology and science fiction*, ed. Gerry Canavan and Kim Stanley Roberts (Middletown, CT: Wesleyan University Press, 2014), 48. See also Brian Stableford, "Science fiction and ecology," in *A companion to science fiction*, ed. David Seed (Oxford: Blackwell, 2005); and Darko Suvin, *Metamorphoses of science fiction* (New Haven, CT: Yale University Press, 1979).

121. See Anna Tsing, "The buck, the bull, and the dream of the stag: some unexpected weeds of the Anthropocene," *Suomen Antropologi: Journal of the Finnish Anthropological Society* 42 (1) (2017): 3–21.

122. Articulations of this perspective include Daniel Hannan, interview on the BBC *Politics Today* (16 September 2020), and Matt Ridley, "Why wealth and wildlife go hand in hand," *The Times* (31 October 2016). An interesting point to reflect on is the degree of similarity between elements of climate change skepticism and biodiversity decline skepticism. On the former see Matthew Lockwood, "Right-wing populism and the climate change agenda: exploring the linkages," *Environmental Politics* 27 (4) (2018): 712–732.

123. Ghosh, *The great derangement*, 7.

124. Ghosh, *The great derangement*, 8.

125. Ursula K. Heise, "Science fiction and the time scales of the Anthropocene," *ELH* 86 (2) (2019): 275–304.

126. Agustina Bazterrica, *Tender is the flesh*, trans. Sarah Moses (London: Pushkin Press, 2020 [2017]).

127. Bazterrica, *Tender is the flesh*, 100.

128. Claude Lévi-Strauss, "Siamo tutti cannibali," *La Repubblica* (10 October 1993).

129. See Timothy Morton, *Hyperobjects: philosophy and ecology after the end of the world* (Minneapolis: University of Minnesota Press, 2013).

130. Jeff VanderMeer, *Southern Reach trilogy* (New York: Farrar, Straus, and Giroux, 2014). See also Kaisa Kortekallio, "Becoming-instrument: thinking with Jeff VanderMeer's

Annihilation and Timothy Morton's *Hyperobjects*," in *Reconfiguring human, nonhuman and post-human in literature and culture*, ed. Sanna Karkulehto, Aino-Kaisa Koistinen, and Essi Varis (London: Routledge, 2019), 57–75; David Tompkins, "Weird ecology: on the *Southern Reach* trilogy," *Los Angeles Review of Books* (30 September 2014); and Gry Ulstein, "Brave new weird: anthropocene monsters in Jeff VanderMeer's 'The Southern Reach,'" in *Concentric: Literary and Cultural Studies* 43 (1) (2017): 71–96.

131. See Clark, *Ecocriticism on the edge.*

132. Roger Caillois, "Mimicry and legendary psychasthenia," trans. John Shepley, *October* 31 (1984): 17–32.

133. Caillois, "Mimicry and legendary psychasthenia."

134. See Axelle Karera, "Blackness and the pitfalls of Anthropocene ethics," *Critical Philosophy of Race* 7 (1) (2019): 45.

135. Michael Shellenberger and Ted Nordhaus, eds., *Love your monsters: postenvironmentalism and the Anthropocene* (Oakland, CA: Breakthrough Institute, 2011).

136. Fredric Jameson, "Future city," *New Left Review* 21 (2003): 76.

137. Lauren Beukes, *Zoo city* (London: Penguin, 2018 [2010]).

138. Heise, "Science fiction and the time scales of the Anthropocene," 301. On environmental crisis and shifting modes of literary representation see also Lawrence Buell, *The future of environmental criticism: environmental crisis and literary imagination* (Malden, MA: John Wiley, 2005).

139. Jameson, "Future city."

140. See Andrew Strombeck, "The network and the archive: the specter of imperial management in William Gibson's *Neuromancer*," *Science Fiction Studies* 37 (2) (2010): 275–295. An example of this shift in cultural representations of the future city might be what the novelist and critic Samuel R. Delaney refers to as the "Junk City" emerging in the wake of the postmodern critical realignment of the 1980s. See Samuel R. Delany and R. M. P., "On 'Triton' and other matters: an interview with Samuel R. Delany," *Science Fiction Studies* 17 (3) (1990): 295–324.

141. John Thornhill, "William Gibson—the prophet of cyberspace talks AI and climate collapse," *Financial Times* (13 February 2020). Gibson's original description of cyberspace is elaborated in his novel *Neuromancer* (New York: Ace Books, 1984).

142. Rachel Hutchinson, "Fukasaku Kinji and Kojima Hideo replay Hiroshima: atomic imagery and cross-media memory," *Japanese Studies* 39 (2) (2019): 169–189.

143. See, for example, Johanna Zylinska, *The end of man: a feminist counterapocalypse* (Minnneapolis: University of Minnesota Press, 2018).

144. See, for example, Michael Hauskeller, "My brain, my mind, and I: some philosophical assumptions of mind-uploading," *International Journal of Machine Consciousness* 4 (1) (2012): 187–200.

145. See Jill Bennett, *Empathic vision: affect, trauma, and contemporary art* (Stanford, CA: Stanford University Press, 2005).

146. Jennifer Fay, *Inhospitable world: cinema in the time of the Anthropocene* (New York: Oxford University Press, 2018), 3.

147. Fay, *Inhospitable world*, 4.

148. Fay, *Inhospitable world*, 4. See also Mark B. N. Hansen, "Our predictive condition; or, prediction in the wild," in *The nonhuman turn*, ed. Richard Grusin (Minneapolis: University of Minnesota Press), 101–138.

149. See, for example, Wallace-Wells, *The uninhabitable earth*.

150. See, for example, Mark Elvin, *The retreat of the elephants: an environmental history of China* (New Haven, CT: Yale University Press, 2004); Alan Mikhail, *Nature and empire in Ottoman Egypt: an environmental history* (Cambridge: Cambridge University Press, 2011); and J. R. McNeill, *The mountains of the Mediterranean world: an environmental history* (Cambridge: Cambridge University Press, 2003).

151. Chakrabarty, "The climate of history," 219.

152. In Tamil-language environmental book fairs, a number of key environmental works such as Rachel Carson's *Silent spring* are now available in translation, and there are also many local-language guides to birds and other wildlife.

153. Ross Exo Adams, "Natura urbans, natura urbanata: ecological urbanism, circulation, and the immunization of nature," *Environment and Planning D: Society and Space* 32 (1) (2014): 12–29.

154. See, for example, Derek Woods, "Scale critique for the Anthropocene," *Minnesota Review* 83 (2014): 133–142.

155. Chris Otter, "The technosphere: a new concept for urban studies," *Urban History* 44 (1) (2017): 145–154.

156. Davies, *The birth of the Anthropocene*.

EPILOGUE

1. The comparison with de Chirico's landscapes was made by the artist Grayson Perry during an interview for Channel 4 News, UK (23 April 2020).

2. James Evans, "Ecology in the urban century: power, place, and the abstraction of nature," in *Grounding urban natures: histories and futures of urban ecologies*, ed. Henrik Ernstson and Sverker Sörlin (Cambridge, MA: MIT Press, 2019), 315.

3. Christoph Kueffer, "Plant sciences for the Anthropocene: what can we learn from research in urban areas?," *Plants, People, Planet* (2020): 286–289.

4. See Michele Acuto, Susan Parnell, and Karen C. Seto, "Building a global urban science," *Nature Sustainability* 1 (1) (2018): 2–4; Hillary Angelo and David Wachsmuth, "Why does everyone think cities can save the planet?," *Urban Studies* 57 (11) (2020): 2201–2221; and Marit Rosol, Vincent Béal, and Samuel Mössner, "Greenest cities? The (post-) politics of new urban environmental regimes," *Environment and Planning A: Economy and Space* 49 (8) (2017): 1710–1718.

5. See, for example, Gary Grant, *Ecosystem services come to town: greening cities by working with nature* (Oxford: Wiley-Blackwell, 2012), and Robert McDonald and Timothy Beatley, *Biophilic cities for an urban century: why nature is essential for the success of cities* (Cham: Palgrave / Springer Nature, 2021). The recently created "biophilic cities network" that is committed to "natureful cities" enjoys support from across a range of disciplines.

6. Timon McPhearson, Steward T. A. Pickett, Nancy B. Grimm, Jari Niemelä, Marina Alberti, Thomas Elmqvist, Christiane Weber, Dagmar Haase, Jürgen Breuste, and Salman Qureshi, "Advancing urban ecology toward a science of cities," *BioScience* 66 (3) (2016): 198–212.

7. See in particular Gerald E. Frug, "The city as a legal concept," *Harvard Law Review* 93 (1980): 1057–1154.

8. See Stephen R. Kellert and Edward O. Wilson, eds., *The biophilia hypothesis* (Washington, DC: Island Press, 1993).

9. See Lorraine Daston, *Against nature* (Cambridge, MA: MIT Press, 2018), 57.

10. See, for example, Bram Büscher, Robert Fletcher, Dan Brockington, Chris Sandbrook, William M. Adams, Lisa Campbell, Catherine Corson, et al., "Half-earth or whole earth? Radical ideas for conservation, and their implications," *Oryx* 51 (3) (2017): 407–410; and Rosaleen Duffy, "Waging a war to save biodiversity: the rise of militarized conservation," *International Affairs* 90 (4) (2014): 819–834.

11. For a detailed exposition of the "planetary urbanization" thesis see Neil Brenner, "Theses on urbanization," *Public Culture* 25 (1) (69) (2013): 85–114.

12. See Sebastian Conrad, *What is global history?* (Princeton, NJ: Princeton University Press, 2016).

13. I am indebted to Brian McGrath for making this interesting distinction between "ecology" and "ecologies" at a panel discussion held at the Brussels Ecosystems symposium hosted by the metrolab, Brussels (18 October 2018).

14. Sara Ahmed, *Living a feminist life* (Durham, NC: Duke University Press, 2016).

15. On the incorporation of nonhuman agency in urban planning and design see also Andrea Mubi Brighenti, "The vegetative city," *Culture, Theory and Critique* 59 (3) (2018): 215–231; Creighton Connolly, "From resilience to multi-species flourishing: (re)imagining urban-environmental governance in Penang, Malaysia," *Urban Studies* 57 (7) (2020): 1485–1501; Christian Hunold, "Urban greening and human-wildlife relations in Philadelphia: from animal control to multispecies coexistence?," *Environmental Values* 29 (1) (2020): 67–87; Jonathan Metzger, "Cultivating torment: the cosmopolitics of more-than-human urban planning," *City* 20 (4) (2016): 581–601; and Donna Houston, Jean Hillier, Diana MacCallum, Wendy Steele, and Jason Byrne, "Make kin, not cities! Multispecies entanglements and 'becoming-world' in planning theory," *Planning Theory* 17 (2) (2018): 190–212. On posthuman modes of cohabitation see, for example, Yamini Narayanan and Sumanth Bindumadhav, "'Posthuman cosmopolitanism' for the Anthropocene in India: urbanism and human-snake relations in the Kali Yuga," *Geoforum* 106 (2019): 402–410; and Thom Van Dooren, *The wake of crows: living and dying in shared worlds* (New York: Columbia University Press, 2019).

16. See Matthew Gandy, "Entropy by design: Gilles Clément, Parc Henri Matisse and the limits to avant-garde urbanism," *International Journal of Urban and Regional Research* 37 (1) (2013): 259–278.

17. See, for example, Katy Fulfer, "The capabilities approach to justice and the flourishing of nonsentient life," *Ethics and the Environment* 18 (1) (2013): 19–42; and Anders Schinkel, "Martha Nussbaum on animal rights," *Ethics and the Environment* 13 (1) (2008): 41–69.

18. Jennifer Wolch, "Zoöpolis," *Capitalism, Nature, Socialism* 7 (2) (1996): 21–47.

19. The term "biosphere" emerges around the same time as that of "ecology" and is introduced by the Austrian geologist Eduard Suess in the mid-1870s. On the international dimensions to urban biodiversity policy see also J. A. Puppim de Oliveira, Osman Balaban, Christopher N. H. Doll, Raquel Moreno-Peñaranda, Alexandros Gasparatos, Deljana Iossifova, and Aki Suwa, "Cities and biodiversity: perspectives and governance challenges for implementing the Convention on Biological Diversity (CBD) at the city level," *Biological Conservation* 144 (5) (2011): 1302–1313.

20. Cara Clancy, "London's National Park City," *Ecos* 38 (6) (19 December 2017).

21. Diana Mejía, "Humedales: to build or protect?," *Bogotá Post* (21 March 2016) and https://cerrosdebogota.org/ (accessed 12 January 2021). I would like to thank Nida Rehman for her insights into urban planning in Bogotá.

22. For an articulation of the differences between "applied" and "epistemological" variants of European cosmopolitanism see Ulrich Beck, "The truth of others: a cosmopolitan approach," *Common Knowledge* 10 (3) (2004): 430–449; and Bruno Latour, "Whose cosmos,

which cosmopolitics? Comments on the peace terms of Ulrich Beck," *Common Knowledge* 10 (3) (2004): 450–462.

23. See Ursula K. Heise, *Sense of place and sense of planet: the environmental imagination of the global* (Oxford: Oxford University Press, 2008).

24. Ernest Ingersoll, *Wild neighbors: out-door studies in the United States* (New York: Macmillan, 1902 [1897]), 120.

25. See, for example, Gurminder K. Bhambra, "Whither Europe? Postcolonial versus neocolonial cosmopolitanism," *Interventions* 18 (2) (2016): 187–202; and Tariq Jazeel, "Spatializing difference beyond cosmopolitanism: rethinking planetary futures," *Theory, Culture and Society* 28 (5) (2011): 75–97.

26. Isabelle Stengers, *Cosmopolitics II*, trans. Robert Bononno (Minneapolis: University of Minnesota Press, 2011 [2003]), 355–356.

FILMOGRAPHY

0°00 navigation part 1: a journey across England. Dir.: Simon Faithfull, 2009.

Airborne. Dir.: Shaunak Sen, 2020.

Calx ruderalis subsp. *Istanbulensis*. Dir.: Elif Kendir-Beraha, Aslıhan Demirtaş, and Ali Mahmut Demirel, 2021.

Cockroach. Dir.: Ai Wei Wei, 2020.

Fish tank. Dir.: Andrea Arnold, 2009.

Foxes. Dir.: Lorcan Finnegan, 2011.

Katzelmacher. Dir.: Rainer Werner Fassbinder, 1969.

Killer of sheep. Dir.: Charles Burnett, 1978.

London orbital. Dir.: Chris Petit and Iain Sinclair, 2002.

Natura urbana: the Brachen of Berlin. Dir.: Matthew Gandy, 2017.

Nightingales in Berlin. Dir.: Ville Tanttu, 2019.

Der Pflanzenprofessor. Mit Herbert Sukopp durch Berlin. Dir.: Heiderose Häsler, 2008.

Rantİstanbul. Dir.: Devrim CK, Alper Şen, 2021.

Robinson in ruins. Dir.: Patrick Keiller, 2010.

La sociologie est un sport de combat. Dir.: Pierre Carles, 2001.

Terrain vague. Dir.: Marcel Carné, 1960.

Le transect. Dir.: Paul-Armand Gette, 1974.

Twelve monkeys. Dir.: Terry Gilliam, 1996.
Urban ark Los Angeles. Dir.: Stefan Wanigatunga, 2017.
Wild. Dir.: Nicolette Krebitz, 2016.

BIBLIOGRAPHY

Abadie, Jean-Claude, Camila Andrade, Nathalie Machon, and Emmanuelle Porcher. "On the use of parataxonomy in biodiversity monitoring: a case study on wild flora." *Biodiversity and Conservation* 17 (14) (2008): 3485–3500.

Acampora, Christa Davis. *Contesting Nietzsche*. Chicago: University of Chicago Press, 2013.

Acampora, Ralph. "Nietzsche's feral philosophy: thinking through an animal imaginary." In *A Nietzschean bestiary: becoming animal beyond docile and brutal*, ed. Christa Davis Acampora and Ralph R. Acampora, 1–16. Lanham, MD: Rowman and Littlefield, 2004.

Acampora, Ralph. "Oikos and domus: on constructive co-habitation with other creatures." *Philosophy and Geography* 7 (2) (2004): 219–235.

Acosta, Oscar Zeta. *The revolt of the cockroach people*. San Francisco: Straight Arrow Press, 1973.

Acuto, Michele, Susan Parnell, and Karen C. Seto. "Building a global urban science." *Nature Sustainability* 1, no. 1 (2018): 2–4.

Adams, Ross Exo. "Natura urbans, natura urbanata: ecological urbanism, circulation, and the immunization of nature." *Environment and Planning D: Society and Space* 32 (1) (2014): 12–19.

Adorno, Theodor W. *Aesthetic theory*. Trans. Robert Hullot-Kentor. London: Continuum, 1997 [1970].

Adorno, Theodor W. *Aspects of the new right-wing extremism*. Trans. Wieland Hoban. Cambridge: Polity, 2020 [1967].

Adorno, Theodor W. "The idea of natural history." *Telos* 1984 [1932] (60): 111–124.

Agard-Jones, Vanessa. "Spray." *Somatosphere* (27 May 2014).

Agyeman, Julian. "Black people in a white landscape: social and environmental justice." *Built Environment* (1990) 16 (3): 232–236.

Agyemen, Julian. "Entering cosmopolis: crossing over, hybridity, conciliation and the intercultural city ecosystem." *Minding Nature* 7 (1) (2014): 20–25.

Ahmad, Zarin. "Delhi's meatscapes: cultural politics of meat in a globalizing city." *IIM Kozhikode Society and Management Review* 3 (1) (2014): 21–31.

Ahmed, Sara. *Living a feminist life*. Durham, NC: Duke University Press, 2016.

Ahmed, Sara. *Queer phenomenology: orientation, objects, others*. Durham, NC: Duke University Press, 2006.

Alagona, Peter S. *After the grizzly: endangered species and the politics of place in California*. Berkeley: University of California Press, 2013.

Alberti, Marina. "Eco-evolutionary dynamics in an urbanizing planet." *Trends in Ecology and Evolution* 30 (2) (2015): 114–126.

Alberti, Marina. "Measuring urban sustainability." *Environmental Impact Assessment Review* 16 (4) (1996): 381–424.

Alberti, Marina, Cristian Correa, John M. Marzluff, Andrew P. Hendry, Eric P. Palkovacs, Kiyoko M. Gotanda, Victoria M. Hunt, Travis M. Apgar, and Yuyu Zhou. "Global urban signatures of phenotypic change in animal and plant populations." *Proceedings of the National Academy of Sciences* 114 (34) (2017): 8951–8956.

Alberti, Marina, and John M. Marzluff. "Ecological resilience in urban ecosystems: linking urban patterns to human and ecological functions." *Urban Ecosystems* 7 (2004): 241–265.

Alberti, Marina, John Marzluff, and Victoria M. Hunt. "Urban driven phenotypic changes: empirical observations and theoretical implications for eco-evolutionary feedback." *Philosophical Transactions of the Royal Society B: Biological Sciences* 372 (1712) (2017): 20160029.

Alberti, Marina, John M. Marzluff, Eric Shulenberger, Gordon Bradley, Clare Ryan, and Craig Zumbrunnen. "Integrating humans into ecology: opportunities and challenges for studying urban ecosystems." *BioScience* 53 (12) (2003): 1169–1179.

Albuquerque, Maria de Fatima P. Militão, Wayner V. de Souza, Antônio da Cruz G. Mendes, Tereza M. Lyra, Ricardo A. A. Ximenes, Thália V. B. Araújo, Cynthia Braga, Demócrito B. Miranda-Filho, Celina M. T. Martelli, and Laura C. Rodrigues. "Pyriproxyfen and the microcephaly epidemic in Brazil—an ecological approach to explore the hypothesis of their association." *Memórias do Instituto Oswaldo Cruz* 111 (12) (2016): 774–776.

Ali, Hazrat, Ezzat Khan, and Muhammad Anwar Sajad. "Phytoremediation of heavy metals—concepts and applications." *Chemosphere* 91 (7) (2013): 869–881.

Ali, S. Harris, and Roger Keil. "Global cities and the spread of infectious disease: the case of severe acute respiratory syndrome (SARS) in Toronto, Canada." *Urban Studies* 43 (3) (2006): 491–509.

Allen, Heather K., Justin Donato, Helena Huimi Wang, Karen A. Cloud-Hansen, Julian Davies, and Jo Handelsman. "Call of the wild: antibiotic resistance genes in natural environments." *Nature Reviews Microbiology* 8 (4) (2010): 251–259.

Aloi, Giovanni. "Sorely visible: plants, roots, and national identity." *Plants, People, Planet* 1 (3) (2019): 204–211.

Alves, Maria Thereza. *The long road to Xico. El largo camino a Xico, 1991–2015*. Berlin: Sternberg Press, 2017.

Anderson, Kay. "Culture and nature at the Adelaide Zoo: at the frontiers of 'human' geography." *Transactions of the Institute of British Geographers* (1995): 275–294.

Angelo, Hillary, and David Wachsmuth. "Why does everyone think cities can save the planet?" *Urban Studies* 57 (11) (2020): 2201–2221.

Angilletta, Michael J., S. Wilson Robbie Jr., Amanda C. Niehaus, Michael W. Sears, Carlos A. Navas, and Pedro L. Ribeiro. "Urban physiology: city ants possess high heat tolerance." *PLoS One* 2 (2) (2007): e258.

Arabindoo, Pushpa. "Unprecedented natures? An anatomy of the Chennai floods." *City* 20 (6) (2016): 800–821.

Araujo, Ricardo Vieira, Marcos Roberto Albertini, André Luis Costa-da-Silva, Lincoln Suesdek, Nathália Cristina Soares Franceschi, Nancy Marçal Bastos, Gizelda Katz, et al. "São Paulo urban heat islands have a higher incidence of dengue than other urban areas." *Brazilian Journal of Infectious Diseases* 19 (2) (2015): 146–155.

Arboleda, Martín. *Planetary mine: territories of extraction under late capitalism*. London: Verso, 2020.

Arrhenius, Thordis. "Preservation and protest: counterculture and protest in 1970s Sweden." *Future Anterior* (Winter 2010): 106–123.

Asafu-Adjaye, John, Linus Blomquist, Stewart Brand, Barry W. Brook, Ruth DeFries, Erle Ellis, Christopher Foreman, et al. *An ecomodernist manifesto*. Oakland, CA: Breakthrough Institute, 2015.

Attaway, William. *Blood on the forge*. New York: New York Review Book, 2005 [1941].

Austin, Kevin. "Botanical processes in urban derelict spaces." PhD thesis, University of Birmingham, 2002.

Ayres, Robert U. "Industrial metabolism." *Technology and Environment* (February 1989): 23–49.

Baccini, Peter. "A city's metabolism: towards the sustainable development of urban systems." *Journal of Urban Technology* 4 (2) (1997): 27–39.

Baccini, Peter, and Paul H. Brunner. *Metabolism of the anthroposphere: analysis, evaluation, design.* Cambridge, MA: MIT Press, 2012.

Baiser, Benjamin, Julian D. Olden, Sydne Record, Julie L. Lockwood, and Michael L. McKinney. "Pattern and process of biotic homogenization in the New Pangaea." *Proceedings of the Royal Society B: Biological Sciences* 279 (1748) (2012): 4772–4777.

Baka, Jennifer. "The political construction of wasteland: governmentality, land acquisition and social inequality in South India." *Development and Change* 44 (2) (2013): 409–428.

Balée, William. "The research program of historical ecology." *Annual Review of Anthropology* 35 (1) (2006): 75–98.

Ballard, J. G. *Memories of the space age.* Sauk City, WI: Arkham House, 1988 [1982].

Barchetta, Lucilla. *La rivolta del verde. Nature e rovine a Torino.* Milan: Agenzia X, 2021.

Barles, Sabine. "Écologie territoriale et métabolisme urbain: quelques enjeux de la transition socioécologique," *Revue d'Economie Régionale Urbaine* 5 (2017): 819–836.

Barles, Sabine. *La ville délétère: médecins et ingénieurs dans l'espace urbain XVIIIe–XIXe siècle.* Seyssel: Champ Vallon, 1999.

Barnett, Anthony. *The lure of greatness: England's Brexit and America's Trump.* London: Unbound, 2017.

Barry, Andrew, and Georgina Born. "Interdisciplinarity: reconfigurations of the social and natural sciences." In *Interdisciplinarity: reconfigurations of the social and natural sciences,* ed. Andrew Barry and Georgina Born, 1–56. London: Routledge, 2013.

Barua, Maan. "Nonhuman labour, encounter value, spectacular accumulation: the geographies of a lively commodity." *Transactions of the Institute of British Geographers* 42 (2) (2017): 274–288.

Barua, Maan, and Anindya Sinha. "Animating the urban: an ethological and geographical conversation." *Social and Cultural Geography* 20 (8) (2019): 1160–1180.

Bateman, Phillip W., and Patricia A. Fleming. "Big city life: carnivores in urban environments." *Journal of Zoology* 287 (1) (2012): 1–23.

Battistoni, Alyssa. "Bringing in the work of nature: from natural capital to hybrid labor," *Political Theory* 45 (1) (2017): 5–31.

Baviskar, Amita. "Urban nature and its publics: shades of green in the remaking of Delhi." In *Grounding urban natures: histories and futures of urban ecologies*, ed. Henrik Ernstson and Sverker Sörlin, 233–246. Cambridge, MA: MIT Press, 2019.

Bazterrica, Agustina. *Tender is the flesh*. Trans. Sarah Moses. London: Pushkin Press, 2020 [2017].

Beames, I. "The spread of the fox in the London area." *Ecologist* 2 (2) (1972): 25–26.

Beck, Alan M. *The ecology of stray dogs: a study of free-ranging urban animals*. West Lafayette, IN: Purdue University Press, 1973.

Beck, Ulrich. "The truth of others: a cosmopolitan approach." *Common Knowledge* 10 (3) (2004): 430–449.

Beebe, William. *Unseen life of New York: as a naturalist sees it*. New York: Duell, Sloan and Pearce, 1953.

Beisel, Uli, Ann H. Kelly, and Noémi Tousignant. "Knowing insects: hosts, vectors and companions of science." *Science as Culture* 22 (1) (2013): 1–15.

Benecke, Mark. "A brief history of forensic entomology." *Forensic Science International* 120 (1–2) (2001): 2–14.

Benjamin, Walter. *Beroliniana*. Munich: Koehler and Amelang, 2001 [1932–1938].

Benjamin, Walter. "The Paris of the Second Empire in Baudelaire." Trans. Howard Eiland. In Walter Benjamin, *The writer of modern life: essays on Charles Baudelaire*, ed. Michael W. Jennings, 46–133. Cambridge, MA: Belknap Press, 2006 [1938].

Bennett, Elena M., Martin Solan, Reinette Biggs, Timon McPhearson, Albert V. Norström, Per Olsson, Laura Pereira, et al. "Bright spots: seeds of a good Anthropocene." *Frontiers in Ecology and the Environment* 14 (8) (2016): 441–448.

Bennett, Jane. "The force of things: steps toward an ecology of matter." *Political Theory* 32 (3) (2004): 347–372.

Bennett, Jill. *Empathic vision: affect, trauma, and contemporary art*. Stanford, CA: Stanford University Press, 2005.

Benson, Etienne. "The urbanization of the eastern gray squirrel in the United States." *Journal of American History* 100 (3) (2013): 691–710.

Bergson, Henri. *Creative evolution*. Trans. Arthur Mitchell, ed. Keith Ansell-Pearson, Michael Kolkman, and Michael Vaughan. Basingstoke: Palgrave-Macmillan, 2007 [1907].

Berleant, Arnold. *The aesthetics of environment*. Philadelphia: Temple University Press, 1992.

Berlin, Bernd. *Ethnobiological classification: principles of categorization of plants and animals in traditional societies.* Princeton, NJ: Princeton University Press, 1992.

Berry, Brian J. L., and John D. Kasarda. *Contemporary urban ecology.* New York: Macmillan, 1977.

Berthier, Alizé, Philippe Clergeau, and Richard Raymond. "De la belle exotique à la belle invasive: perceptions et appréciations de la Perruche à collier (*Psittacula krameri*) dans la métropole parisienne." *Annales de géographie* 716 (2017): 408–434.

Bertone, Matthew A., Misha Leong, Keith M. Bayless, Tara L. F. Malow, Robert R. Dunn, and Michelle D. Trautwein. "Arthropods of the great indoors: characterizing diversity inside urban and suburban homes." *PeerJ* 4 (2016): e1582.

Bertram, Georg W. "Two conceptions of second nature." *Open Philosophy* 3 (1) (2020): 68–80.

Beukes, Lauren. *Zoo City.* London: Penguin, 2018 [2010].

Bewell, Alan. *Natures in translation: romanticism and colonial natural history.* Baltimore: Johns Hopkins University Press, 2017.

Bhambra, Gurminder K. "Whither Europe? Postcolonial versus neocolonial cosmopolitanism." *Interventions* 18 (2) (2016): 187–202.

Bhaskar, Roy. *A realist theory of science.* London: Verso, 1997 [1975].

Bhattacharyya, Debjani. *Empire and ecology in the Bengal Delta: the making of Calcutta.* Cambridge: Cambridge University Press, 2018.

Bhuwania, Anuj. *Courting the people: public interest litigation in post-emergency India.* Cambridge: Cambridge University Press, 2017.

Biedermann, Alex. "The role of the subjectivist position in the probabilization of forensic science." *Journal of Forensic Science and Medicine* 1 (2) (2015): 140–148.

Biehler, Dawn Day. "Permeable homes: a historical political ecology of insects and pesticides in US public housing." *Geoforum* 40 (6) (2009): 1014–1023.

Biehler, Dawn Day. *Pests in the city: flies, bedbugs, cockroaches, and rats.* Seattle: University of Washington Press, 2013.

Bird, Adrian. "Perceptions of epigenetics." *Nature* 447 (7143) (2007): 396–398.

Biro, Andrew, ed. *Critical ecologies: The Frankfurt School and contemporary environmental crises.* Toronto: University of Toronto Press, 2011.

Björklund, Mats, Iker Ruiz, and Juan Carlos Senar. "Genetic differentiation in the urban habitat: the great tits (*Parus major*) of the parks of Barcelona city." *Biological Journal of the Linnaean Society* 99 (1) (2010): 9–19.

Blanc, Nathalie. *Les nouvelles esthétiques urbains*. Paris: Armand Colin, 2012.

Blanc, Nathalie. *Vers une esthétique environmentale*. Paris: Éditions Quæ, 2008.

Blanchette, Alex. *Porkopolis: American animality, standardized life, and the factory farm*. Durham, NC: Duke University Press, 2020.

Blomley, Nicholas K. *Law, space, and the geographies of power*. New York: Guilford Press, 1994.

Blumer, George. "Some remarks on the early history of trichinosis (1822–1866)." *Yale Journal of Biology and Medicine* 11 (6) (1939): 581–588.

Boase, Clive. "Bed bugs (Hemiptera: Cimicidae): an evidence-based analysis of the current situation." In *Proceedings of the Sixth International Conference on Urban Pests*, 7–14. Veszprém, Hungary: OOK-Press, 2008.

Böhme, Hartmut. *Aussichten der Natur: Naturästhetik in Wechselwirkung von Natur und Kultur*. Berlin: Matthes & Seitz, 2017.

Bois, Yve-Alain, Michel Feher, Hal Foster, and Eyal Weizman. "On forensic architecture: a conversation with Eyal Weizman." *October* 156 (2016): 116–140.

Boisseron, Bénédicte. *Afro-dog: blackness and the animal question*. New York: Columbia University Press, 2018.

Bon, François. *Paysage fer*. Lagrasse: Verdier, 2000.

Bonnet, Edmond. *Petite flore parisienne*. Paris: Librairie F. Savy, 1883.

Bonneuil, Christophe, and Jean-Baptiste Fressoz. *The shock of the Anthropocene*. Trans. David Fernbach. London: Verso, 2016 [2013].

Bonney, Rick, Caren B. Cooper, Janis L. Dickinson, Steve Kelling, Tina Phillips, Kenneth V. Rosenberg, and Jennifer Shirk. "Citizen science: a developing tool for expanding science knowledge and scientific literacy." *BioScience* 59 (11) (2009): 977–984.

Borges, Jorge Luis. *Collected fictions*. Trans. Andrew Hurley. London: Penguin, 2000.

Borut, Jacob. "Struggles for spaces: where could Jews spend free time in Nazi Germany?" *Leo Baeck Institute Year Book* 56 (1) (2011): 307–350.

Bourdieu, Pierre. "The force of law: toward a sociology of the juridical field." *Hastings Law Journal* 38 (5) (1987): 814–853.

Bourdieu, Pierre. *On television*. Trans. Priscilla Parkhurst Ferguson. New York: New Press, 1998 [1996].

Boyd, David R. "Recognizing the rights of nature: lofty rhetoric or legal revolution." *Natural Resources and Environment* 32 (4) (2018): 13–17.

Boyd, William. "Making meat: science, technology, and American poultry production." *Technology and Culture* 42 (4) (2001): 631–664.

Braczkowski, Alexander R., Christopher J. O'Bryan, Martin J. Stringer, James E. M. Watson, Hugh P. Possingham, and Hawthorne L. Beyer. "Leopards provide public health benefits in Mumbai, India." *Frontiers in Ecology and the Environment* 16 (3) (2018): 176–182.

Brantz, Dorothee. "Animal bodies, human health, and the reform of slaughterhouses in nineteenth-century Berlin." In *Meat, modernity and the rise of the slaughterhouse*, ed. Paula Young Lee, 71–85. Durham: University of New Hampshire Press, 2008.

Brantz, Dorothee. "The urban politics of nature: two centuries of green spaces in Berlin, 1800–2014." In *Green Landscapes in the European City, 1750 to 2010*, ed. Peter Clark, Marjaana Niemi, and Catharina Nolin, 141–159. London: Routledge, 2017.

Brantz, Dorothee, and Sonja Dümpelmann, eds. *Greening the city: urban landscapes in the twentieth century*. Charlottesville: University of Virginia Press, 2011.

Bratton, Benjamin H. *The stack: on software and sovereignty*. Cambridge, MA: MIT Press, 2015.

Braun, Bruce. "Environmental issues: writing a more-than-human urban geography." *Progress in Human Geography* 29 (5) (2005): 635–650.

Braun, Bruce. "A new urban dispositif? Governing life in an age of climate change." *Environment and Planning D: Society and Space* 32 (1) (2014): 49–64.

Braverman, Irus. "Anticipating endangerment: the biopolitics of threatened species lists." *BioSocieties* 12 (2017): 132–157.

Braverman, Irus. "En-listing life: red is the color of threatened species lists." In *Critical animal geographies: politics, intersections, and hierarchies in a multispecies world*, ed. Kathryn Gillespie and Rosemary-Claire Collard, 184–202. London: Routledge/Earthscan, 2015.

Braverman, Irus. "Law's underdog: a call for more-than-human legalities." *Annual Review of Law and Social Science* 14 (2018): 127–144.

Braverman, Irus. *Wild life: the institution of nature*. Stanford, CA: Stanford University Press, 2015.

Brenner, Neil. "Theses on urbanization." *Public Culture* 25 (1) (69) (2013): 85–114.

Bresnihan, Patrick. "John Clare and the manifold commons." *Environmental Humanities* 3 (1) (2013): 71–91.

Brighenti, Andrea Mubi. "The vegetative city." *Culture, Theory and Critique* 59 (3) (2018): 215–231.

Brinkley, Catherine, and Domenic Vitiello. "From farm to nuisance: animal agriculture and the rise of planning regulation." *Journal of Planning History* 13 (2) (2014): 113–135.

Brito-Henriques, Eduardo. "Arruinamento e regeneração do espaço edificado na metrópole do século XXI: o caso de Lisboa." *EURE* 43 (128) (2017): 251–272.

Brito-Henriques, Eduardo, Daniel Paiva, and Pablo Costa. "Cyborg urbanization beyond the human: the construction and ruination of the Matinha gasworks site." *Urban Geography* 40 (10) (2019): 1596–1615.

Broich, Jacqueline Maria, and Daniel Ritter. *Die Stadtbrache als "terrain vague": Geschichte und Theorie eines unbestimmten Zwischenraums in Literatur, Kino und Architektur.* Bielefeld: transcript, 2017.

Brownlow, Alec. "An archaeology of fear and environmental change in Philadelphia." *Geoforum* 37 (2) (2006): 227–245.

Bruno, Giuliana. *Streetwalking on a ruined map: cultural theory and the city films of Elvira Notari.* Princeton, NJ: Princeton University Press, 1993.

Bruyninckx, Joeri. *Listening in the field: recording and the science of birdsong.* Cambridge, MA: MIT Press, 2018.

Bryant, Taimie L. "Sacrificing the sacrifice of animals: legal personhood for animals, the status of animals as property, and the presumed primacy of humans." *Rutgers Law Journal* 39 (2007): 247–330.

Bryson, Michael A. "Empty lots and secret places: Leonard Dubkin's exploration of urban nature in Chicago." *Interdisciplinary Studies in Literature and Environment* 18 (1) (2011): 47–66.

Buell, Lawrence. *The future of environmental criticism: environmental crisis and literary imagination.* Malden, MA: Blackwell, 2005.

Bulkeley, Harriet, and Vanesa Castán Broto. "Government by experiment? Global cities and the governing of climate change." *Transactions of the Institute of British Geographers* 38 (3) (2013): 361–375.

Bullard, Robert D. *Dumping in Dixie: race, class, and environmental quality.* Boulder, CO: Westview Press, 2000.

Buller, Henry. "Reconfiguring wild spaces: the porous boundaries of wild animal geographies." In *Routledge handbook of human-animal studies*, ed. Garry Marvin and Susan McHugh, 233–245. London: Routledge, 2014.

Buller, Henry. "Safe from the wolf: biosecurity, biodiversity, and competing philosophies of nature." *Environment and Planning A* 40 (7) (2008): 1583–1597.

Burckhardt, Lucius. "documenta urbana: sichtbar machen." In *Landschafts-theoretische Aquarelle und Spaziergangs-wissenschaft*, ed. Noah Regenass, Markus Ritter, and Martin Schmitz, 225–254. Berlin: Martin Schmitz, 2017 [1982].

Burckhardt, Lucius. "Wissenschaft ohne Fragen." In *Transect and some other attitudes towards landscape*. Ipswich: European Visual Arts Centre, c. 1988.

Büscher, Bram, Robert Fletcher, Dan Brockington, Chris Sandbrook, William M. Adams, Lisa Campbell, Catherine Corson, et al. "Half-earth or whole earth? Radical ideas for conservation, and their implications." *Oryx* 51 (3) (2017): 407–410.

Byrne, Jason. "When green is White: The cultural politics of race, nature and social exclusion in a Los Angeles urban national park." *Geoforum* 43 (3) (2012): 595–611.

Byrne, Jason, Jennifer Wolch, and Jin Zhang. "Planning for environmental justice in an urban national park." *Journal of Environmental Planning and Management* 52 (3) (2009): 365–392.

Byrne, Katharine, and Richard A. Nichols. "*Culex pipiens* in London Underground tunnels: differentiation between surface and subterranean populations." *Heredity* 82 (1) (1999): 7–15.

Caillois, Roger. "Mimicry and legendary psychasthenia." Trans. John Shepley. *October* 31 (1984): 17–32.

Callicott, J. Baird. "The land aesthetic." *Renewable Resources Journal* 10 (1992): 12–17.

Cameron, Laura. "Histories of disturbance." *Radical History Review* 74 (1999): 5–24.

Campbell, Michael O'Neal. "An animal geography of avian ecology in Glasgow." *Applied Geography* 27 (2) (2007): 78–88.

Campkin, Ben. "Terrors by night: bedbug infestations in London." In *Urban constellations*, ed. Matthew Gandy, 139–144. London: jovis, 2011.

Careri, Francesco. *Walkscapes: walking as aesthetic practice*. Ames, IA: Culicidae Architectural Press, 2017.

Carey, Peta. "Making possums pay." *New Zealand Geographic* 63 (May–June 2003).

Carlson, Allen. "Appreciating art and appreciating nature." In *Landscape, natural beauty and the arts*, ed. Salim Kemal and Ivan Gaskell, 199–227. Cambridge: Cambridge University Press, 1993.

Carlson, Allen. "Nature, aesthetic appreciation, and knowledge." *Journal of Aesthetics and Art Criticism* 53 (4) (1995): 393–400.

Cassidy, Angela, and Brett Mills. "'Fox tots attack shock': urban foxes, mass media and boundary-breaching." *Environmental Communication: A Journal of Nature and Culture* 6 (4) (2012): 494–511.

Castellano, Katey. "Moles, molehills, and common right in John Clare's poetry." *Studies in Romanticism* 56 (2) (2017): 157–176.

Castellano, Katey. "Multispecies work in John Clare's 'Birds nesting' poems." In *Palgrave Advances in John Clare Studies*, ed. Simon Kövesi and Erin Lafford, 179–197. Cham: Palgrave Macmillan/Springer Nature, 2020.

Castoriadis, Cornelius. "Radical imagination and the social instituting imaginary." In *The Castoriadis reader*, ed. David A. Curtis. Oxford: Blackwell, 1997.

Castree, Noel. "Environmental issues: relational ontologies and hybrid politics." *Progress in Human Geography* 27 (2) (2003): 203–211.

Cavin, Joëlle Salomon, and Christian A. Kull. "Invasion ecology goes to town: from disdain to sympathy." *Biological Invasions* 19 (12) (2017): 3471–3487.

Celermajer, Danielle, Sria Chatterjee, Alasdair Cochrane, Stefanie Fishel, Astrida Neimanis, Anne O'Brien, Susan Reid, Krithika Srinivasan, David Schlosberg, and Anik Waldow. "Justice through a multispecies lens." *Contemporary Political Theory* (March 2020): 1–38.

Chakrabarty, Dipesh. "The climate of history: four theses." *Critical Inquiry* 35 (2) (2009): 197–222.

Chan, Ying-kit. "No room to swing a cat? Animal treatment and urban space in Singapore." *Southeast Asian Studies* 5 (2016): 305–329.

Cheptou, Pierre-Olivier, O. Carrue, Soraya Rouifed, and Amélie M. Cantarel. "Rapid evolution of seed dispersal in an urban environment in the weed *Crepis sancta*." *Proceedings of the National Academy of Sciences* 105 (10) (2008): 3796–3799.

Chiesura, Anna. "The role of urban parks for the sustainable city." *Landscape and Urban Planning* 68 (1) (2004): 129–138.

Choy, Tim. *Ecologies of comparison: an ethnography of endangerment in Hong Kong*. Durham, NC: Duke University Press, 2011.

Christophers, Brett. "Revisiting the urbanization of capital." *Annals of the Association of American Geographers* 101 (6) (2011): 1347–1364.

Claborn, John. "From black Marxism to industrial ecosystem: racial and ecological crisis in William Attaway's *Blood on the forge*." *MFS Modern Fiction Studies* 55 (3) (2009): 566–595.

Clancy, Cara. "London's National Park City." *Ecos* 38 (6) (19 December 2017).

Clancy, Cara. "Wild entanglements: exploring the visions and dilemmas of 'renaturing' urban Britain." PhD dissertation, University of Plymouth, 2019.

Clancy, Cara, and Kim Ward. "Auto-rewilding in post-industrial cities: the case of inland cormorants in urban Britain." *Conservation and Society* 18 (2) (2020): 126–136.

Clark, Nigel. "'Botanizing on the asphalt'? The complex life of cosmopolitan bodies." *Body and Society* 6 (3–4) (2000): 12–33.

Clark, Nigel. "The demon-seed: bioinvasion as the unsettling of environmental cosmopolitanism." *Theory, Culture and Society* 19 (1–2) (2002): 101–125.

Clark, Nigel, and Yasmin Gunaratnam. "Earthing the Anthropos? From 'socializing the Anthropocene' to geologizing the social." *European Journal of Social Theory* 20 (1) (2017): 146–163.

Clark, Timothy. *Ecocriticism on the edge: the Anthropocene as a threshold concept*. London: Bloomsbury, 2015.

Clément, Gilles. *Manifeste du tiers paysage*. Paris: Éditions Sujet/Objet, 2004.

Clements, Frederic Edward. *Research methods in ecology*. Lincoln, NE: University Publishing Company, 1905.

Cohn, Jeffrey P. "Citizen science: can volunteers do real research?" *BioScience* 58 (3) (2008): 192–197.

Colebrook, Claire. "On the very possibility of queer theory." In *Deleuze and queer theory*, ed. Chrysanthi Nigianni and Merl Storr, 11–23. Edinburgh: Edinburgh University Press, 2009.

Coleman, Loren. "Alligators-in-the-sewers: a journalistic origin." *Journal of American Folklore* 92 (365) (1979): 335–338.

Collard, Rosemary-Claire. "Cougar-human entanglements and the biopolitical un/making of safe space." *Environment and Planning D: Society and Space* 30 (1) (2012): 23–42.

Collard, Rosemary-Claire. "Putting animals back together, taking commodities apart." *Annals of the Association of American Geographers* 104 (1) (2014): 151–165.

Collin, Michèle. "Nouvelles urbanités des friches." *Multitudes* 6 (2001): 148–155.

Collins, James P., Ann Kinzig, Nancy B. Grimm, William F. Fagan, Diane Hope, Jianguo Wu, and Elizabeth T. Borer. "A new urban ecology: modeling human communities as integral parts of ecosystems poses special problems for the development and testing of ecological theory." *American Scientist* 88 (5) (2000): 416–425.

Collins, Patricia Hill. *Intersectionality*. Cambridge: Polity Press, 2016.

Combs, Matthew, Kaylee A. Byers, Bruno M. Ghersi, Michael J. Blum, Adalgisa Caccone, Federico Costa, Chelsea G. Himsworth, Jonathan L. Richardson, and Jason Munshi-South. "Urban rat races: spatial population genomics of brown rats (*Rattus norvegicus*) compared across multiple cities." *Proceedings of the Royal Society B: Biological Sciences* 285 (1880) (2018): 20180245.

Connolly, Creighton. "From resilience to multi-species flourishing: (re)imagining urban-environmental governance in Penang, Malaysia." *Urban Studies* 57 (7) (2020): 1485–1501.

Connolly, Creighton, Roger Keil, and S. Harris Ali. "Extended urbanisation and the spatialities of infectious disease: demographic change, infrastructure and governance." *Urban Studies* 58 (2) (2021): 245–263.

Conrad, Cathy C., and Krista G. Hilchey. "A review of citizen science and community-based environmental monitoring: issues and opportunities." *Environmental Monitoring and Assessment* 176 (1–4) (2011): 273–291.

Conrad, Sebastian. *Deutsche Kolonialgeschichte*. Munich: C. H. Beck, 2008.

Conrad, Sebastian. *What is global history?* Princeton, NJ: Princeton University Press, 2016.

Conrads, L. A. "Observations of meteorological urban effects. The heat island of Utrecht." PhD thesis, Rijksuniversiteit Utrecht, 1975.

Cook, Laurence M., Bruce S. Grant, Ilik J. Saccheri, and Jim Mallet. "Selective bird predation on the peppered moth: the last experiment of Michael Majerus." *Biology Letters* 8 (4) (2012): 609–612.

Cooper, Melinda E. *Life as surplus: biotechnology and capitalism in the neoliberal era*. Seattle: University of Washington Press, 2011.

Cope, Jennifer, Raoult C. Ratard, Vincent R. Hill, Theresa Sokol, Jonathan Jake Causey, Jonathan S. Yoder, Gayatri Mirani, et al. "The first association of a primary amebic meningoencephalitis death with culturable *Naegleria fowleri* in tap water from a US treated public drinking water system." *Clinical Infectious Diseases* 60 (8) (2015): e36–e42.

Cosgrove, Denis. "Prospect, perspective and the evolution of the landscape idea." *Transactions of the Institute of British Geographers* 10 (1) (1985): 45–62.

Cotoi, Calin. "We should have asked what year we were in! Wastelands and wilderness in the Văcăreşti Park." *Antipode* 53 (2021).

Cronon, William. *Nature's metropolis: Chicago and the great west*. New York: W. W. Norton, 1991.

Crosby, Alfred W. *Ecological imperialism: the biological expansion of Europe, 900–1900*. Cambridge: Cambridge University Press, 2004.

Cuvier, Georges. *Essay on the theory of the earth*. 2nd ed., trans. Robert Kerr. Cambridge: Cambridge University Press, 2009 [1813].

Dagenais, Danielle. "The garden of movement: ecological rhetoric in support of gardening practice." *Studies in the History of Gardens and Designed Landscapes* 24 (4) (2004): 313–340.

Dagg, Anne I. "Wildlife in an urban area." *Naturaliste Canadienne* 97 (1970): 201–212.

Daley, Robert. *The World Beneath the City*. Philadelphia: Lippincott, 1959.

Daniel, Terry C. "Whither scenic beauty? Visual landscape quality in the 21st century." *Landscape and Urban Planning* 54 (2001): 267–281.

Danneels, Koenraad. "From sociobiology to urban metabolism: the interaction of urbanism, science and politics in Brussels (1900–1978)." PhD thesis, University of Antwerp and KU Leuven, 2021.

Dansereau, Pierre. "Les dimensions écologiques de l'espace urbain." *Cahiers de géographie du Québec* 31 (84) (1987): 333–395.

Daston, Lorraine. *Against nature.* Cambridge, MA: MIT Press, 2019.

Daston, Lorraine, and Peter Galison. *Objectivity.* 2nd ed. New York: Zone Books, 2010 [2007].

Davey, Stuart. "Role of wildlife in an urban environment." In *Transactions of the Thirty-Second North American Wildlife and Natural Resources Conference*, ed. J. B. Trefethen, 50–59. Washington, DC: Wildlife Management Institute, 1967.

Davies, Jeremy. *The birth of the Anthropocene.* Berkeley: University of California Press, 2015.

Davies, Thom. "Toxic space and time: slow violence, necropolitics, and petrochemical pollution." *Annals of the American Association of Geographers* 108 (6) (2018): 1537–1553.

Davis, Lucy. "Zones of contagion: the Singapore body politic and the body of the street-cat." In *Considering animals: contemporary studies in human-animal relations*, ed. Carol Freeman, Elizabeth Leane, and Yvette Watt, 183–198. London: Routledge, 2011.

Davis, Mike. *Ecology of fear: Los Angeles and the imagination of disaster.* New York: Metropolitan Books, 1998.

Davis, Mike. *The monster at our door: the global threat of avian flu.* New York: New Press, 2005.

D'Costa, Vanessa M., Emma Griffiths, and Gerard D. Wright. "Expanding the soil antibiotic resistome: exploring environmental diversity." *Current Opinion in Microbiology* 10 (5) (2007): 481–489.

Deakin, Richard. *Flora of the Colosseum of Rome; or illustrations and descriptions of four hundred and twenty plants growing spontaneously upon the ruins of the Colosseum of Rome.* London: Groombridge and sons, 1855.

Dean, Dennis. *James Hutton and the history of geology.* Ithaca, NY: Cornell University Press, 2019.

de Araújo, Thalia Velho Barreto, Ricardo Arraes de Alencar Ximenes, Demócrito de Barros Miranda-Filho, Wayner Vieira Souza, Ulisses Ramos Montarroyos, Ana Paula Lopes de Melo, Sandra Valongueiro, et al. "Association between microcephaly, Zika virus infection, and other risk factors in Brazil: final report of a case-control study." *Lancet Infectious Diseases* 18 (3) (2018): 328–336.

DeCandia, Alexandra L., Carol S. Henger, Amelia Krause, Linda J. Gormezano, Mark Weckel, Christopher Nagy, Jason Munshi-South, and Bridgett M. von Holdt. "Genetics of urban colonization: neutral and adaptive variation in coyotes (*Canis latrans*) inhabiting the New York metropolitan area." *Journal of Urban Ecology* 5 (1) (2019).

de la Bellacasa, María Puig. *Matters of care: speculative ethics in more than human worlds.* Minneapolis: University of Minnesota Press, 2017.

Delany, Samuel R. "On 'Triton' and other matters: an interview with Samuel R. Delany." *Science Fiction Studies* 17 (3) (1990): 295–324.

Deleuze, Gilles. "Bergson's conception of difference." In *The new Bergson*, ed. John Mullarkey, 42–65. Manchester: Manchester University Press, 1999 [1956].

Del Tredici, Peter. "Spontaneous urban vegetation: reflections of change in a globalizing world." *Nature and Culture* 5 (3) (2010): 299–315.

DeLuca, Kevin, and Anne Demo. "Imagining nature and erasing class and race: Carleton Watkins, John Muir, and the construction of wilderness." *Environmental History* 6 (4) (2001): 541–560.

Demos, T. J. *Against the Anthropocene: visual culture and environment today.* Berlin: Sternberg, 2017.

Demos, T. J. "The politics of sustainability: art and ecology." In *Radical nature: art and architecture for a changing planet, 1969–2009*, ed. Francesco Manacorda and Ariella Yedgar, 17–30. London: Koenig/Barbican, 2009.

Demos, T. J. "Rights of nature: the art and politics of earth jurisprudence." Gallery notes for exhibition held at the Nottingham Contemporary, 2015.

Denizen, Seth. "The flora of bombed areas (an allegorical key)." In *The botanical city*, ed. Matthew Gandy and Sandra Jasper, 38–45. Berlin: jovis, 2020.

d'Eramo, Marco. *The pig and the skyscraper. Chicago: a history of our future.* Trans. Graeme Thomson. London: Verso, 2002.

Descola, Philippe. *Beyond nature and culture.* Chicago: University of Chicago Press, 2013.

Descola, Philippe. "The difficult art of composing worlds (and of replying to objections)." *HAU: Journal of Ethnographic Theory* 4 (3) (2014): 431–443.

Descola, Philippe. *The ecology of others.* Trans. Geneviève Godbout and Benjamin P. Luley. Chicago: Prickly Paradigm Press, 2013 [2007].

Desimini, Jill. "Notions of nature and a model for managed urban wilds." In *Terrain vague: interstices at the edge of the pale*, ed. Patrick Barron and Manuela Mariani, 173–186. London: Routledge, 2014.

de Solà-Morales Rubió, Ignasi. "Terrain vague." In *Anyplace*, ed. Cynthia Davidson, 118–123. Cambridge, MA: MIT Press, 1993.

Deutsche, Rosalyn. *Evictions: art and spatial politics*. Cambridge, MA: MIT Press, 1996.

Devos, Yannick, Luc Vrydaghs, Ann Degraeve, and Sylvianne Modrie. "Unravelling urban stratigraphy. The study of Brussels'(Belgium) Dark Earth. An archaeopedological perspective." *Medieval and Modern Matters* 2 (2011): 51–76.

de Vries, Herman. "Terrain vague." In *No art—no city! Stadtutopien in der zeitgenössischen kunst*, ed. Florian Matzner, Hans-Joachim Manske, and Rose Pfister, 151–153. Ostfildern: Hatje Cantz, 2003.

Dickinson, Janis L., Benjamin Zuckerberg, and David N. Bonter. "Citizen science as an ecological research tool: challenges and benefits." *Annual Review of Ecology, Evolution, and Systematics* 41 (2010): 149–172.

Dikeç, Mustafa. *Badlands of the republic: space, politics, and urban policy*. Oxford: Blackwell, 2007.

Di Luca, Marco, Luciano Toma, Daniela Boccolini, Francesco Severini, Giuseppe La Rosa, Giada Minelli, Gioia Bongiorno, et al. "Ecological distribution and CQ11 genetic structure of *Culex pipiens* complex (Diptera: Culicidae) in Italy." *PLoS One* 11 (1) (2016): e0146476.

Di Palma, Vittoria. *Wasteland: a history*. New Haven, CT: Yale University Press, 2014.

Doherty, Jacob. "Filthy flourishing: para-sites, animal infrastructure, and the waste frontier in Kampala." *Current Anthropology* 60 (S20) (2019): S321–S332.

Donald, Diana. "'Beastly Sights': the treatment of animals as a moral theme in representations of London c. 1820–1850." *Art History* 22 (4) (1999): 514–544.

Donaldson, Sue, and Will Kymlicka. *Zoopolis: a political theory of animal rights*. Oxford: Oxford University Press, 2011.

Dotson, Kristie. "Tracking epistemic violence, tracking practices of silencing." *Hypatia* 26 (2) (2011): 236–257.

Dotson, Kristie, and Kyle Whyte. "Environmental justice, unknowability and unqualified affectability." *Ethics and the Environment* 18 (2) (2013): 55–79.

Draus, Paul, and Juliette Roddy. "Weeds, pheasants and wild dogs: resituating the ecological paradigm in postindustrial Detroit." *International Journal of Urban and Regional Research* 42 (5) 2018: 807–827.

Drenthen, Martin. "The return of the wild in the Anthropocene: wolf resurgence in the Netherlands." *Ethics, Policy and Environment* 18 (3) (2015): 318–337.

Drenthen, Martin, Jozef Keulartz, and James Proctor, eds. *New visions of nature: complexity and authenticity*. Dordrecht: Springer, 2009.

Drouin, Jean-Marc, and Bernadette Bensaude-Vincent. "Nature for the people." In *Cultures of natural history*, ed. Nick Jardine, Jim A. Secord, and Emma C. Spary, 408–425. Cambridge: Cambridge University Press, 1996.

Drouin, Jean-Marc. "Paul Jovet: les concepts de l'écologie végétale à l'épreuve de la ville." In *Sauvages dans la ville. Actes du colloque, organisé pour le centenaire de la naissance de Paul Jovet*, ed. Bernadette Lizet, Anne-Élisabeth Wolf, and John Celecia, 75–90. Paris: JATBA/Publications scientifiques du MNHN, 1997.

Duany, Andrés, and Emily Talen. "Transect planning." *Journal of the American Planning Association* 68 (3) (2002): 245–266.

Dubkin, Leonard. *The natural history of a yard*. Chicago: Henry Regnery, 1955.

Dudley, Nigel. *Authenticity in nature: making choices about the naturalness of ecosystems*. London: Earthscan, 2011.

Duffy, Rosaleen. "Waging a war to save biodiversity: the rise of militarized conservation." *International Affairs* 90 (4) (2014): 819–834.

Duijzings, Ger. "Dictators, dogs, and survival in a post-totalitarian city." In *Urban constellations*, ed. Matthew Gandy, 145–148. Berlin: jovis, 2011.

Duvigneaud, Paul. "Étude écologique de l'écosystème urbain bruxellois: 1. L'écosystème 'urbs.'" *Mémoires de la Société Royale de Botanique de Belgique* 6 (1974): 5–35.

Dwyer, John F., Herbert W. Schroeder, and Paul H. Gobster. "The deep significance of urban trees and forests." In *The ecological city: preserving and restoring urban biodiversity*, ed. Rutherford H. Platt, Rowan A. Rowntree, and Pamela C. Muick, 137–150. Amherst: University of Massachusetts Press, 1994.

Ebner, Paulus. "Nützen und Schützen. Städtischer Tierschutz im 19. Jahrhundert." In *Umwelt Stadt. Geschichte des Natur- und Lebensraumes Wien*, ed. Karl Brunner and Petra Schneider, 433–437. Vienna: Böhlau, 2005.

Eckersley, Robyn. "Geopolitan democracy in the Anthropocene." *Political Studies* 65 (4) (2017): 983–999.

Eddington, John. "Early London botanists: '. . . in the fieldes aboute London, plentuously . . . .'" *London Naturalist* 90 (2011): 21–45.

Edensor, Tim. *Industrial ruins: spaces, aesthetics and materiality*. London: Berg, 2005.

Edwards, Michael. "London for sale." In *Urban Constellations*, ed. Matthew Gandy, 54–57. Berlin: jovis, 2011.

Ekers, Michael, and Scott Prudham. "The socioecological fix: fixed capital, metabolism, and hegemony." *Annals of the American Association of Geographers* 108 (1) (2017): 17–34.

Elkin, Lauren. *Flâneuse: women walk the city in Paris, New York, Tokyo, Venice, and London*. London: Chatto and Windus, 2016.

Ellis, Erle C., Jed O. Kaplan, Dorian Q. Fuller, Steve Vavrus, Kees Klein Goldewijk, and Peter H. Verburg. "Used planet: a global history." *Proceedings of the National Academy of Sciences* 110 (20) (2013): 7978–7985.

Ellis, Erle C., and Navin Ramankutty. "Putting people in the map: anthropogenic biomes of the world." *Frontiers in Ecology and the Environment* 6 (8) (2008): 439–447.

Ellis, Rebecca, and Claire Waterton. "Caught between the cartographic and the ethnographic imagination: the whereabouts of amateurs, professionals, and nature in knowing biodiversity." *Environment and Planning D: Society and Space* 23 (2005): 673–693.

Ellul, Jacques. *The technological society.* New York: Alfred A. Knopf, 1964.

Elvin, Mark. *The retreat of the elephants: an environmental history of China.* New Haven, CT: Yale University Press, 2004.

Emel, Jody, and Harvey Neo. *Political ecology of meat.* London: Routledge, 2015.

Endlicher, Wilfried. *Einführung in die Stadtökologie.* Stuttgart: Ulmer, 2012.

Ernstson, Henrik. "Urban plants and colonial durabilities." In *The botanical city,* ed. Matthew Gandy and Sandra Jasper, 71–81. Berlin: jovis, 2020.

Ernstson, Henrik, and Sverker Sörlin, eds. *Grounding urban natures: histories and futures of urban ecologies.* Cambridge, MA: MIT Press, 2019.

Ernstson, Henrik, Sverker Sörlin, and Thomas Elmqvist. "Social movements and ecosystem services—the role of social network structure in protecting and managing urban green areas in Stockholm." *Ecology and Society* 13 (2) (2008).

Ernstson, Henrik, and Erik Swyngedouw, eds. *Urban political ecology in the anthro-obscene: interruptions and possibilities.* London: Routledge, 2018.

Ernwein, Marion. "Bringing urban parks to life: the more-than-human politics of urban ecological work." *Annals of the American Association of Geographers* 111 (2) (2020): 559–576.

Ernwein, Marion. *Les natures de la ville néolibérale. Une écologie politique du végétal urbain.* Grenoble: UGA Éditions, 2019.

Escobar, Arturo. "After nature: steps to an antiessentialist political ecology." *Current Anthropology* 40 (1) (1999): 1–30.

Escobar, Arturo. *Designs for the pluriverse: radical interdependence, autonomy, and the making of worlds.* Durham, NC: Duke University Press, 2018.

Eser, Ute. *Der Naturschutz und das Fremde: ökologische und normative Grundlagen der Umweltethik.* Frankfurt: Campus, 1999.

Esposito, Roberto. *Bíos: politics and philosophy.* Trans. T. Campbell. Minneapolis: University of Minnesota Press, 2008 [2004].

Evans, Alfred B. "Protests and civil society in Russia: the struggle for the Khimki forest." *Communist and Post-Communist Studies* 45 (3–4) (2012): 233–242.

Evans, David, ed. *The art of walking*. London: Black Dog, 2012.

Evans, James. "Ecology in the urban century: power, place, and the abstraction of nature." In *Grounding urban natures: histories and futures of urban ecologies*, ed. Henrik Ernstson and Sverker Sörlin, 303–322. Cambridge, MA: MIT Press, 2019.

Evans, James. "Resilience, ecology and adaptation in the experimental city." *Transactions of the Institute of British Geographers* 36 (2011): 223–237.

Evans, Jennifer. "Life among the ruins: sex, space, and subculture in zero hour Berlin." In *Berlin: divided city, 1945–1989*, ed. Philip Broadbent and Sabine Hake, 11–22. New York: Berghahn, 2010.

Evans, Karl L., Kevin J. Gaston, Alain C. Frantz, Michelle Simeoni, Stewart P. Sharp, Andrew McGowan, Deborah A. Dawson, K. Walasz, et al. "Independent colonization of multiple urban centres by a formerly forest specialist bird species." *Proceedings of the Royal Society B* 276 (2009): 2403–2410.

Evans, Richard J. *Death in Hamburg: society and politics in the cholera years, 1830–1910*. Oxford: Clarendon Press, 1987.

Ewald, Klaus C. "The neglect of aesthetics in landscape planning in Switzerland." *Landscape and Urban Planning* 54 (2011): 255–266.

Eyre, M., M. Luff, and J. Woodward. "Beetles (Coleoptera) on brownfield sites in England: an important conservation resource?" *Journal of Insect Conservation* 7 (4) (2003): 223–231.

Falchi, Fabio. "The new world atlas of artificial night sky brightness." *Science Advances* 2 (2016): e1600377.

Falck, Zachary J. S. *Weeds: an environmental history of metropolitan America*. Pittsburgh: University of Pittsburgh Press, 2010.

Farrier, David. *Anthropocene poetics: deep time, sacrifice zones, and extinction*. Minneapolis: University of Minnesota Press, 2019.

Favre, Brian. "Is there a need for a new, an ecological, understanding of legal animal rights?" *Journal of Human Rights and the Environment* 11 (2) (2020): 297–319.

Fay, Jennifer. *Inhospitable world: cinema in the time of the Anthropocene*. New York: Oxford University Press, 2018.

Feagan, Robert B., and Michael Ripmeester. "Contesting natural(ized) lawns: a geography of private green space in the Niagara region." *Urban Geography* 20 (7) (1999): 617–634.

Feagin, Joe R. "The secondary circuit of capital: office construction in Houston, Texas." *International Journal of Urban and Regional Research* 11 (2) (1987): 172–192.

Feenberg, Andrew. *Technosystem: the social life of reason.* Cambridge, MA: Harvard University Press, 2017.

Feenberg, Andrew. *Transforming technology: a critical theory revisited.* Oxford: Oxford University Press, 2002.

Fehn, Klaus. "'Germanisch-deutsche Kulturlandschaft'—Historische Geographie und NS-Forschung." *Petermanns geographische Mitteilungen* 146 (6) (2002): 64–69.

Feinstein, Julie. *Field guide to urban wildlife.* Mechanicsburg, PA: Stackpole, 2011.

Fenton, James. "Native or alien." *Ecos* 7 (2) (1986): 20–23.

Filoche, Sébastien, Gérard Arnal, and Jacques Moret. *La biodiversité du département de la Seine-Saint-Denis. Atlas de la flore sauvage.* Mèze: Biotope; Paris: Muséum National d'Histoire Naturelle, 2006.

Finney, Carolyn. *Black faces, white spaces: reimagining the relationship of African Americans to the great outdoors.* Chapel Hill: University of North Carolina Press, 2014.

Fischer, Leonie K., Verena Rodorff, Moritz von der Lippe, and Ingo Kowarik. "Drivers of biodiversity patterns in parks of a growing South American megacity." *Urban Ecosystems* 19 (2016): 1–19.

Fischer-Kowalski, Marina, and Walter Hüttler. "Society's metabolism: the intellectual history of materials flow analysis, part II, 1970–1998." *Journal of Industrial Ecology* 2 (4) (1998): 107–136.

Fischer-Kowalski, Marina, and Jan Rotmans. "Conceptualizing, observing, and influencing social-ecological transitions." *Ecology and Society* 14 (2) (2009): 3.

Fischer-Kowalski, Marina, and Helga Weisz. "The archipelago of social ecology and the island of the Vienna school." In *Social ecology: society-nature relations across time and space,* ed. Helmut Haberl, Marina Fischer-Kowalski, Fridolin Krausmann, and Verena Winiwarter, 3–28. Basel: Springer, 2016.

Fisher, Dana R., and William R. Freudenburg. "Ecological modernization and its critics: assessing the past and looking toward the future." *Society and Natural Resources* 14 (8) (2001): 701–709.

Fitter, Richard S. R. *London's natural history.* London: Collins, 1945.

Fitz-Henry, Erin. "The natural contract: from Lévi-Strauss to the Ecuadorian constitutional court." *Oceania* 82 (3) (2012): 264–277.

Flinders, Matthew, and Matthew Wood. "Ethnographic insights into competing forms of co-production: a case study of the politics of street trees in a northern English city." *Social Policy and Administration* 53 (2) (2019): 279–294.

Forman, Richard T. T. *Land mosaics: the ecology of landscapes and regions.* Cambridge: Cambridge University Press, 1995.

Forsberg, Kevin J., Alejandro Reyes, Bin Wang, Elizabeth M. Selleck, Morten O. A. Sommer, and Gautam Dantas. "The shared antibiotic resistome of soil bacteria and human pathogens." *Science* 337 (6098) (2012): 1107–1111.

Foster, Cheryl. "The narrative and the ambient in environmental aesthetics." *Journal of Aesthetics and Art Criticism* 56 (2) (1998): 127–137.

Foster, John Bellamy. "Marx's theory of metabolic rift: classical foundations for environmental sociology." *American Journal of Sociology* 105 (2) (1999): 366–405.

Francis, Robert A., Jamie Lorimer, and Mike Raco. "Urban ecosystems as 'natural' homes for biogeographical boundary crossings." *Transactions of the Institute of British Geographers* 37 (2) (2012): 183–190.

Franck, Karen A., and Quentin Stevens, eds. *Loose space: possibility and diversity in urban life.* London: Routledge, 2007.

Frank, Susanne. "Rückkehr der Natur. Die Neuerfindung von Natur und Landschaft in der Emscherzone." *EMSCHERplayer* (October 2010).

Franz, Martin, Orhan Güles, and Gisela Prey. "Place-making and 'green' reuses of brownfields in the Ruhr." *Tijdschrift voor Economische en Sociale Geografie* 99 (3) (2008): 316–328.

Fricker, Miranda. *Epistemic injustice: power and the ethics of knowing.* New York: Oxford University Press, 2007.

Frisby, David. "The metropolis as text: Otto Wagner and Vienna's 'second renaissance.'" *Culture, Theory and Critique* 40 (1) (1997): 1–16.

Frost, Mark. "Entering the 'circles of vitality': beauty, sympathy, and fellowship." In *Vital beauty: reclaiming aesthetics in the tangle of technology and nature,* ed. Joke Brouwer, Arne Mulder, and Lars Spuybroek, 132–153. Rotterdam: V2_Publishing, 2012.

Frug, Gerald E. "The city as a legal concept." *Harvard Law Review* 93 (1980): 1057–1154.

Frutos, Roger, Jordi Serra-Cobo, Tianmu Chen, and Christian A. Devaux. "COVID-19: time to exonerate the pangolin from the transmission of SARS-CoV-2 to humans." *Infection, Genetics and Evolution* 84 (2020): e104493.

Fulfer, Katy. "The capabilities approach to justice and the flourishing of nonsentient life." *Ethics and the Environment* 18 (1) (2013): 19–42.

Gabrys, Jennifer. "Sensing lichens: from ecological microcosms to environmental subjects." *Third Text* 32 (2–3) (2018): 350–367.

Gaertner, Mirijam, Brendon M. H. Larson, Ulrike M. Irlich, Patricia M. Holmes, Louise Stafford, Brian W. van Wilgen, and David M. Richardson. "Managing invasive species in cities: a framework from Cape Town, South Africa." *Landscape and Urban Planning* 151 (2016): 1–9.

Gajendran, V. "Chennai's peri-urban: accumulation of capital and environmental exploitation." *Environment and Urbanization Asia* 7 (1) (2016): 1–19.

Gandy, Matthew. "At a tangent: delineating a new ecological imaginary in Berlin's Park am Gleisdreieck." *Architectural Design* 90 (1) (2020): 106–113.

Gandy, Matthew. "Borrowed light: a journey through Weimar Berlin." In *The fabric of space: water, modernity, and the urban imagination.* Cambridge, MA: MIT Press, 2014.

Gandy, Matthew. *Concrete and clay: reworking nature in New York City.* Cambridge, MA: MIT Press, 2002.

Gandy, Matthew. "Entropy by design: Gilles Clément, Parc Henri Matisse and the limits to avant-garde urbanism." *International Journal of Urban and Regional Research* 37 (1) (2013): 259–278.

Gandy, Matthew. "The fly that tried to save the world: saproxylic geographies and other-than-human ecologies." *Transactions of the Institute of British Geographers* 44 (2) (2019): 392–406.

Gandy, Matthew. "Interstitial landscapes: reflections on a Berlin corner." In *Urban constellations*, ed. Matthew Gandy, 149–152. Berlin: jovis, 2011.

Gandy, Matthew. "Negative luminescence." *Annals of the American Association of Geographers* 107 (5) (2017): 1090–1107.

Gandy, Matthew. "Queer ecology: nature, sexuality and heterotopic alliances." *Environment and Planning D: Society and Space* 30 (2012): 727–747.

Gandy, Matthew. "Unintentional landscapes." *Journal of Landscape Research* 41 (4) (2016): 433–440.

Gandy, Matthew. "Where does the city end?" *Architectural Design* 82 (1) (2012): 128–132.

Gao, L., Yali Shi, Wenhui Li, Jiemin Liu, and Yaqi Cai. "Occurrence and distribution of antibiotics in urban soil in Beijing and Shanghai, China." *Environmental Science and Pollution Research* 22 (15) (2015): 11360–11371.

Garber, Steven D. *The urban naturalist.* New York: John Wiley, 1987.

Garbolino, Paolo. "The scientification of forensic practice." In *New challenges to philosophy of science*, ed. Hanne Andersen, Dennis Dieks, Wenceslao J. Gonzalez, Thomas Uebel, and Gregory Wheeler, 287–297. Dordrecht: Springer, 2013.

Gaston, Kevin J. "Urban ecology." In *Urban ecology*, ed. Kevin J. Gaston, 1–9. Cambridge: Cambridge University Press, 2010.

Gaziano, Emanuel. "Ecological metaphors as scientific boundary work: innovation and authority in interwar sociology and biology." *American Journal of Sociology* 101 (4) (1996): 874–907.

Gee, Henry. *In search of deep time: beyond the fossil record to a new history of life.* Ithaca, NY: Cornell University Press, 2000.

Gehrt, Stanley D., Justin L. Brown, and Chris Anchor. "Is the urban coyote a misanthropic synanthrope? The case from Chicago." *Cities and the Environment (CATE)* 4 (1) (2011): 3.

Gehrt, Stanley D., and Max McGraw. "Ecology of coyotes in urban landscapes." *Wildlife Damage Management Conferences* 63 (2007): 303–311.

Gerhardt, Karen E., Xiao-Dong Huang, Bernard R. Glick, and Bruce M. Greenberg. "Phytoremediation and rhizoremediation of organic soil contaminants: potential and challenges." *Plant Science* 176 (1) (2009): 20–30.

Ghertner, D. Asher. "Analysis of new legal discourse behind Delhi's slum demolitions." *Economic and Political Weekly* 43 (20) (2008): 57–66.

Ghosh, Amitav. *The great derangement: climate change and the unthinkable.* Chicago: University of Chicago Press, 2016.

Gibson, William. *Neuromancer.* New York: Ace Books, 1984.

Gidwani, Vinay, and Rajyashree N. Reddy. "The afterlives of 'waste': notes from India for a minor history of capitalist surplus." *Antipode* 43 (5) (2011): 1625–1658.

Gieryn, Thomas F. "Boundary-work and the demarcation of science from non-science: strains and interests in professional ideologies of scientists." *American Sociological Review* 48 (6) (1983): 781–795.

Gieryn, Thomas F. *Cultural boundaries of science: credibility on the line.* Chicago: University of Chicago Press, 1999.

Gilbert, Marius, Scott H. Newman, John Y. Takekawa, Leo Loth, Chandrashekhar Biradar, Diann J. Prosser, Sivananinthaperumal Balachandran, et al. "Flying over an infected landscape: distribution of highly pathogenic avian influenza H5N1 risk in South Asia and satellite tracking of wild waterfowl." *EcoHealth* 7 (4) (2010): 448–458.

Gilbert, Oliver. *The ecology of urban habitats.* London: Chapman and Hall, 1991 [1989].

Gilbert, Oliver. *The flowering of the cities: the natural flora of "urban commons."* Peterborough: English Nature, 1992.

Gilbert, Oliver. "Wild figs by the river Don, Sheffield." *Watsonia* 18 (1990): 84–85.

Gill, Don, and Penelope Bonnett. *Nature in the urban landscape: a study of city ecosystems.* Baltimore: York Press, 1973.

Gilloch, Graeme. *Myth and metropolis: Walter Benjamin and the city.* Cambridge: Polity, 1996.

Ginn, Franklin. "Sticky lives: slugs, detachment and more-than-human ethics in the garden." *Transactions of the Institute of British Geographers* 39 (4) (2014): 532–544.

Gioielli, Robert. *Environmental activism and the urban crisis: Baltimore, St. Louis, Chicago.* Philadelphia: Temple University Press, 2014.

Gliwicz, Joanna, J. Goszczynski, and Maciej Luniak. "Characteristic features of animal populations under synurbization—the case of the Blackbird and of the Striped Field Mouse." *Memorabilia Zoologica* 49 (1994).

Gobster, Paul H. "Visions of restoration: conflict and compatibility in urban park restoration." *Landscape and Urban Planning* 56 (1–2) (2001): 35–51.

Gobster, Paul H., Joan I. Nassauer, Terry C. Daniel, and Gary Fry. "The shared landscape: what does aesthetics have to do with ecology?" *Landscape Ecology* 22 (7) (2007): 959–972.

Gómez-Barris, Macarena. *The extractive zone: social ecologies and decolonial perspectives.* Durham, NC: Duke University Press, 2017.

Goode, David. *Nature in towns and cities.* London: Collins, 2014.

Goodness, Julie, and Pippin M. L. Anderson. "Local assessment of Cape Town: navigating the management complexities of urbanization, biodiversity, and ecosystem services in the Cape Floristic Region." In *Urbanization, biodiversity and ecosystem services: challenges and opportunities: a global assessment*, ed. Thomas Elmqvist et al., 461–484. Dordrecht: Springer, 2013.

Gopal, Divya, Moritz von der Lippe, and Ingo Kowarik. "Sacred sites, biodiversity and urbanization in an Indian megacity." *Urban Ecosystems* 22 (2019): 161–172.

Gopal, Divya, Harini Nagendra, and Michael Manthey. "Vegetation in Bangalore's slums: composition, species distribution, density, diversity, and history." *Environmental Management* 55 (2015): 1390–1401.

Gopinath, Gayatri. *Unruly visions: the aesthetic practices of queer diaspora.* Durham, NC: Duke University Press, 2018.

Gordillo, Gastón R. *Rubble: the afterlife of destruction.* Durham, NC: Duke University Press, 2014.

Görg, Christof. *Gesellschaftliche Naturverhältnisse.* Münster: Westfälisches Dampfboot, 1999.

Görg, Christof. "Societal relationships with nature: a dialectical approach to environmental politics." In *Critical ecologies: the Frankfurt School and contemporary environmental crisis*, ed. Andrew Biro, 43–72. Toronto: University of Toronto Press, 2011.

Gottlieb, Scott. "West Nile virus detected in mosquitoes in Central Park." *Bulletin of the World Health Organization* 78 (9) (2000): 1168.

Grant, Gary. *Ecosystem services come to town: greening cities by working with nature.* Oxford: Wiley-Blackwell, 2012.

Gregor, Thomas, Dirk Bönsel, Indra Starke-Ottich, and Georg Zizka. "Drivers of floristic change in large cities—a case study of Frankfurt/Main (Germany)." *Landscape and Urban Planning* 104 (2) (2012): 230–237.

Gröning, Gert, and Joachim Wolschke-Bulmahn. "Politics, planning and the protection of nature: political abuse of early ecological ideas in Germany, 1933–45." *Planning Perspectives* 2 (2) (1987): 127–148.

Gros, Frédéric. *A philosophy of walking*. Trans. John Howe. London: Verso, 2014 [2009].

Grosz, Elizabeth. "Bergson, Deleuze, and the becoming of unbecoming." *Parallax* 11 (2) (2005): 4–13.

Grove, J. Morgan, Mary L. Cadenasso, Steward T. Pickett, Gary E. Machlis, and William R. Burch. *The Baltimore school of urban ecology: space, scale, and time for the study of cities*. New Haven, CT: Yale University Press, 2015.

Guattari, Félix. *The three ecologies*. Trans. Ian Pindar and Paul Sutton. London: Athlone, 2000 [1989].

Gubler, Duane J. "Dengue, urbanization and globalization: the unholy trinity of the 21st century." *Tropical Medicine and Health* 39 (4) (2011): 3–11.

Gulsrud, Natalie, and Henriette Steiner. "When urban greening becomes an accumulation strategy: exploring the ecological, social and economic calculus of the High Line." *Journal of Landscape Architecture* 14 (3) (2019): 82–87.

Gururani, Shubhra. "'When land becomes gold': changing political ecology of the commons in a rural-urban frontier." In *Land rights, biodiversity conservation and justice: rethinking parks and people*, ed. Sharlene Mollett and Thembela Kepe, 107–125. London: Routledge, 2018.

Hacking, Ian. *The emergence of probability: a philosophical study of early ideas about probability, induction and statistical inference*. Cambridge: Cambridge University Press, 2006 [1975].

Hajer, Maarten. "Policy without polity? Policy analysis and the institutional void." *Policy Sciences* 36 (2) (2003): 175–195.

Häkli, Jouni. "Culture and politics of nature in the city: the case of Berlin's 'green wedge.'" *Capitalism Nature Socialism* 7 (2) (1996): 125–138.

Hall, Stuart. "When was 'the post-colonial'? Thinking at the limit." In *The postcolonial question*, ed. Iain Chambers and Lidia Curti, 248–266. London: Routledge, 2002.

Hallmann, Caspar A., Martin Sorg, Eelke Jongejans, Henk Siepel, Nick Hofland, Heinz Schwan, Werner Stenmans, et al. "More than 75 percent decline over 27 years in total flying insect biomass in protected areas." *PLoS One* 12 (10) (2007): e0185809.

Hansen, Mark B. N. "Our predictive condition; or, prediction in the wild." In *The nonhuman turn*, ed. Richard Grusin, 101–138. Minneapolis: University of Minnesota Press.

Harari, Yuval Noah. *Sapiens: a brief history of humankind*. New York: Random House, 2014.

Hard, Gerhard. "Die Natur, die Stadt und die Ökologie. Reflexionen über 'Stadtnatur' und 'Stadtökologie.'" In *Dimensionen geographischen Denkens*, 341–370. Osnabrück: V & R unipress, 2003[1994].

Hard, Gerhard. *Ruderalvegetation. Ökologie und Ethnoökologie, Ästhetik und "Schutz."* Notizbuch 49 der Kasseler Schule. Kassel: Arbeitsgemeinschaft Freiraum und Vegetation, 1998.

Hard, Gerhard. *Spuren und Spurenleser. Zur Theorie und Ästhetik des Spurenlesens in der Vegetation und anderswo.* Osnabrück: Universitätsverlag Rasch, 1995.

Hard, Gerhard. "Vegetationsgeographie und sozialökologie einer Stadt." *Geographische Zeitung* 75 (1985): 125–144.

Harding, Sandra. "'Strong objectivity': a response to the new objectivity question." *Synthese* 104 (3) (1995): 331–349.

Hare, Nathan. "Black ecology." *Black Scholar* 1 (6) (1970): 2–8.

Harms, Erik. "Knowing into oblivion: clearing wastelands and imagining emptiness in Vietnamese New Urban Zones." *Singapore Journal of Tropical Geography* 35 (3) (2014): 312–327.

Harris, Stephen, and Jeremy M. V. Rayner. "Urban fox (*Vulpes vulpes*) population estimates and habitat requirements in several British cities." *Journal of Animal Ecology* 55 (2) (1986): 575–591.

Harvey, David. *Consciousness and the urban experience: studies in the history and theory of capitalist urbanization.* Vol. 1. Baltimore: Johns Hopkins University Press, 1985.

Harvey, David. *The limits to capital.* Oxford: Blackwell, 1982.

Hauser, Susanne. *Metamorphosen des Abfalls, Konzepte für alte Industrieareale.* Frankfurt: Campus, 2001.

Hauskeller, Michael. "My brain, my mind, and I: some philosophical assumptions of mind-uploading." *International Journal of Machine Consciousness* 4 (1) (2012): 187–200.

Hawksworth, David L., and Francis Rose. *Lichens as pollution monitors.* London: Edward Arnold, 1976.

Hawley, Amos H. "Ecology and human ecology." *Social Forces* 22 (4) (1944): 398–405.

Hawley, Amos H. *Human ecology: a theoretical essay.* Chicago: University of Chicago Press, 1986.

Heald, O. J. N., C. Fraticelli, S. E. Cox, M. C. A. Stevens, S. C. Faulkner, T. M. Blackburn, and S. C. Le Comber. "Understanding the origins of the ring-necked parakeet in the UK." *Journal of Zoology* 312 (2020) 1–11.

Hebbert, Michael, and Vladimir Jankovic. "Cities and climate change: the precedents and why they matter." *Urban Studies* 50 (7) (2013): 1332–1347.

Heink, Ulrich, and Ingo Kowarik. "What are indicators? On the definition of indicators in ecology and environmental planning." *Ecological Indicators* 10 (3) (2010): 584–593.

Heinrich, Gudrun. "Rot-Grün in Berlin 1989–1990." In *Die Grünen. Wie sie wurden, was sie sind*, ed. Joachim Raschke, 809–822. Cologne: Bund-Verlag, 1993.

Heise, Ursula K. *Imagining extinction: the cultural meanings of endangered species*. Chicago: University of Chicago Press, 2016.

Heise, Ursula K. "Science fiction and the time scales of the Anthropocene." *ELH* 86 (2) (2019): 275–304.

Heise, Ursula K. *Sense of place and sense of planet: the environmental imagination of the global*. Oxford: Oxford University Press, 2008.

Heynen, Nik, Maria Kaïka, and Erik Swyngedouw, eds. *In the nature of cities: urban political ecology and the politics of urban metabolism*. London: Routledge, 2006.

Higgins, Lila, and Gregory B. Pauly. *Wild LA: explore the amazing nature in and around Los Angeles*. Portland, OR: Timber Press/Los Angeles County Natural History Museum, 2019.

Higgs, Eric. *Nature by design: people, natural process, and ecological restoration*. Cambridge, MA: MIT Press, 2003.

Hinchliffe, Steve. "Reconstituting nature conservation: towards a careful political ecology." *Geoforum* 39 (1) (2008): 88–97.

Hinchliffe, Steve, Matthew B. Kearns, Monica Degen, and Sarah Whatmore. "Urban wild things: a cosmopolitical experiment." *Environment and Planning D: Society and Space* 23 (5) (2005): 643–658.

Hinchliffe, Steve, and Sarah Whatmore. "Living cities: towards a politics of conviviality." *Science as Culture* 15 (2) (2006): 123–138.

Ho, Pui-yin. *Making Hong Kong: a history of urban development*. Cheltenham: Edward Elgar, 2018.

Hoag, Colin, Filippo Bertoni, and Nils Bubandt. "Wasteland ecologies: undomestication and multispecies gains on an Anthropocene dumping ground." *Journal of Ethnobiology* 38 (1) (2018): 88–104.

Hogg, John. "On the ballast-flora of the coasts of Durham and Northumberland." *Annals and magazine of natural history: zoology, botany, and geology* 19 (3) (1867): 38–43.

Hohler, Franz. *Die Rückeroberung*. Munich: btb, 2012 [1982].

Holm, Nicholas. "Consider the possum: foes, anti-animals, and colonists in paradise." *Animal Studies Journal* 4 (1) (2015): 32–56.

Holm, Nicholas. "Consider the squirrel: freaks, vermin, and value in the ruin(s) of nature." *Cultural Critique* 80 (2012): 56–95.

Höppner, Hans, and Hans Preuss. *Flora des westfälisch-rheinischen Industriegebietes unter Einschluß der Rheinischen Bucht*. Dortmund: Friedrich Wilhelm Ruhfus, 1926.

Hörl, Erich. "Introduction to general ecology: the ecologization of thinking." In *General ecology: the new ecological paradigm*, ed. Erich Hörl and James Edward Burton, 1–74. London: Bloomsbury, 2017.

Hough, Michael. *City form and natural process: towards a new vernacular*. Toronto: McGraw-Hill, 1984.

Houston, Donna, Jean Hillier, Diana MacCallum, Wendy Steele, and Jason Byrne. "Make kin, not cities! Multispecies entanglements and 'becoming-world' in planning theory." *Planning Theory* 17 (2) (2018): 190–212.

Hovorka, Alice. "Trans-species urban theory: chickens in an African city." *Cultural Geographies* 15 (1) (2008): 95–117.

Howell, Philip. "Between wild and domestic, animal and human, life and death: the problem of the stray in the Victorian city." In *Animal history in the modern city: exploring liminality*, ed. Clemens Wischermann, Aline Steinbrecher, and Philip Howell, 145–160. London: Bloomsbury, 2019.

Howlett, Rory J., and Michael E. N. Majerus. "The understanding of industrial melanism in the peppered moth (*Biston betularia*) (Lepidoptera: Geometridae)." *Biological Journal of the Linnaean Society* 30 (1) (1987): 31–44.

Hudson, Michael. "From Marx to Goldman Sachs: the fictions of fictitious capital, and the financialization of industry." *Critique* 38 (3) (2010): 419–444.

Hudson, W. H. *Birds in London*. London: Longmans, Green, 1898.

Hülsberg, Werner. *The German Greens: a social and political profile*. Trans. Gus Fagan. London: Verso, 1988.

Hunold, Christian. "Urban greening and human-wildlife relations in Philadelphia: from animal control to multispecies coexistence?" *Environmental Values* 29 (1) (2020): 67–87.

Hurley, Andrew. *Environmental inequalities: class, race, and industrial pollution in Gary, Indiana, 1945–1980*. Chapel Hill: University of North Carolina Press, 1995.

Hurrell, Julio Alberto, and María Lelia Pochettino. "Urban ethnobotany: theoretical and methodological contributions." In *Methods and techniques in ethnobiology and ethnoecology*, ed. Ulysses Paulino Albuquerque et al., 293–309. New York: Springer/Humana Press, 2014.

Hutchinson, Rachel. "Fukasaku Kinji and Kojima Hideo replay Hiroshima: atomic imagery and cross-media memory." *Japanese Studies* 39 (2) (2019): 169–189.

Huyssen, Andreas. *Twilight memories: marking time in a culture of amnesia*. London: Routledge, 2012.

Huyssen, Andreas. "The voids of Berlin." *Critical Inquiry* 24 (1) (1997): 57–81.

Ineichen, Stefan, and Max Ruckstuhl. *Stadtfauna Zürich*. Bern: Haupt Verlag, 2012.

Ingersoll, Ernest. *Wild neighbors: out-door studies in the United States*. New York: Macmillan, 1902 [1897].

Ingold, Tim. *Lines*. London: Routledge, 2007.

Irvine, Richard D. G., and Mina Gorji. "John Clare in the Anthropocene." *Cambridge Journal of Anthropology* 31 (1) (2013): 119–132.

Irwin, Alan. *Citizen science: a study of people, expertise and sustainable development.* London: Routledge, 1995.

Jaffe, Rivke. "Political animals: an interspecies approach to urban inequalities." Paper presented to the Berkeley Black Geographies Symposium, University of California, Berkeley, 12–13 March 2020.

James, Simon P. "Protecting nature for the sake of human beings." *Ratio* (2) (2016): 213–227.

Jameson, Fredric. "The brick and the balloon: architecture, idealism and land speculation." *New Left Review* 228 (March 1998): 25–46.

Jameson, Fredric. "Future city." *New Left Review* 21 (2003): 65–79.

Jameson, Fredric. *The seeds of time.* New York: Columbia University Press, 1993.

Jasper, Sandra. "Acoustic ecologies: architecture, nature, and modernist experimentation in West Berlin." *Annals of the American Association of Geographers* 110 (4) (2020): 1114–1133.

Jazeel, Tariq. "Spatializing difference beyond cosmopolitanism: rethinking planetary futures." *Theory, Culture and Society* 28 (5) (2011): 75–97.

Jefferies, Richard. *Nature near London.* London: John Clare Books, 1980 [1893].

Jerolmack, Colin. "How pigeons became rats: the cultural-spatial logic of problem animals." *Social Problems* 55 (1) (2008): 72–94.

Jim, Chi Yung. "Old masonry walls as ruderal habitats for biodiversity conservation and enhancement in urban Hong Kong." In *Urban biodiversity and design*, ed. Norbert Müller, Peter Werner, and John G. Kelcey, 323–347. Hoboken, NJ: Wiley-Blackwell, 2010.

Johnson, Thomas. *Botanical journeys in Kent and Hampstead: a facsimile reprint with introduction and translation of his Iter plantarum 1629 [and] Descriptio itineris plantarum 1632.* Ed. John S. L. Gilmour. Pittsburgh: Hunt Botanical Library, 1972.

Johnston, Catherine. "Beyond the clearing: towards a dwelt animal geography." *Progress in Human Geography* 32 (5) (2008): 633–649.

Jones, Bryony A., Delia Grace, Richard Kock, Silvia Alonso, Jonathan Rushton, Mohammed Y. Said, Declan McKeever, et al. "Zoonosis emergence linked to agricultural intensification and environmental change." *Proceedings of the National Academy of Sciences* 110 (21) (2013): 8399–8404.

Jorgensen, Anna, and Marian Tylecote. "Ambivalent landscapes—wilderness in the urban interstices." *Landscape Research* 32 (4) (2007): 443–462.

Jost, Lou. "Partitioning diversity into independent alpha and beta components." *Ecology* 88 (10) (2007): 2427–2439.

Jovet, Paul. "Evolution des groupements rudéraux 'parisiens.'" *Bulletin de la Société Botanique de France* 87 (1940): 305–312.

Kadas, Gyongyver. "Rare invertebrates colonising green roofs in London." *Urban Habitats* 4 (1) (2006): 66–86.

Kaplan, Marion A. *Between dignity and despair: Jewish life in Nazi Germany*. Oxford: Oxford University Press, 1998.

Karera, Axelle. "Blackness and the pitfalls of Anthropocene ethics." *Critical Philosophy of Race* 7 (1) (2019): 32–56.

Katzschner, Tania. "Cape flats nature: rethinking urban ecologies." In *Contested ecologies: dialogues in the South on nature and knowledge*, ed. Lesley Green, 202–206. Cape Town: HSRC Press, 2013.

Kaup, Brent Z. "Pathogenic metabolisms: a rift and the Zika Virus in Mato Grosso, Brazil." *Antipode* 53 (2) (2021): 567–586.

Kaviraj, Sudipta. "Filth and the public sphere: concepts and practices about waste in Calcutta." *Public Culture* 10 (1) (1997): 83–113.

Kean, Hilda. "Traces and representations: animal pasts in London's present." *London Journal* 36 (1) (2011): 54–71.

Keck, Frédéric. *Avian reservoirs: virus hunters and birdwatchers in Chinese sentinel posts*. Durham, NC: Duke University Press, 2020.

Keck, Frédéric. "Livestock revolution and ghostly apparitions: South China as a sentinel territory for influenza pandemics." *Current Anthropology* 60 (S20) (2019): S251–S259.

Keck, Frédéric. "Sentinels for the environment: birdwatchers in Taiwan and Hong Kong." *China Perspectives* 2 (2015): 43–52.

Keeney, Elizabeth B. *The botanizers: amateur scientists in nineteenth-century America*. Chapel Hill: University of North Carolina Press, 1992.

Kegel, Bernhard. *Tiere in der Stadt. Eine Naturgeschichte*. Cologne: DuMont, 2013.

Keil, Andreas. "Use and perception of post-industrial urban landscapes in the Ruhr." In *Wild urban woodlands: new perspectives for urban forestry*, ed. Ingo Kowarik and Stefan Körner, 117–130. Berlin: Springer, 2005.

Keil, Roger. "Urban political ecology." *Urban Geography* 24 (8) (2003): 723–738.

Kellert, Stephen R., and Edward O. Wilson, eds. *The biophilia hypothesis*. Washington, DC: Island Press, 1993.

Kelly, Ann H., and Javier Lezaun. "Urban mosquitoes, situational publics, and the pursuit of interspecies separation in Dar es Salaam." *American Ethnologist* 41 (2) (2014): 368–383.

Keppel, Gunnar, Kimberly P. Van Niel, Grant W. Wardell-Johnson, Colin J. Yates, Margaret Byrne, Ladislav Mucina, Antonius G. T. Schut, Stephen D. Hopper, and Steven E. Franklin. "Refugia: identifying and understanding safe havens for biodiversity under climate change." *Global Ecology and Biogeography* 21 (4) (2012): 393–404.

Khan, A. G., C. Kuek, T. M. Chaudhry, C. S. Khoo, and W. J. Hayes. "Role of plants, mycorrhizae and phytochelators in heavy metal contaminated land remediation." *Chemosphere* 41 (1–2) (2000): 197–207.

Kim, Claire Jean. *Dangerous crossings: race, species, nature in a multicultural age.* Cambridge: Cambridge University Press, 2015.

Kinchy, Abby J., and Simona L. Perry. "Can volunteers pick up the slack? Efforts to remedy knowledge gaps about the watershed impacts of Marcellus Shale gas development." *Duke Environmental Law and Policy Forum* 22 (2) (2012): 303–339.

King, Tiffany Lethabo. "Humans involved: lurking in the lines of posthumanist flight." *Critical Ethnic Studies* 3 (1) (2017): 162–185.

Kingsland, Sharon E. *The evolution of American ecology 1890–2000.* Baltimore: Johns Hopkins University Press, 2005.

Kinney, Rebecca J. *Beautiful wasteland: the rise of Detroit as America's postindustrial frontier.* Minneapolis: University of Minnesota Press, 2016.

Kitchin, Rob, Justin Gleeson, Karen Keaveney, and Cian O'Callaghan. "A haunted landscape: housing and ghost estates in post-Celtic Tiger Ireland." *National Institute for Regional and Spatial Analysis Working Paper Series* 59 (1) (2010).

Kitchin, Rob, Cian O'Callaghan, and Justin Gleeson. "The new ruins of Ireland? Unfinished estates in the post-Celtic Tiger era." *International Journal of Urban and Regional Research* 38 (3) (2014): 1069–1080.

Kiuchi, Kumiko. "Tokyo ecology: the Akabane Nature Observation Park." In *The botanical city,* ed. Matthew Gandy and Sandra Jasper, 253–265. Berlin: jovis, 2020.

Klausnitzer, Bernhard. *Ökologie der Großstadtfauna.* Stuttgart: Gustav Fischer, 1987.

Knight, David, ed. *The evolution debate 1813–1870.* Vol. 1: Georges Cuvier, *Essay on the theory of the earth.* London: Routledge, 2003.

Kohler, Matthias. "Aerial architecture." *Log* 25 (2012): 23–30.

Kohler, Robert E. *All creatures: naturalists, collectors, and biodiversity, 1850–1950.* Princeton, NJ: Princeton University Press, 2006.

Kohler, Robert E. *Landscapes and labscapes: exploring the lab-field border in biology.* Chicago: University of Chicago Press, 2002.

Kolinsky, Eva. "Non-German minorities in contemporary German society." In *Turkish culture in German society today*, ed. David Horrocks and Eva Kolinsky, 71–111. Providence, RI: Berghahn, 1996.

Kollin, Susan. "Not yet another world: ecopolitics and urban natures in Jonathan Lethem's *Chronic city*." *Literature Interpretation Theory* 26 (4) (2015): 255–275.

Kortekallio, Kaisa. "Becoming-instrument: thinking with Jeff VanderMeer's *Annihilation* and Timothy Morton's *Hyperobjects*." In *Reconfiguring human, nonhuman and posthuman in literature and culture*, ed. Sanna Karkulehto, Aino-Kaisa Koistinen, and Essi Varis, 57–75. London: Routledge, 2019.

Kowarik, Ingo. *Biologische Invasionen. Neophyten und Neozoen in Mitteleuropa*. Stuttgart: Ulmer, 2003.

Kowarik, Ingo. "Novel urban ecosystems, biodiversity, and conservation." *Environmental Pollution* 159 (8–9) (2011): 1974–1983.

Kowarik, Ingo. "Unkraut oder Urwald? Natur der Vierten Art auf dem Gleisdreieck." In *Dokumentation Gleisdreieck Morgen. Sechs Ideen für einen Park*, ed. Bundesgartenschau, 45–55. Berlin: Bezirksamt Kreuzberg, 1995 [1991].

Kowarik, Ingo. "Wild urban woodlands: towards a conceptual framework." In *Wild urban woodlands: new perspectives for urban forestry*, ed. Ingo Kowarik and Stefan Körner, 1–32. Berlin: Springer, 2005.

Kowarik, Ingo, Sascha Buchholz, Moritz von der Lippe, and Birgit Seitz. "Biodiversity functions of urban cemeteries: evidence from one of the largest Jewish cemeteries in Europe." *Urban Forestry and Urban Greening* 19 (2016): 68–78.

Kowarik, Ingo, Leonie K. Fischer, Ina Säumel, Moritz von der Lippe, Frauke Weber, and Janneke Westermann. "Plants in urban settings: from patterns to mechanisms and ecosystem services." In *Perspectives in urban ecology: ecosystems and interactions between humans and nature in the metropolis of Berlin*, ed. Wilfried Endlicher, 135–166. Heidelberg: Springer, 2011.

Kowarik, Ingo, and Ina Säumel. "Biological flora of Central Europe: *Ailanthus altissima* (Mill.) Swingle." *Perspectives in Plant Ecology, Evolution and Systematics* 8 (2007): 207–237.

Kracauer, Siegfried. "Aus dem Fenster gesehen." In *Straßen in Berlin und anderswo*, 50–52. Berlin: Das Arsenal, 1987 [1931].

Krasny, Marianne E., Cecilia Lundholm, Soul Shava, Eunju Lee, and Hiromi Kobori. "Urban landscapes as learning arenas for biodiversity and ecosystem services management." In *Urbanization, biodiversity and ecosystem services: challenges and opportunities*, ed. Thomas Elmqvist et al., 629–664. Dordrecht: Springer, 2013.

Kratzer, Albert. "Das Klima der Städte." *Geographische Zeitschrift* 41 (9) (1935): 321–339.

Kueffer, Christoph. "Plant sciences for the Anthropocene: what can we learn from research in urban areas?" *Plants, People, Planet* 2 (4) (2020): 286–289.

Kühn, Norbert. "Intentions for the unintentional: spontaneous vegetation as the basis for innovative planting design in urban areas." *Journal of Landscape Architecture* (Autumn 2006): 46–53.

Kühn, Rudolf. *Die Strassenbäume*. Hannover: B. Patzer, 1961.

Kukla, Rebecca. "Objectivity and perspective in empirical knowledge." *Episteme* 3 (1) (2006): 80–95.

Kumar, Mukul, K. Saravanan, and Nityanand Jayaraman. "Mapping the coastal commons: fisherfolk and the politics of coastal urbanisation in Chennai." *Economic and Political Weekly* 48 (2014): 46–53.

Kumar, Nishant, Urvi Gupta, Harsha Malhotra, Yadvendradev V. Jhala, Qamar Qureshi, Andrew G. Gosler, and Fabrizio Sergio. "The population density of an urban raptor is inextricably tied to human cultural practices." *Proceedings of the Royal Society B* 286 (1900) (2019): 20182932.

Kunick, Wolfram. "Zonierung des Stadtgebietes von Berlin (West). Ergebnisse floristischer Untersuchungen." *Landschaftsentwicklung und Umweltforschung* 14 (1982): 1–164.

Kunkel, Benjamin. "The capitalocene." *London Review of Books* (2 March 2017).

Kuzmin, Ivan V., Brooke Bozick, Sarah A. Guagliardo, Rebekah Kunkel, Joshua R. Shak, Suxiang Tong, and Charles E. Rupprecht. "Bats, emerging infectious diseases, and the rabies paradigm revisited." *Emerging Health Threats Journal* 4 (1) (2011): e7159.

Kwa, Chunglin. "The visual grasp of the fragmented landscape: plant geographers vs. plant sociologists." *Historical Studies in the Natural Sciences* 48 (2) (2018): 180–222.

Lachmund, Jens. "The city as ecosystem: Paul Duvigneaud and the ecological study of Brussels." In *Spatializing the history of ecology: sites, journeys, mappings*, ed. Raf de Bont and Jens Lachmund, 141–162. London: Routledge, 2017.

Lachmund, Jens. "Exploring the city of rubble: botanical fieldwork in bombed cities in Germany after World War II." *Osiris* 18 (2003): 234–254.

Lachmund, Jens. *Greening Berlin: the co-production of science, politics, and urban nature*. Cambridge, MA: MIT Press, 2013.

Lacoste, Yves. *La géographie ça sert d'abord à faire la guerre*. Paris: François Maspero, 1976.

Lacoste, Yves. "An illustration of geographical warfare: bombing of the dikes on the Red River, North Vietnam." In *Radical geography: alternative viewpoints on contemporary social issues*, ed. Richard Peet, 244–261. London: Methuen, 1978 [1972].

LaDeau, Shannon L., Paul T. Leisnham, Dawn Day Biehler, and Danielle Bodner. "Higher mosquito production in low-income neighborhoods of Baltimore and Washington, DC: understanding ecological drivers and mosquito-borne disease risk in temperate cities." *International Journal of Environmental Research and Public Health* 10 (4) (2013): 1505–1526.

Laichmann, Michaela. "Arbeitsvieh und Schoßtier. Hunde im mittelalterlichen und früh-neuzeitlichen Wien." In *Umwelt Stadt. Geschichte des Natur-und Lebensraumes Wien*, ed. Karl Brunner and Petra Schneider, 410–417. Vienna: Böhlau, 2005.

Lakoff, Andrew. "The indicator species: tracking ecosystem collapse in arid California." *Public Culture* 28 (2016): 237–259.

Lampugnani, Vittorio Magnago, Konstanze Sylva Domhardt, and Rainer Schützeichel, eds. *Enzyklopädie zum gestalteten Raum: Im Spannungsfeld zwischen Stadt und Landschaft.* Zurich: gta Verlag, 2014.

Landecker, Hannah. "Antibiotic resistance and the biology of history." *Body and Society* 22 (4) (2016): 19–52.

Landecker, Hannah, and Aaron Panofsky. "From social structure to gene regulation, and back: a critical introduction to environmental epigenetics for sociology." *Annual Review of Sociology* 39 (2013): 333–357.

Latour, Bruno. "Whose cosmos, which cosmopolitics? Comments on the peace terms of Ulrich Beck." *Common Knowledge* 10 (3) (2004): 450–462.

Laundon, Jack R. "London's lichens." *London Naturalist* 49 (1970): 20–68.

Laurie, Ian C., ed. *Nature in cities: the natural environment in the design and development of green space.* Chichester: John Wiley, 1979.

Lave, Rebecca. "The future of environmental expertise." *Annals of the Association of American Geographers* 105 (2) (2015): 244–252.

Lepawski, Josh. *Reassembling rubbish: worlding electronic waste.* Cambridge, MA: MIT Press, 2018.

Lerner, Steve. *Diamond: a struggle for environmental justice in Louisiana's chemical corridor.* Cambridge, MA: MIT Press, 2006.

Lethem, Jonathan. *Chronic city.* London: Faber and Faber, 2010.

Lévesque, Luc. "Montréal, l'informe urbanité des terrains vagues: pour une gestion créatrice du mobilier urbain." *Annales de la Recherche Urbaine* 85 (1) (1999): 47–57.

Levy, Sharon. "The new top dog." *Nature* 485 (2012): 296–297.

Lewis, Simon L., and Mark A. Maslin. "Defining the Anthropocene." *Nature* 519 (7542) (2015): 171–180.

Liebig, Justus von. *Letters on modern agriculture*. Ed. John Blyth. London: Walton and Maberly, 1859.

Livingston, Julie, and Jasbir K. Puar. "Interspecies." *Social Text* 106 (29) (1) (2011): 1–14.

Lizet, Bernadette. "Du terrain vague à la friche paysagée." *Ethnologie Française* 40 (4) (2010): 597–608.

Lizet, Bernadette, Anne-Élisabeth Wolf, and John Celecia, eds. *Sauvages dans la ville. Actes du colloque, organisé pour le centenaire de la naissance de Paul Jovet*. Paris: JATBA/Publications scientifiques du MNHN, 1997.

Lockwood, Matthew. "Right-wing populism and the climate change agenda: exploring the linkages." *Environmental Politics* 27 (4) (2018): 712–732.

Lora-Wainwright, Anna. *Resigned activism: living with pollution in rural China*. Cambridge, MA: MIT Press, 2017.

Lorde, Audre. "Outside." *American Poetry Review* 6 (1) (January–February 1977).

Lorimer, Jamie. "Living roofs and brownfield wildlife: towards a fluid biogeography of UK nature conservation." *Environment and Planning A* 40 (2008): 2042–2060.

Lorimer, Jamie. "Nonhuman charisma." *Environment and Planning D: Society and Space* 25 (2007): 911–932.

Lorimer, Jamie, and Clemens Driessen. "Wild experiments at the Oostvaardersplassen: rethinking environmentalism in the Anthropocene." *Transactions of the Institute of British Geographers* 39 (2) (2014): 169–181.

Loughran, Kevin. "Race and the construction of city and nature." *Environment and Planning A* 49 (9) (2017): 1948–1967.

Loughran, Kevin. "Urban parks and urban problems: an historical perspective on green space development as a cultural fix." *Urban Studies* 57 (11) (2020): 2321–2338.

Lourenço, José, Maricelia Maia de Lima, Nuno Rodrigues Faria, Andrew Walker, Moritz U. G. Kraemer, Christian Julian Villabona-Arenas, Ben Lambert, et al. "Epidemiological and ecological determinants of Zika virus transmission in an urban setting." *Elife* 6 (2017): e29820.

Luhmann, Niklas. *Social systems*. Trans. John Bednarz and Dirk Baecker. Stanford, CA: Stanford University Press, 1995 [1984].

Luke, Timothy W. "Reconstructing social theory and the Anthropocene." *European Journal of Social Theory* 20 (1) (2017): 80–94.

Luniak, Maciej. "The birds of the park habitats in Warsaw." *Acta ornithologica* 18 (6) (1981): 335–370.

Luniak, Maciej. "Synurbization—adaptation of animal wildlife to urban development." In *Proceedings of the 4th International Urban Wildlife Symposium*, ed. William W. Shaw, Lisa K. Harris, and Larry Van Druff, 50–55. Tucson: University of Arizona, 2004.

Luther, Erin. "Tales of cruelty and belonging: in search of an ethic for urban human-wildlife relations." *Animal Studies Journal* 2 (1) (2013): 35–54.

Lutz, Frank Eugene. *A lot of insects: entomology in a suburban garden*. New York: Putnam and Sons, 1941.

Lyotard, Jean-François. *The postmodern condition: a report on knowledge*. Trans. Geoffrey Bennington and Brian Massumi. Minneapolis: University of Minnesota Press, 1984, [1979].

Mabey, Richard. *The unofficial countryside*. London: Collins, 1973.

Mabey, Richard. *Weeds*. London: Profile, 2010.

Macaulay, Rose. *The world my wilderness*. London: Collins, 1950.

Mace, Georgina M. "The role of taxonomy in species conservation." *Philosophical Transactions: Biological Sciences* 359 (1444) (2004): 711–719.

Macgregor, Callum J., Darren M. Evans, Richard Fox, and Michael J. O. Pocock. "The dark side of street lighting: impacts on moths and evidence for the disruption of nocturnal pollen transport." *Global Change Biology* 23 (2) (2017): 697–707.

Mah, Alice, and Xinhong Wang. "Accumulated injuries of environmental injustice: living and working with petrochemical pollution in Nanjing, China." *Annals of the American Association of Geographers* 109 (6) (2019): 1961–1977.

Malm, Andreas. *Fossil capital: the rise of steam power and the roots of global warming*. London: Verso, 2016.

Mandel, Ernest. *Late capitalism*. Trans. Joris De Bres. London: Verso, 1978 [1972].

Mandel, Ruth. *Cosmopolitan anxieties: Turkish challenges to citizenship and belonging in Germany*. Durham, NC: Duke University Press, 2008.

Marchand, Bernard, and Joëlle Salomon Cavin. "Anti-urban ideologies and planning in France and Switzerland: Jean-François Gravier and Armin Meili." *Planning Perspectives* 22 (1) (2007): 29–53.

Marchand, Yves, and Richard Meffre. *The ruins of Detroit*. Göttingen: Steidl; London: Thames and Hudson, 2010.

Marder, Michael. "The life of plants and the limits of empathy." *Dialogue: Canadian Philosophical Review / Revue Canadienne de Philosophie* 5 (2) (2012): 259–273.

Marris, Emma. *Rambunctious garden: saving nature in a post-wild world*. New York: Bloomsbury, 2011.

Martin, Patricia Yancey, John R. Reynolds, and Shelley Keith. "Gender bias and feminist consciousness among judges and attorneys: a standpoint theory analysis." *Signs: Journal of Women in Culture and Society* 27 (3) (2002): 665–701.

Martinez-Alier, Joan. "Urban 'unsustainability' and environmental conflict." In *The human sustainable city: challenges and perspectives from the Habitat Agenda*, ed. Luigi Fusco Girard, Bruno Forte, Maria Cerreta, Pasquale De Toro, and Fabiana Forte, 89–105. London: Routledge, 2019 [2003].

Martínez Estrada, Ezequeil. *X-ray of the Pampa*. Trans. Thomas F. McGann. Austin: University of Texas Press, 1971 [1933].

Marx, Karl. *Capital*. Vol. 1. London: Lawrence & Wishart, 1977 [1887].

Marx, Karl. "Feuerbach. Opposition of the materialist and idealist outlook." In Karl Marx and Friedrich Engels, *The German ideology*, part one, trans. W. Lough, 39–95. London: Lawrence & Wishart, 1965 [1844].

Marzluff, John M. "A decadal review of urban ornithology and a prospectus for the future." *Ibis* 159 (1) (2017): 1–13.

Marzluff, John M., Eric Shulenberger, Wilfried Endlicher, Marina Alberti, Gordon Bradley, Clare Ryan, Craig ZumBrunnen, and Ute Simon, eds. *An international perspective on the interaction between humans and nature*. New York: Springer, 2008.

Maspero, François. *Roissy Express: a journey through the Paris suburbs*. Trans. Paul Jones. London: Verso, 1994 [1990].

Mathey, Juliane, and Dieter Rink. "Urban wastelands—a chance for biodiversity in cities? Ecological aspects, social perceptions and acceptance of wilderness by residents." In *Urban biodiversity and design*, ed. Norbert Müller, Peter Werner, and John G. Kelcey, 406–424. Hoboken, NJ: Wiley-Blackwell, 2010.

Matthews, Anne. *Wild nights: the nature of New York City*. New York: North Point Press, 2001.

Maturana, Humberto R., and Francisco G. Varela. *Autopoietic systems: a characterization of the living organization*. Urbana-Champaign: University of Illinois Press, 1975.

Mawdsley, Emma. "India's middle classes and the environment." *Development and Change* 35 (1) (2004): 79–103.

Mazanik, Anna. "'Shiny shoes' for the city: the public abattoir and the reform of meat supply in imperial Moscow." *Urban History* 45 (2) (2018): 214–232.

Mazumdar, Subhendu, Dipankar Ghose, and Goutam Kumar Saha. "Foraging strategies of Black Kites (*Milvus migrans govinda*) in urban garbage dumps." *Journal of Ethology* 34 (3) (2016): 243–247.

Mbembe, Achille. *Critique of black reason*. Trans. Laurent Dubois. Durham, NC: Duke University Press, 2017.

Mbembe, Achille. "Provincializing France." *Public Culture* 23 (1) (2011): 85–111.

McCarthy, Thomas. "Private irony and public decency: Richard Rorty's new pragmatism." *Critical Inquiry* 16 (2) (1990): 355–370.

McCauley, Douglas J. "Selling out on nature." *Nature* 443 (2006): 27–28.

McDonald, Robert, and Timothy Beatley. *Biophilic cities for an urban century: why nature is essential for the success of cities*. Cham: Palgrave / Springer Nature, 2021.

McDonnell, Mark J. "The history of urban ecology: an ecologist's perspective." In *Urban ecology: patterns, processes and applications*, ed. Jari Niemelä, Jürgen H. Breuste, Thomas Elmqvist, Glenn Guntenspergen, Philip James, and Nancy E. McIntyre, 5–13. Oxford: Oxford University Press, 2011.

McDonough, Tom. "The crimes of the flâneur." *October* 102 (2002): 101–122.

McHarg, Ian L. *Design with nature*. New York: American Museum of Natural History, 1969.

McKinney, Michael L. "Urbanisation as a major cause of biotic homogenisation." *Biological Conservation* 127 (2006): 247–260.

McNeill, J. R. *The mountains of the Mediterranean world: an environmental history*. Cambridge: Cambridge University Press, 2003.

McNeill, J. R., and Peter Engelke. *The great acceleration: an environmental history of the Anthropocene since 1945*. Cambridge, MA: Harvard University Press, 2014.

McNeuer, Catherine. *Taming Manhattan: environmental battles in the antebellum city*. Cambridge, MA: Harvard University Press, 2014.

McNew, Sabrina, Daniel Beck, Ingrid Sadler-Riggleman, Sarah A. Knutie, Jennifer A. H. Koop, Dale H. Clayton, and Michael K. Skinner. "Epigenetic variation between urban and rural populations of Darwin's finches." *BMC Evolutionary Biology* 17 (1) (2017): 183.

McPhearson, Timon, Steward T. A. Pickett, Nancy B. Grimm, Jari Niemelä, Marina Alberti, Thomas Elmqvist, Christiane Weber, Dagmar Haase, Jürgen Breuste, and Salman Qureshi. "Advancing urban ecology toward a science of cities." *BioScience* 66 (3) (2016): 198–212.

McShane, Clay, and Joel A. Tarr. *The horse in the city: living machines in the nineteenth century*. Baltimore: Johns Hopkins University Press, 2007.

Mégnin, Jean-Pierre. *Le faune des cadavres: application de l'entomologie à la médecine légale*. Langes: Klincksieck, 2015 [1894].

Mellor, David, ed. *A paradise lost: the neo-romantic imagination in Britain 1935–1955*. London: Lund Humphries and Barbican Art Gallery, 1987.

Mellor, Leo. "Words from the bombsites: debris, modernism and literary salvage." *Critical Quarterly* 46 (4) (2004): 77–90.

Menachery, Vineet D., Boyd L. Yount, Amy C. Sims, Kari Debbink, Sudhakar S. Agnihothram, Lisa E. Gralinski, Rachel L. Graham, et al. "SARS-like WIV1-CoV poised for human emergence." *Proceedings of the National Academy of Sciences* 113 (11) (2016): 3048–3053.

Mendieta, Eduardo. "The biotechnological *Scala Naturae* and interspecies cosmopolitanism: Patricia Piccinini, Jane Alexander, and Guillermo Gómez-Peña." In *Biopower: Foucault and beyond*, ed. Vernon W. Cisney and Nicolae Morar, 158–182. Chicago: University of Chicago Press, 2015.

Mendieta, Eduardo. "Interspecies cosmopolitanism: towards a discourse ethics grounding of animal rights." *Philosophy Today* 54 (2010): 208–216.

Merleau-Ponty, Maurice. *Nature: course notes from the Collège de France*. Compiled by D. Ségland. Trans. R. Vallier. Evanston, IL: Northwestern University Press, 2003 [1956–1960].

Merleau-Ponty, Maurice. *Phenomenology of perception*. Trans. C. Smith. London: Routledge and Kegan Paul, 1962 [1945].

Metken, Günter. "*Ailanthus altissima* in Wedding." In Paul-Armand Gette, *Exotik als Banalität / De l'exotisme en tant que banalité*, i–iii. Berlin: DAAD, 1980.

Metken, Günter. "Gettes parallele Wissenschaft." In Paul-Armand Gette, *Arbeiten 1959–1979*. Munich: Städtische Galerie im Lenbachhaus, 1979.

Metken, Günter. *Spurensicherung*. Cologne: DuMont, 1977.

Metzger, Jonathan. "Cultivating torment: the cosmopolitics of more-than-human urban planning." *City* 20 (4) (2016): 581–601.

Middeldorp, Nick, and Philippe Le Billon. "Deadly environmental governance: authoritarianism, eco-populism, and the repression of environmental and land defenders." *Annals of the American Association of Geographers* 109 (2) (2019): 324–337.

Mikhail, Alan. *Nature and empire in Ottoman Egypt: an environmental history*. Cambridge: Cambridge University Press, 2011.

Millington, Nate. "Post-industrial imaginaries: nature, representation and ruin in Detroit, Michigan." *International Journal of Urban and Regional Research* 37 (1) (2013): 279–296.

Misrach, Richard, and Kate Orff. *Petrochemical America*. New York: Aperture, 2012.

Mitchell, Robert. *Experimental life: vitalism in romantic science and literature*. Baltimore: Johns Hopkins University Press, 2013.

Mitchell, Timothy. *Carbon democracy: political power in the age of oil*. London: Verso, 2011.

Mohai, Paul, and Bunyan Bryant. "Is there a 'race' effect on concern for environmental quality?" *Public Opinion Quarterly* (December 1998): 475–505.

Mohanty, Chandra Talpade. "Under Western eyes: feminist scholarship and colonial discourses." *Feminist Review* 30 (1988): 61–88.

Montgomery, Alesia. *Greening the Black urban regime: the culture and commerce of sustainability in Detroit*. Detroit: Wayne State University Press, 2020.

Moore, Jason W., ed. *Anthropocene or Capitalocene? Nature, history, and the crisis of capitalism*. Oakland, CA: PM Press, 2016.

Moore, Jason W. *Capitalism in the web of life: ecology and the accumulation of capital*. London: Verso, 2015.

Mora, Camilo, Derek P. Tittensor, Sina Adl, Alastair G. B. Simpson, and Boris Worm. "How many species are there on Earth and in the ocean?" *PLoS Biology* 9 (8) (2011): e1001127.

Morrison, Blake. *South of the river*. London: Chatto & Windus, 2007.

Morton, Timothy. *Hyperobjects: philosophy and ecology after the end of the world*. Minneapolis: University of Minnesota Press, 2013.

Mosyakin, Sergei L., and Oksana G. Yavorska. "The nonnative flora of the Kiev (Kyiv) urban area, Ukraine: a checklist and brief analysis." *Urban Habitats* 1 (1) (2002): 45–65.

Mugerauer, Robert. "Toward a theory of integrated urban ecology: complementing Pickett et al." *Ecology and Society* 15 (4) (2010).

Müller, Jakob C., Jesko Partecke, Ben J. Hatchwell, Kevin J. Gaston, and Karl L. Evans. "Candidate gene polymorphisms for behavioural adaptations during urbanization in blackbirds." *Molecular Ecology* 22 (2013): 3629–3637.

Müller, Norbert. "Most frequently occurring vascular plants and the role of non-native species in urban areas—a comparison of selected cities of the old and new worlds." In *Urban Biodiversity and Design*, ed. Norbert Müller, Peter Werner, and John G. Kelcey, 227–242. Hoboken, NJ: Wiley-Blackwell, 2010.

Müller-Stoll, Wolfgang. *Der Pflanzenwelt Brandenburgs*. Berlin-Kleinmachnow: Gartenverlag, 1955.

Muratet, Audrey, Nathalie Machon, Frédéric Jiguet, Jacques Moret, and Emmanuelle Porcher. "The role of urban structures in the distribution of wasteland flora in the Greater Paris Area, France." *Ecosystems* 10 (2007): 661–671.

Murphy, Kathleen S. "Collecting slave traders: James Petiver, natural history, and the British slave trade." *William and Mary Quarterly* 70 (4) (2013): 637–670.

Murphy, Michelle. "Alterlife and decolonial chemical relations." *Cultural Anthropology* 32 (4) (2017): 494–503.

Nading, Alex M. *Mosquito trails: ecology, health, and the politics of entanglement.* Berkeley: University of California Press, 2014.

Naegeli, O., and A. Thellung. "Die Flora Kantons Zürich. I Teil: Die Ruderal- und Adventivflora des Kantons Zürich." *Vierteljahrsschrift der Naturforschenden Gesellschaft in Zürich* 50 (1905): 225–305.

Nagendra, Harini. *Nature in the city: Bengaluru in the past, present, and future.* Delhi: Oxford University Press, 2016.

Nagy, Kelsi, and Phillip David Johnson II, eds. *Trash animals: how we live with nature's filthy, feral, invasive, and unwanted species.* Minneapolis: University of Minnesota Press, 2013.

Napoletano, Brian M., John Bellamy Foster, Brett Clark, Pedro S. Urquijo, Michael K. McCall, and Jaime Paneque-Gálvez. "Making space in critical environmental geography for the metabolic rift." *Annals of the American Association of Geographers* 109 (6) (2019): 1811–1828.

Narayanan, Yamini. "Animating caste: visceral geographies of pigs, caste, and violent nationalisms in Chennai city." *Urban Geography* 42 (2021).

Narayanan, Yamini. "Street dogs at the intersection of colonialism and informality: 'subaltern animism' as a posthuman critique of Indian cities." *Environment and Planning D: Society and Space* 35 (3) (2017): 475–494.

Narayanan, Yamini, and Sumanth Bindumadhav. "'Posthuman cosmopolitanism' for the Anthropocene in India: urbanism and human-snake relations in the Kali Yuga." *Geoforum* 106 (2019): 402–410.

Nash, T. H., and C. Gries. "Lichens as indicators of air pollution." In *Air pollution,* ed. Otto Hutzinger, 1–29. Berlin: Springer, 1991.

Needham, Andrew. *Power lines: Pheonix and the making of the modern Southwest.* Princeton, NJ: Princeton University Press, 2014.

Nesbitt, Kate. *Theorizing a new agenda for architecture: an anthology of architectural theory 1965–1995.* Princeton, NJ: Princeton Architectural Press, 1996.

Neumann, Roderick P. "Life zones: the rise and decline of a theory of the geographic distribution of species." In *Spatializing the history of ecology: sites, journeys, mappings,* ed. Raf de Bont and Jens Lachmund, 37–55. London: Routledge, 2017.

Newcombe, Ken, Jetse D. Kalma, and Alan R. Aston. "The metabolism of a city: the case of Hong Kong." *Ambio* (1978): 3–15.

Newell, Joshua P., and Laura A. Henry. "The state of environmental protection in the Russian Federation: a review of the post-Soviet era." *Eurasian Geography and Economics* 57 (6) (2016): 779–801.

Newman, Andrew. *Landscape of discontent: urban sustainability in immigrant Paris*. Minneapolis: University of Minnesota Press, 2015.

Neyrat, Frédéric. *The unconstructable earth: an ecology of separation*. Trans. Drew S. Burk. New York: Fordham University Press, 2019 [2016].

Ngai, Sianne. *Our aesthetic categories: zany, cute, interesting*. Cambridge, MA: Harvard University Press, 2012.

Niewöhner, Jörg. "Epigenetics: embedded bodies and the molecularisation of biography and milieu." *BioSocieties* 6 (3) (2011): 279–298.

Nixon, Rob. *Slow violence and the environmentalism of the poor*. Cambridge, MA: Harvard University Press, 2011.

Nohl, Werner. "Sustainable landscape use and aesthetic perception—preliminary reflections on future landscape aesthetics." *Landscape and Urban Planning* 54 (1–4) (2001): 223–237.

Norris, Margot. *Beasts of the modern imagination: Darwin, Nietzsche, Kafka, Ernst, and Lawrence*. Baltimore: Johns Hopkins University Press, 1985.

Northridge, Jennifer, et al. "The role of housing type and housing quality in urban children with asthma." *Journal of Urban Health* 87 (2) (2010): 211–224.

Numata, Makoto, ed. *Tokyo project: interdisciplinary studies of urban ecosystems in the metropolis of Tokyo*. Chiba: Chiba University, 1977.

Nurka, Camille. "Animal techne: transing posthumanism." *TSQ: Transgender Studies Quarterly* 2 (2) (2015): 209–226.

Nyhart, Lynn K. "Natural history and the 'new' biology." In *Cultures of natural history*, ed. Nicholas Jardine, James A. Secord, and Emma C. Spary, 426–443. Cambridge: Cambridge University Press, 1996.

Nylander, William. "Les lichens du Jardin du Luxembourg." *Bulletin de la Société botanique de France* 13 (7) (1866): 364–371.

Nyman, Marcus. "Food, meaning-making and ontological uncertainty: exploring 'urban foraging' and productive landscapes in London." *Geoforum* 99 (2019): 170–180.

O'Connor, James R. *Natural causes: essays in ecological Marxism*. London: Guilford, 1998.

O'Connor, Terry. *Animals as neighbors: the past and present of commensal species*. East Lansing: Michigan State University Press, 2009.

Olden, Julian D. "Biotic homogenization: a new research agenda for conservation biogeography." *Journal of Biogeography* 33 (2006): 2027–2039.

Ordish, George. *The living house*. London: Rupert Hart-Davis, 1960.

Orff, Kate. *Toward an urban ecology*. New York: Monacelli Press, 2016.

Orwell, George. *Down and out in Paris and London*. Harmondsworth: Penguin, 1940 [1933].

Otter, Chris. "Technosphere." In *Concepts of urban-environmental history*, ed. Sebastian Haumann, Martin Knoll, and Detlev Mares, 21–32. Bielefeld: transcript, 2020.

Otter, Chris. "The technosphere: a new concept for urban studies." *Urban History* 44 (1) (2017): 145–154.

Ottich, Indra, Dirk Bönsel, Thomas Gregor, Andreas Malten, Georg Zizka, Uwe Barth, Kurt Baumann, and Klaus Hoppe. *Natur vor der Haustür—Stadtnatur in Frankfurt am Main*. Stuttgart: E. Schweizerbart'sche Verlagsbuchhandlung, 2009.

Owen, Jennifer. *The ecology of a garden: the first fifteen years*. Cambridge: Cambridge University Press, 1991.

Owen, Jennifer. *Wildlife of a garden: a thirty-year study*. London: Royal Horticultural Society, 2010.

Özkaynak, Begüm, Cem İskender Aydın, Pınar Ertör-Akyazı, and Irmak Ertör. "The Gezi Park resistance from an environmental justice and social metabolism perspective." *Capitalism Nature Socialism* 26 (1) (2015): 99–114.

Pachirat, Timothy. *Every twelve seconds: industrialized slaughter and the politics of sight*. New Haven, CT: Yale University Press, 2011.

Page, Michael. "Evolution and apocalypse in the Golden Age." In *Green planets: ecology and science fiction*, ed. Gerry Canavan and Kim Stanley Roberts, 40–55. Middletown, CT: Wesleyan University Press, 2014.

Paiva, Daniel, and Eduardo Brito-Henriques. "A podcast on urban ruins, or the aural weaving of theory and field." *Cultural Geographies* 26 (4) (2019): 535–540.

Paixão, Enny S., Maria Gloria Teixeira, and Laura C. Rodrigues. "Zika, chikungunya and dengue: the causes and threats of new and re-emerging arboviral diseases." *BMJ Global Health* 3 (1) (2018): e000530.

Palmer, Clare. *Animal ethics in context*. New York: Columbia University Press, 2010.

Palmer, Clare. "Placing animals in urban environmental ethics." *Journal of Social Philosophy* 34 (1) (2003): 64–78.

Panagiotakopulu, Eva, and Paul C. Buckland. "A thousand bites—insect introductions and late Holocene environments." *Quaternary Science Reviews* 156 (2017): 23–35.

Parikka, Jussi. *The Anthrobscene*. Minneapolis: University of Minnesota Press, 2014.

Parikka, Jussi. "Deep times of planetary trouble." *Cultural Politics* 12 (3) (2016): 279–292.

Park, Chris C. *Chernobyl: the long shadow*. London: Routledge, 1989.

Park, Lisa Sun-Hee, and David N. Pellow. *The slums of Aspen: immigrants vs. the environment in America's Eden*. New York: New York University Press, 2011.

Park, Robert E., Ernest W. Burgess, and R. D. McKenzie, eds. *The City*. Chicago: University of Chicago Press, 1925.

Pascal, Blaise. *Expériences nouvelles touchant le vuide*. Paris: Pierre Margat, 1647.

Patchin, Paige Marie. "Thresholds of empire: women, biosecurity, and the Zika chemical vector program in Puerto Rico." *Annals of the American Association of Geographers* 110 (4) (2020): 967–982.

Patrick, Darren. "Queering the urban forest: invasions, mutualisms, and eco-political creativity with the tree of heaven (*Ailanthus altissima*)." In *Urban forests, trees, and greenspace*, ed. L. Anders Sandberg, Adrina Bardekjian, and Sadia Butt, 191–206. London: Routledge, 2015.

Pellegrini, Patricia, and Sandrine Baudry. "Streets as new places to bring together both humans and plants: examples from Paris and Montpellier (France)." *Social and Cultural Geography* 15 (8) (2014): 871–900.

Pennock, Hanna. "Natural history museum security." In *The future of natural history museums*, ed. Eric Dorfman, 49–64. London: Routledge, 2018.

Peppler, A. "Die Temperaturverhältnisse von Karlsruhe an heissen Sommertagen." *Deutsches meteorologisches Jahrbuch Baden* 61 (1929): 59–60.

Perkins, Chas. E. "Ballast plants in Boston and vicinity." *Botanical Gazette* 8 (3) (1883): 188–190.

Phillips, Richard. "Georges Perec's experimental fieldwork; Perecquian fieldwork." *Social and Cultural Geography* 19 (2) (2018): 171–191.

Philo, Chris. "Animals, geography and the city: notes on inclusions and exclusions." *Environment and Planning D: Society and Space* 13 (1995): 655–681.

Pick, Daniel. *War machine: the rationalization of slaughter in the modern age*. New Haven, CT: Yale University Press, 1993.

Pickering, Andrew. *The mangle of practice: time, agency, and science*. Chicago: University of Chicago Press, 1995.

Pickett, Steward T. A., and Mary L. Cadenasso. "Advancing urban ecological studies: frameworks, concepts, and results from the Baltimore Ecosystem Study." *Austral Ecology* 31 (2) (2006): 114–125.

Pickett, Steward T. A., Mary L. Cadenasso, Matthew E. Baker, Lawrence E. Band, Christopher G. Boone, Geoffrey L. Buckley, Peter M. Groffman et al. "Theoretical perspectives of the Baltimore Ecosystem Study: conceptual evolution in a social-ecological research project." *BioScience* 70 (4) (2020): 297–314.

Pickett, Steward T. A., Mary L. Cadenasso, and J. Morgan Grove. "Resilient cities: meaning, models, and metaphor for integrating the ecological, socio-economic, and planning realms." *Landscape and Urban Planning* 69 (4) (2004): 369–384.

Pieroni, Andrea, and Ina Vanderbroek. *Traveling cultures and plants: the ethnobiology and ethnopharmacy of human migration.* New York: Berghahn, 2007.

Pietri, Pedro. *Puerto Rican obituary.* New York: Monthly Review Press, 1973.

Pincetl, Stephanie. "Cities as novel biomes: recognizing urban ecosystem services as anthropogenic." *Frontiers in Ecology and Evolution* 3 (2015).

Plater, Zygmunt J. B. "From the beginning, a fundamental shift of paradigms: a theory and short history of environmental law." *Loyola of Los Angeles Law Review* 27 (3) (1994): 981–1008.

Poe, Melissa R., Joyce LeCompte, Rebecca McLain, and Patrick Hurley. "Urban foraging and the relational ecologies of belonging." *Social and Cultural Geography* 15 (8) (2014): 901–919.

Polgovsky Ezcurra, Mara. "The flight of seeds." In *The botanical city*, ed. Matthew Gandy and Sandra Jasper, 122–130. Berlin: jovis, 2020.

Pongratz, Julia, C. H. Reick, T. Raddatz, and Martin Claussen. "Effects of anthropogenic land cover change on the carbon cycle of the last millennium." *Global Biogeochemical Cycles* 23 (4) (2009): GB4001.

Poppendieck, Hans-Helmut, Gisela Bertram, and Barbara Engelschall, eds. *Der botanische Wanderführer für Hamburg und Umgebung.* Hamburg and Munich: Dörling and Galitz, 2016.

Potter, Michael F. "The history of bed bug management—with lessons from the past." *American Entomologist* 57 (1) (2011): 12–25.

Primack, Richard, Hiromi Kobori, and Seiwa Mori. "Dragonfly pond restoration promotes conservation awareness in Japan." *Conservation Biology* 14 (5) (2000): 1553–1554.

Prior, Jonathan, and Kim J. Ward. "Rethinking rewilding: a response to Jørgensen." *Geoforum* 69 (2016): 132–135.

Pugliese, Joseph. *Biopolitics of the more-than-human: forensic ecologies of violence.* Durham, NC: Duke University Press, 2020.

Puppim de Oliveira, J. A., Osman Balaban, Christopher N. H. Doll, Raquel Moreno-Peñaranda, Alexandros Gasparatos, Deljana Iossifova, and Aki Suwa. "Cities and biodiversity: perspectives and governance challenges for implementing the Convention on Biological Diversity (CBD) at the city level." *Biological Conservation* 144 (5) (2011): 1302–1313.

385

Pyšek, Petr. "Alien and native species in Central European urban floras: a quantitative comparison." *Journal of Biogeography* 25 (1998): 155–163.

Rademacher, Anne M., and Kalyanakrishnan Sivaramakrishnan, eds. *Ecologies of urbanism in India: metropolitan civility and sustainability.* Hong Kong: Hong Kong University Press, 2013.

Radkau, Joachim, and Frank Uekötter, eds. *Naturschutz und Nationalsozialismus.* Frankfurt: Campus Verlag, 2003.

Rajamani, Lavanya. "Public interest environmental litigation in India: exploring issues of access, participation, equity, effectiveness and sustainability." *Journal of Environmental Law* 19 (3) (2007): 293–321.

Raman, Bhavani. "Sovereignty, property and land development: the East India Company in Madras." *Journal of the Economic and Social History of the Orient* 61 (5–6) (2018): 976–1004.

Ramirez, Kelly S., Jonathan W. Leff, Albert Barberán, Scott Thomas Bates, Jason Betley, Thomas W. Crowther, Eugene F. Kelly, et al. "Biogeographic patterns in below-ground diversity in New York City's Central Park are similar to those observed globally." *Proceedings of the Royal Society B: Biological Sciences* 281 (1795) (2014): 20141988.

Rasmussen, Kim Su. "Foucault's genealogy of racism." *Theory, Culture and Society* 28 (5) (2011): 34–51.

Räthzel, Nora. "Germany: one race, one nation." *Race and Class* 32 (3) (1990): 31–48.

Rebele, Franz. "Urban ecology and special features of urban ecosystems." *Global Ecology and Biogeography Letters* 4 (6) (1994): 173–187.

Rees, Ronald. "The taste for mountain scenery." *History Today* 25 (1975): 305–312.

Rehman, Nida. "Epidemiological landscapes: the spaces and politics of mosquito control in Lahore." PhD thesis, University of Cambridge, 2020.

Reinhardt, Klaus. *Bedbug.* London: Reaktion, 2018.

Reumer, Jelle. *Wildlife in Rotterdam: nature in the city.* Trans. Anthony Runia. Rotterdam: Natuurhistorisch Museum Rotterdam, 2014.

Reuß, Jürgen von. "Freiflächenpolitik als Sozialpolitik." In *Martin Wagner 1885–1975. Wohnungsbau und Weltstadtplanung. Die Rationalisierung des Glücks,* ed. Klaus Homann, Martin Kieren, and Ludovica Scarpa, 49–65. Berlin: Akademie der Künste, 1987.

Reznick, David N. *The "Origin" then and now: an interpretive guide to the "Origin of species."* Princeton, NJ: Princeton University Press, 2010.

Richter, Steffen, and Andreas Rötzer, eds. *Dritte Natur: Technik Kapital Umwelt.* Berlin: Matthes & Seitz, 2018.

Riechelmann, Cord. *Wilde Tiere in der Großstadt*. Berlin: Nicolai, 2004.

Rieley, John O., and Susan E. Page. "Survey, mapping, and evaluation of green space in the Federal Territory of Kuala Lumpur, Malaysia." In *Urban ecology as the basis of urban planning*, ed. Herbert Sukopp, Makoto Numata, and A. Huber, 173–183. The Hague: SPB Academic Publishing, 1995.

Riley, Seth P. D., Brian L. Cypher, and Stanley D. Gehrt. *Urban carnivores: ecology, conflict, and conservation*. Baltimore: Johns Hopkins University Press, 2010.

Rindi, Fabio. "Diversity, distribution and ecology of green algae and cyanobacteria in urban habitats." In *Algae and cyanobacteria in extreme environments*, ed. Joseph Seckbach, 619–638. Dordrecht: Springer.

Rindi, Fabio, and Michael D. Guiry. "Composition and distribution of subaerial algal assemblages in Galway City, western Ireland." *Cryptogamie. Algologie* 24 (3) (2003): 245–267.

Rink, Dieter. "Wilderness: the nature of urban shrinkage?" *Nature and Culture* 4 (2009): 275–292.

Ritvo, Harriet. *The animal estate: the English and other creatures in the Victorian Age*. London: Penguin, 1987.

Robbins, Paul, and Joanne Sharp. "Turfgrass subjects: the political economy of urban monoculture." In *In the nature of cities: urban political ecology and the politics of urban metabolism*, ed. Nik Heynen, Maria Kaïka, and Erik Swyngedouw, 110–128. London: Routledge, 2006.

Robertson, Morgan M. "The nature that capital can see: science, state, and market in the commodification of ecosystem services." *Environment and Planning D: Society and Space* 24 (3) (2006): 367–387.

Rodríguez-Díaz, Carlos E., Adriana Garriga-López, Souhail M. Malavé-Rivera, and Ricardo L. Vargas-Molina. "Zika virus epidemic in Puerto Rico: health justice too long delayed." *International Journal of Infectious Diseases* 65 (2017): 144–147.

Rorty, Richard. *Philosophy and the mirror of nature*. Princeton, NJ: Princeton University Press, 1979.

Rose, Morag. "Confessions of an anarcho-flâneuse, or psychogeography the Mancunian way." In *Walking inside out: contemporary British psychogeography*, ed. Tina Richardson. London: Rowman & Littlefield, 2015.

Rosengren, Mathilda. "There's life in dead wood: tracing a more-than-human urbanity in the spontaneous nature of Gothenburg." In *The botanical city*, ed. Matthew Gandy and Sandra Jasper, 229–236. Berlin: jovis, 2020.

Rosenzweig, Michael L. *Species diversity in space and time*. Cambridge: Cambridge University Press, 1995.

Rosol, Marit, Vincent Béal, and Samuel Mössner. "Greenest cities? The (post-) politics of new urban environmental regimes." *Environment and Planning A: Economy and Space* 49 (8) (2017): 1710–1718.

Ross, Andrew. *Bird on fire: lessons from the world's least sustainable city.* Oxford: Oxford University Press, 2011.

Rothenberg, David. *Nightingales in Berlin: searching for the perfect sound.* Chicago: University of Chicago Press, 2019.

Rotheram, Ian D. "'The city that hates trees': standing up to the Sheffield street tree slaughter." *Ecos* 39 (3) (13 July 2018).

Rublowsky, John. *Nature in the city.* New York: Basic Books, 1967.

Ruff, Allan. "Holland and the ecological landscape." *Garden History* 30 (2) (2002): 239–251.

Rupprecht, Christoph D. D., and Jason A. Byrne. "Informal urban greenspace: a typology and trilingual systematic review of its role for urban residents and trends in the literature." *Urban Forestry and Urban Greening* 13 (4) (2014): 597–611.

Rutherford, Stephanie. "The Anthropocene's animal? Coywolves as feral cotravelers." *Environment and Planning E: Nature and Space* 1 (1–2) (2018): 206–223.

Sadler, Jon, Adam Bates, James Hale, and Philip James. "Bringing cities alive: the importance of urban green spaces for people and biodiversity." In *Urban ecology*, ed. Kevin J. Gaston, 230–261. Cambridge: Cambridge University Press, 2010.

Safransky, Sara. "Greening the urban frontier: race, property, and resettlement in Detroit." *Geoforum* 56 (2014): 237–248.

Saint-Laurent, Diane. "Approches biogéographiques de la nature en ville: parcs, espaces verts et friches." *Cahiers de Géographie du Québec* 44 (2000): 147–166.

Saito, Yuriko. "The aesthetics of unscenic nature." *Journal of Aesthetics and Art Criticism* 56 (2) (1998): 101–111.

Saldanha, Arun. "A date with destiny: racial capitalism and the beginnings of the Anthropocene." *Environment and Planning D: Society and Space* 38 (1) (2020): 12–34.

Šálek, Martin, Lucie Drahníková, and Emil Tkadlec. "Changes in home range sizes and population densities of carnivore species along the natural to urban habitat gradient." *Mammal Review* 45 (1) (2015): 1–14.

Samuelson, Ash E., Richard J. Gill, Mark J. F. Brown, and Ellouise Leadbeater. "Lower bumblebee colony reproductive success in agricultural compared with urban environments." *Proceedings of the Royal Society B: Biological Sciences* 285 (1881) (2018): 20180807.

Sanders, Dirk, Enric Frago, Rachel Kehoe, Christophe Patterson, and Kevin J. Gaston. "A meta-analysis of biological impacts of artificial light at night." *Nature Ecology and Evolution* 5 (1) (2021): 74–81.

Sanderson, Eric W. *Mannahatta: a natural history of New York City*. New York: Abrams, 2009.

Santner, Eric L. *On creaturely life: Rilke, Benjamin, Sebald*. Chicago: University of Chicago Press, 2006.

Santos, Boaventura de Sousa. "Law: a map of misreading. Toward a postmodern conception of law." *Journal of Law and Society* 14 (3) (1987): 297–302.

Santos, Boaventura de Sousa, and César A. Rodríguez-Garavito, eds. *Law and globalization from below: towards a cosmopolitan legality*. Cambridge: Cambridge University Press, 2005.

Sarasin, Philipp. "Zweierlei Rassismus? Die Selektion des Fremden als Problem in Michel Foucaults Verbindung von Biopolitik und Rassismus." In *Biopolitik und Rassismus*, ed. Martin Stingelin, 55–79. Frankfurt: Suhrkamp, 2003.

Sarpong, Sampson B., et al. "Socioeconomic status and race as risk factors for cockroach allergen exposure and sensitization in children with asthma." *Journal of Allergy and Clinical Immunology* 97 (6) (1996): 1393–1401.

Savard, Jean-Pierre, Philippe Clergeau, and Gwenaelle Mennechez. "Biodiversity concepts and urban ecosystems." *Landscape and Urban Planning* 48 (3–4) (2000): 131–142.

Savransky, Martin. "The pluralistic problematic: William James and the pragmatics of the pluriverse." *Theory, Culture and Society* 38 (2) (2021): 141–159.

Schadek, Ute, Barbara Strauss, Robert Biedermann, and Michael Kleyer. "Plant species richness, vegetation structure and soil resources of urban brownfield sites linked to successional age." *Urban Ecosystems* 12 (2009): 115–126.

Schenk, Winfried. "'Landschaft' und 'Kulturlandschaft'—'getönte' Leitbegriffe für aktuelle Konzepte geographischer Forschung und räumlicher Planung." *Petermanns geographische Mitteilungen* 146 (6) (2002): 6–13.

Schiebinger, Londa. "Gender and natural history." In *Cultures of natural history*, ed. Nicholas Jardine, James A. Secord, and Emma C. Spary, 163–177. Cambridge: Cambridge University Press, 1996.

Schiebinger, Londa. *Plants and empire: colonial bioprospecting in the Atlantic world*. Cambridge, MA: Harvard University Press, 2004.

Schilthuizen, Menno. *Darwin comes to town: how the urban jungle drives evolution*. London: Quercus, 2018.

Schinkel, Anders. "Martha Nussbaum on animal rights." *Ethics and the Environment* 13 (1) (2008): 41–69.

Schivelbusch, Wolfgang. *In a cold crater: cultural and intellectual life in Berlin, 1945–1948*, trans. Kelly Barry. Berkeley: University of California Press, 1998.

Schmidt, Alfred. *The concept of nature in Marx*. Trans. Ben Fowkes. London: Verso, 2014 [1962].

Scholz, Hildemar. "Die Ruderalvegetation Berlins." PhD dissertation, Free University Berlin, 1956.

Scholz, Hildemar. "Die Veränderungen in der Ruderalflora Berlins. Ein Beitrag zur jüngsten Florengeschichte." *Willdenowia* 2 (3) (1960): 379–397.

Schroeder, Susanne, ed. *Skulpturenpark Berlin_Zentrum*. Cologne: Walther König, 2010.

Schultz, Hans-Dietrich. "Deutschlands 'natürliche' Grenzen. 'Mittellage' und 'Mitteleuropa' in der Diskussion der Geographen seit dem Beginn des 19. Jahrhunderts." *Geschichte und Gesellschaft* 15 (2) (1989): 248–281.

Schuppli, Susan. "Dirty pictures." In *Living earth: field notes from the Dark Ecology Project 2014–2016*, ed. Mirna Belina, 189–210. Amsterdam: Sonic Acts Press, 2016.

Schuppli, Susan. *Material witness: media, forensics, evidence*. Cambridge, MA: MIT Press, 2020.

Sears, Paul B. "Human ecology: a problem in synthesis." *Science* 120 (3128) (1954): 959–963.

Sebald, W. G. *Austerlitz*. Trans. Anthea Bell. London: Penguin, 2002 [2001].

Sebald, W. G. *The natural history of destruction*. Trans. Anthea Bell. New York: Modern Library, 2004 [1999].

Sebald, W. G. "Zwischen Geschichte und Naturgeschichte. Über die literarische Beschreibung totaler Zerstörung." In *Campo santo*, 69–100. Frankfurt am Main: Fischer, 2006.

Secord, Anne. "Science in the pub: artisan botanists in early 19th century Lancashire." *History of Science* 32 (3) (1994): 269–315.

Secord, James A. "The crisis of nature." In *Cultures of natural history*, ed. Nicholas Jardine, James A. Secord, and Emma C. Spary, 447–459. Cambridge: Cambridge University Press, 1996.

Şekercioğlu, Çağan H., Sean Anderson, Erol Akçay, Raşit Bilgin, Özgün Emre Can, Gürkan Semiz, Çağatay Tavşanoğlu, et al. "Turkey's globally important biodiversity in crisis." *Biological Conservation* 144 (12) (2011): 2752–2769.

Sepers, Bernice, Krista van den Heuvel, Melanie Lindner, Heidi Viitaniemi, Arild Husby, and Kees van Oers. "Avian ecological epigenetics: pitfalls and promises." *Journal of Ornithology* (2019): 1–21.

Serres, Michel. *The natural contract*. Trans. Elizabeth MacArther and William Paulson. Ann Arbor: University of Michigan Press, 1995.

Shanahan, Danielle F., Richard A. Fuller, Robert Bush, Brenda B. Lin, and Kevin J. Gaston. "The health benefits of urban nature: how much do we need?" *BioScience* 65 (5) (2015): 476–485.

Shellenberger, Michael, and Ted Nordhaus, eds. *Love your monsters: postenvironmentalism and the Anthropocene.* Oakland, CA: Breakthrough Institute, 2011.

Shenstone, J. C. "The flora of London building-sites." *Journal of Botany* 50 (1912): 117–124.

Shochat, Eyal, Paige S. Warren, Stanley H. Faeth, Nancy E. McIntyre, and Diane Hope. "From patterns to emerging processes in mechanistic urban ecology." *Trends in Ecology and Evolution* 21 (4) (2006): 186–191.

Siddig, Ahmed, Aaron Ellison, Alison Ochs, Claudia Villar-Leeman, and Matthew K. Lau. "How do ecologists select and use indicator species to monitor ecological change? Insights from 14 years of publication in *Ecological Indicators.*" *Ecological Indicators* 60 (2016): 223–230.

Simak, Clifford D. *City.* Garden City, NY: Doubleday, 1952.

Simonis, Udo E. "Ecological modernization of industrial society: three strategic elements." WZB Discussion Paper, No. FS II 88-401. Berlin: Wissenschaftszentrum Berlin für Sozialforschung, 1988.

Sinclair, Upton. *The Jungle.* London: Penguin, 1985 [1906].

Sinha, Anindya. "Not in their genes: phenotypic flexibility, behavioural traditions and cultural evolution in wild bonnet macaques." *Journal of Biosciences* 30 (1) (2005): 51–64.

Skogen, Ketil, and Olve Krange. "A wolf at the gate: the anti-carnivore alliance and the symbolic construction of community." *Sociologia Ruralis* 43 (3) (2003): 309–325.

Slabbekoorn, Hans, and Ardie den Boer-Visser. "Cities change the songs of birds." *Current Biology* 16 (23) (2006): 2326–2331.

Slabbekoorn, Hans, and Margriet Peet. "Birds sing at higher pitch in urban noise." *Nature* 424 (2003): 267.

Smith, Betty. *A tree grows in Brooklyn.* Harmondsworth: Penguin, 1951 [1943].

Smith, Neil. "Nature at the millennium: production and re-enchantment." In *Remaking reality: nature at the millennium*, ed. Bruce Braun and Noel Castree, 269–282. London: Routledge, 1998.

Smith, Neil. *Uneven development: nature, capital, and the production of space.* Oxford: Blackwell, 1984.

Sorensen, Willis Conner. *Brethren of the net: American entomology, 1840–1880.* Tuscaloosa: University of Alabama Press, 1995.

Sorkin, Michael. "See you in Disneyland." In *Variations on a theme park: the new American city and the end of public space*, ed. Michael Sorkin, 205–232. New York: Hill and Wang, 1992.

Soulé, Michael. "The onslaught of alien species, and other challenges in the coming decades." *Conservation Biology* 4 (3) (1990): 233–239.

Soulier, Nicolas. *Reconquérir les rues. Exemples à travers le monde et pistes d'actions*. Paris: Ulmer, 2012.

Sparke, Matthew, and Dimitar Anguelov. "H1N1, globalization and the epidemiology of inequality." *Health and Place* 18 (4) (2012): 726–736.

Sperb, Jason. "The end of Detropia: Fordist nostalgia and the ambivalence of poetic ruins in visions of Detroit." *Journal of American Culture* 39 (2) (2016): 212–227.

Spirn, Anne Whiston. *The granite garden: urban nature and human design*. New York: Basic Books, 1984.

Springgay, Stephanie, and Sarah E. Truman. *Walking methodologies in a more-than-human world: WalkingLab*. London: Routledge, 2018.

Srinivasan, Krithika. "The biopolitics of animal being and welfare: dog control and care in the UK and India." *Transactions of the Institute of British Geographers* 38 (1) (2012): 106–119.

Srinivasan, Krithika. "Remaking more-than-human society: thought experiments on street dogs as 'nature.'" *Transactions of the Institute of British Geographers* 44 (2) (2019): 376–391.

Stableford, Brian. "Science fiction and ecology." In *A companion to science fiction*, ed. David Seed, 127–141. Oxford: Blackwell, 2005.

Stace, Clive A., and Michael J. Crawley. *Alien plants*. London: HarperCollins, 2015.

Steiner, Frederick. "Landscape ecological urbanism: origins and trajectories." *Landscape and Urban Planning* 100 (4) (2011): 333–337.

Stengers, Isabelle. *Another science is possible: a manifesto for slow science*. Trans. Stephen Muecke. Cambridge: Polity, 2017 [2013].

Stengers, Isabelle. *Cosmopolitics II*. Trans. Robert Bononno. Minneapolis: University of Minnesota Press, 2011 [2003].

Stevens, Wallace. "Thirteen ways of looking at a blackbird." In *Harmonium* (New York: Alfred A. Knopf, 1931 [1917]).

Stjernfelt, Frederik. "Simple animals and complex biology: von Uexküll's two-fold influence on Cassirer's philosophy." *Synthese* 179 (1) (2011): 169–186.

Stoetzer, Bettina. "*Ailanthus altissima*, or the botanical afterlives of European power." In *The botanical city*, ed. Matthew Gandy and Sandra Jasper, 82–91. Berlin: jovis, 2020.

Stoetzer, Bettina. "Ruderal ecologies: rethinking nature, migration, and the urban landscape in Berlin." *Cultural Anthropology* 33 (2) (2018): 295–323.

Stoler, Ann Laura. "Rethinking colonial categories: European communities and the boundaries of rule." *Comparative Studies in Society and History* 31 (1) (1989): 135–161.

Strombeck, Andrew. "The network and the archive: the specter of imperial management in William Gibson's *Neuromancer*." *Science Fiction Studies* 37 (2) (2010): 275–295.

Strubbe, Diederik, and Erik Matthysen. "Experimental evidence for nest-site competition between invasive ring-necked parakeets (*Psittacula krameri*) and native nuthatches (*Sitta europaea*)." *Biological Conservation* 142 (8) (2009): 1588–1594.

Sukopp, Herbert. "Beitrage zur Ökologie von *Chenopodium botrys* L. I. Verbreitung und Vergesellschaftung." *Verhandlungen des botanischen Vereins der Provinz Brandenburg* 108 (1971): 2–25.

Sukopp, Herbert. "On the early history of urban ecology in Europe." *Preslia, Praha* 74 (2002): 373–393.

Sukopp, Herbert, ed. *Stadtökologie. Das Beispiel Berlins*. Berlin: Dietrich Reimer, 1990.

Sukopp, Herbert, Hans-Peter Blume, and Wolfram Kunick. "The soil, flora and vegetation of Berlin's waste lands." In *Nature in cities: the natural environment in the design and development of urban green space*, ed. Ian Laurie, 115–134. Chichester: Wiley, 1979.

Sumberg, James. "Poultry production in and around Dar es Salaam, Tanzania: competition and complementarity." *Outlook on Agriculture* 27 (3) (1998): 177–185.

Suvin, Darko. *Metamorphoses of Science Fiction*. New Haven, CT: Yale University Press, 1979.

Swyngedouw, Erik. "Into the sea: desalination as hydro-social fix in Spain." *Annals of the Association of American Geographers* 103 (2) (2013): 261–270.

Swyngedouw, Erik. *Social power and the urbanization of water: flows of power*. Oxford: Oxford University Press, 2004.

Swyngedouw, Erik, and Henrik Ernstson. "Interrupting the anthropo-obScene: immuno-biopolitics and depoliticizing ontologies in the Anthropocene." *Theory, Culture and Society* 35 (6) (2018): 3–30.

Takacs, David. *The idea of biodiversity: philosophies of paradise*. Baltimore: Johns Hopkins University Press, 1996.

Tansley, Arthur G. "The use and abuse of vegetational concepts and terms." *Ecology* 16 (3) (1935): 284–307.

Tarlock, A. Dan. "The nonequilibrium paradigm in ecology and the partial unraveling of environmental law." *Loyola of Los Angeles Law Review* 27 (1994): 1121–1144.

Tarr, Joel A. *The search for the ultimate sink: urban pollution in historical perspective*. Akron, OH: University of Akron Press, 1996.

Tavares, Paolo. "Nonhuman rights." In *Forensis: the architecture of public truth*, ed. Eyal Weizman, 553–572. Berlin: Sternberg Press, 2014.

Taylor, Dorceta E. "Minority environmental activism in Britain: from Brixton to the Lake District." *Qualitative Sociology* 16 (3) (1993): 263–295.

Teagle, W. G. "The fox in the London suburbs." *London Naturalist* 46 (1967): 44–68.

Teixeira, Catarina Patoilo, and Cláudia Oliveira Fernandes. "Novel ecosystems: a review of the concept in non-urban and urban contexts." *Landscape Ecology* 35 (1) (2020): 23–39.

Teubner, Gunther. "Rights of non-humans? Electronic agents and animals as new actors in politics and law." *Journal of Law and Society* 33 (4) (2006): 497–521.

Thomas, Chris. *Inheritors of the earth: how nature is thriving in an era of extinction.* London: Allen Lane, 2017.

Till, Karen E. "Interim use at a former death strip? Art, politics and urbanism at Skulpturenpark Berlin_Zentrum." In *After the Wall: Berlin in Germany and Europe*, ed. Marc Silberman, 99–122. Basingstoke: Palgrave Macmillan.

Tironi, Manuel. "Lithic abstractions: geophysical operations against the Anthropocene." *Distinktion: Journal of Social Theory* 20 (3) (2019): 284–300.

Tixier, Nicolas. "Le transect urbain. Pour une écriture corrélée des ambiances et de l'environnement." In *Écologies urbaines. Sur le terrain*, ed. Sabine Barles and Nathalie Blanc, 130–148. Paris: Economica-Anthropos, 2016.

Tolia-Kelly, Divya P. "Landscape, race and memory: biographical mapping of the routes of British Asian landscape values." *Landscape Research* 29 (3) (2004): 277–292.

Tompkins, David. "Weird ecology: on the *Southern Reach* trilogy." *Los Angeles Review of Books* (30 September 2014).

Tournefort, Joseph Pitton de. *Histoire des plantes qui naissent aux environs de Paris, avec leur usage dans la medecine.* Paris: Imprimerie Royale, 1698.

Trepl, Ludwig. "City and ecology." *Capitalism, Nature, Socialism* 7 (2) (1996): 85–94.

Tsing, Anna. "The buck, the bull, and the dream of the stag: some unexpected weeds of the Anthropocene." *Suomen Antropologi: Journal of the Finnish Anthropological Society* 42 (1) (2017): 3–21.

Tsing, Anna. *The mushroom at the end of the world: on the possibility of life in capitalist ruins.* Princeton, NJ: Princeton University Press, 2015.

Tuan, Yi-Fu. *Dominance and affection.* New Haven, CT: Yale University Press, 2003.

Turcu, Catalina. "Re-thinking sustainability indicators: local perspectives of urban sustainability." *Journal of Environmental Planning and Management* 56 (5) (2013): 695–719.

Turnbull, Jonathon. "Checkpoint dogs: photovoicing canine companionship in the Chernobyl exclusion zone." *Anthropology Today* 36 (6) (2020): 21–24.

Udayakumar, Muthulingam, and Thangave Sekar. "Density, diversity and richness of woody plants in urban green spaces: a case study in Chennai metropolitan city." *Urban Forestry and Urban Greening* 11 (4) (2012): 450–459.

Uexküll, Jakob von. *A foray into the worlds of animals and humans.* Trans. Joseph D. O'Neil. Minneapolis: University of Minnesota Press, 2010 [1934].

Ulstein, Gry. "Brave new weird: anthropocene monsters in Jeff VanderMeer's 'The Southern Reach.'" *Concentric: Literary and Cultural Studies* 43 (1) (2017): 71–96.

Uma, Saumya, and Arvind Narrain. "Human rights and its future: some reflections." *Jindal Global Law Review* 9 (2) (2018): 287–297.

Urfi, A. J., Monalisa Sen, A. Kalam, and T. Meganathan. "Counting birds in India: methodologies and trends." *Current Science* (December 2005): 1997–2003.

Valverde, Mariana. "The ethic of diversity: local law and the negotiation of urban norms." *Law and Social Inquiry* 33 (4) (2008): 895–923.

VanderMeer, Jeff. *Southern Reach trilogy.* New York: Farrar, Straus, and Giroux, 2014.

Van Dooren, Thom. *The wake of crows: living and dying in shared worlds.* New York: Columbia University Press, 2019.

Van Stiphout, Maike. *First guide to nature inclusive design.* nextcity.nl, 2019.

Vaquin, Jean-Baptiste. *Atlas de la nature à Paris.* Paris: Atelier Parisien d'urbanisme / Le Passage, 2006.

Varela, Francisco G., Humberto R. Maturana, and Ricardo Uribe. "Autopoiesis: the organization of living systems, its characterization and a model." *Biosystems* 5 (4) (1974): 187–196.

Vasishth, Ashwani, and David C. Sloane. "Returning to ecology: an ecosystem approach to understanding the city." In *From Chicago to LA: making sense of urban theory,* ed. Michael Dear, 343–366. Thousand Oaks, CA: Sage, 2002.

Vasset, Philippe. *Un livre blanc.* Paris: Fayard, 2007.

Vera, Frans W. M. *Grazing ecology and forest history.* Wallingford: CABI, 2000.

Vertovec, Steven. "Berlin multikulti: Germany, 'foreigners' and world openness." *New Community* 22 (3) (1996): 381–399.

Vessel, Matthew F., and Herbert H. Wong. *Natural history of vacant lots.* Berkeley: University of California Press, 1987.

Vialles, Noëlie. *Animal to edible.* Trans. J. A. Underwood. Cambridge: Cambridge University Press, 1994 [1987].

Vicenzotti, Vera. *Der "Zwischenstadt"-Diskurs: eine Analyse zwischen Wildnis, Kulturlandschaft und Stadt*. Bielefeld: transcript, 2014.

Vidal, Fernando, and Nélia Dias, eds. *Endangerment, biodiversity, and culture*. New York: Routledge, 2016.

Vidler, Anthony. *Warped space: art, architecture, and anxiety in modern culture*. Cambridge, MA: MIT Press, 2000.

Voss, Edward, and Anton A. Reznicek. *Field manual of Michigan flora*. Ann Arbor: University of Michigan Press, 2012.

Wagar, W. Warren. "J. G. Ballard and the transvaluation of utopia." *Science Fiction Studies* (1991): 53–70.

Walker, Peter A. "Political ecology: where is the ecology?" *Progress in Human Geography* 29 (1) (2005): 73–82.

Wallace, Kathleen R. "'All things natural are strange': Audre Lorde, urban nature, and cultural place." In *The nature of cities: ecocroticism and urban environments*, ed. Michael Bennett and David W. Teague, 55–76. Tucson: University of Arizona Press, 1999.

Wallace, Robert G. "Breeding influenza: the political virology of offshore farming." *Antipode* 41 (5) (2009): 916–951.

Wallace-Wells, David. *The uninhabitable earth: a story of the future*. London: Penguin, 2019.

Wang, Hua-Feng, Jordi López-Pujol, Laura A. Meyerson, Jiang-Xiao Qiu, Xiao-Ke Wang, and Zhi-Yun Ouyang. "Biological invasions in rapidly urbanizing areas: a case study of Beijing, China." *Biodiversity and Conservation* 20 (11) (2011): 2483–2509.

Waterton, Claire. *Barcoding nature: shifting cultures of taxonomy in an age of biodiversity loss*. London: Routledge, 2017.

Waterton, Claire. "From field to fantasy: clarifying nature, constructing Europe." *Social Studies of Science* 32 (2) (2002): 177–204.

Watts, Michael J. *Silent violence: food, famine, and peasantry in northern Nigeria*. Berkeley: University of California Press, 1983.

Weaver, Scott C. "Urbanization and geographic expansion of zoonotic arboviral diseases: mechanisms and potential strategies for prevention." *Trends in Microbiology* 21 (8) (2013): 360–363.

Webster, Sarah C., Michael E. Byrne, Stacey L. Lance, Cara N. Love, Thomas G. Hinton, Dmitry Shamovich, and James C. Beasley. "Where the wild things are: influence of radiation on the distribution of four mammalian species within the Chernobyl exclusion zone." *Frontiers in Ecology and the Environment* 14 (4) (2016): 185–190.

Weckel, Mark E., Deborah Mack, Christopher Nagy, Roderick Christie, and Anastasia Wincorn. "Using citizen science to map human-coyote interaction in suburban New York, USA." *Journal of Wildlife Management* 74 (5) (2010): 1163–1171.

Weisberg, Zipporah. "The broken promises of monsters: Haraway, animals and the humanist legacy." *Journal for Critical Animal Studies* 7 (2) (2009): 22–62.

Weisman, Alan. *The world without us.* London: Virgin, 2007.

Weizman, Eyal. *Forensic architecture: violence at the threshold of detectability.* New York: Zone Books, 2017.

Weizman, Eyal. "Introduction: forensis." In *Forensis: the architecture of public truth.* Berlin: Sternberg Press, Berlin, 2014.

Weizman, Eyal. "Matters of calculation: the evidence of the Anthropocene." Eyal Weizman in conversation with Heather Davis and Etienne Turpin. In *Architecture in the Anthropocene: encounters among design, deep time, science and philosophy*, ed. Etienne Turpin, 63–81. Ann Arbor, MI: Open Humanities Press, 2013.

Weller, Richard, Zuzanna Drozdz, and Sara Padgett Kjaersgaard. "Hotspot cities: identifying peri-urban conflict zones." *Journal of Landscape Architecture* 14 (1) (2019): 8–19.

Whittaker, Robert H. "Evolution and measurement of species diversity." *Taxon* 21 (2–3) (1972): 213–251.

Whittaker, Robert H. "Gradient analysis of vegetation." *Biological Reviews* 42 (2) (1967): 207–264.

Whyman, Philip B. "Street trees, the private finance initiative and participatory regeneration: policy innovation or incompatible perspectives." *Political Quarterly* 91 (1) (2020): 156–164.

Williams, Mark, Jan Zalasiewicz, Peter K. Haff, Christian Schwägerl, Anthony D. Barnosky, and Erle C. Ellis. "The Anthropocene biosphere." *Anthropocene Review* 2 (3) (2015): 196–219.

Wilson, Elizabeth. "The invisible flâneur." *New Left Review* 191 (1992): 90–110.

Winchell, Kristin M., R. Graham Reynolds, Sofia R. Prado-Irwin, Alberto R. Puente-Rolón, and Liam J. Revell. "Phenotypic shifts in urban areas in the tropical lizard *Anolis cristatellus*." *Evolution* 70 (5) (2016): 1009–1022.

Witcher, Robert E. "On Rome's ecological contribution to British flora and fauna: landscape, legacy and identity." *Landscape History* 34 (2) (2013): 5–26.

Wittig, Rüdiger. "Biodiversity of urban-industrial areas and its evaluation: a critical review." In *Urban Biodiversity and Design*, ed. Norbert Müller, Peter Werner, and John G. Kelcey, 37–55. Hoboken, NJ: Wiley-Blackwell, 2010.

Wolch, Jennifer. "Anima urbis." *Progress in Human Geography* 26 (6) (2002): 721–742.

Wolch, Jennifer. "Zoöpolis." *Capitalism, Nature, Socialism* 7 (2) (1996): 21–48.

Wolch, Jennifer, Alec Brownlow, and Unna Lassiter. "Constructing the animal worlds of inner-city Los Angeles." In *Animal spaces, beastly places: new geographies of human–animal relations*, ed. Chris Philo and Chris Wilbert, 71–97. London: Routledge, 2000.

Wolch, Jennifer, Josh Newell, Mona Seymour, Hilary Bradbury Huang, Kim Reynolds, and Jennifer Mapes. "The forgotten and the future: reclaiming back alleys for a sustainable city." *Environment and Planning A* 42 (12) (2010): 2874–2896.

Wolch, Jennifer, Stephanie Pincetl, and Laura Pulido. "Urban nature and the nature of urbanism." In *From Chicago to LA: making sense of urban theory*, ed. Michael Dear, 369–402. Thousand Oaks, CA: Sage, 2002.

Wolch, Jennifer, Kathleen West, and Thomas E. Gaines. "Transspecies urban theory." *Environment and Planning D: Society and Space* 13 (6) (1995): 735–760.

Wolf, Anne-Élisabeth. "L'herbier parisien de Paul Jovet: première analyse." In *Sauvages dans la ville. Actes du colloque, organisé pour le centenaire de la naissance de Paul Jovet*, ed. Bernadette Lizet, Anne-Élisabeth Wolf, and John Celecia, 35–52. Paris: JATBA/Publications scientifiques du MNHN, 1997.

Wolf, Meike. "Rethinking urban epidemiology: natures, networks, and materialities." *International Journal of Urban and Regional Research* 40 (5) (2016): 958–982.

Wolf, Meike. "Zoonoses: towards an urban epidemiology." In *Urban constellations*, ed. Matthew Gandy, 75–79. Berlin: jovis, 2011.

Wolfe, Cary. *Before the law: humans and other animals in a biopolitical frame*. Chicago: University of Chicago Press, 2013.

Wolford, Wendy. "The Plantationocene: a lusotropical contribution to the theory." *Annals of the American Association of Geographers* (2021).

Woods, Derek. "Scale critique for the Anthropocene." *Minnesota Review* 83 (2014): 133–142.

Woolfson, Esther. *Field notes from a hidden city: an urban nature diary*. London: Granta, 2014.

Woudstra, Jan. "The eco-cathedral: Louis Le Roy's expression of a free landscape architecture." *Die Gartenkunst* 20 (1) (2008): 185–202.

Wu, J. G., and O. L. Loucks. "From balance of nature to hierarchical patch dynamics: a paradigm shift in ecology." *Quarterly Review of Biology* 70 (1995): 439–466.

Yeoh, Brenda S. A. "Cosmopolitanism and its exclusions in Singapore." *Urban Studies* 41 (12) (2004): 2431–2445.

Yunus, Reva. "Unpacking the histories, contours and multiplicity of India's women's movement(s): an interview with Uma Chakravarti." *Journal of Law, Social Justice and Global Development* 23 (2019): 55–73.

Yusoff, Kathryn. "Aesthetics of loss: biodiversity, banal violence and biotic subjects." *Transactions of the Institute of British Geographers* 37 (4) (2012): 578–592.

Yusoff, Kathryn. "Anthropogenesis: origins and endings in the Anthropocene." *Theory, Culture and Society* 33 (2) (2016): 3–28.

Yusoff, Kathryn. *A billion black Anthropocenes or none.* Minneapolis: University of Minnesota Press, 2018.

Yusoff, Kathryn. "Politics of the Anthropocene: formation of the commons as a geologic process." *Antipode* 50 (1) (2018): 255–276.

Zacharias, Frank. "Blühphaseneintritt an Straßenbäumen (insbesondere Tilia x euchlora KOCH) und Temperaturverteilung in Westberlin." PhD dissertation, Technical University, Berlin, 1972.

Zahran, Sammy, Shawn P. McElmurry, Paul E. Kilgore, David Mushinski, Jack Press, Nancy G. Love, Richard C. Sadler, and Michele S. Swanson. "Assessment of the Legionnaires' disease outbreak in Flint, Michigan." *Proceedings of the National Academy of Sciences* 115 (8) (2018): 1730–1739.

Zantop, Susanne. *Colonial fantasies: conquest, family, and nation in precolonial Germany, 1770–1870.* Durham, NC: Duke University Press, 1997.

Zask, Joëlle. *Zoocities. Des animaux sauvages dans la ville.* Paris: Premier Parallèle, 2020.

Zérah, Marie-Hélène. "Conflict between green space preservation and housing needs: the case of the Sanjay Gandhi National Park in Mumbai." *Cities* 24 (2) (2007): 122–132.

Zerbe, Stefan, Il-Ki Choi, and Ingo Kowarik. "Characteristics and habitats of non-native plant species in the city of Chonju, southern Korea." *Ecological Research* 19 (2004): 91–98.

Zerbe, Stefan, Ute Maurer, Solveig Schmitz, and Herbert Sukopp. "Biodiversity in Berlin and its potential for nature conservation." *Landscape and Urban Planning* 62 (3) (2003): 139–148.

Zhang, Amy. "Circularity and enclosures: metabolizing waste with the black soldier fly." *Cultural Anthropology* 35 (1) (2020): 74–103.

Zhang, Shen. "The forest and urban greening in Shanghai." In *Urban ecology: plants and plant communities in urban environments*, ed. Herbert Sukopp, Slavomil Hejný, and Ingo Kowarik, 141–154. The Hague: SPB Academic Publishing, 1990.

Zhao, Yimin. "Space as method: field sites and encounters in Beijing's green belts." *City* 21 (2) (2017): 190–206.

Zimmer, Anna. "Urban political ecology: theoretical concepts, challenges, and suggested future directions." *Erdkunde* 64 (4) (2010): 343–354.

Zimmerer, Karl S. "Human geography and the 'new ecology': the prospect and promise of integration." *Annals of the Association of American Geographers* 84 (1) (1994): 108–125.

Zimmerer, Karl S. "The reworking of conservation geographies: non-equilibrium landscapes and nature-society hybrids." *Annals of the Association of American Geographers* 90 (2) (2000): 356–369.

Zitouni, Benedikte. *Agglomérer: une anatomie de l'extension bruxelloise (1828–1915)*. Brussels: University of Brussels Press, 2010.

Zylinska, Joanna. *The end of man: a feminist counterapocalypse*. Minneapolis: University of Minnesota Press, 2018.

INDEX

Note: Illustrations are indicated by page numbers in italics.

urban metabolism, vii, 13, *21*, 23, 30, 91, 211–213, 266 (n. 70)
urban microbiomes, 9, 65, 70, 98
urban migration, 45, 60, 133, 134, 149, 150, 219
urban nature (definition), 13–14. *See also* nature
urban parataxonomy, 176
urban planning, 12, 15, 22, 87, 88, 132, 170, 177, 179, 181, 191, 207, 253
urban political ecology, viii, 18, 29–31, 33, 61, 212, 242, 243, 245, 257
urban refugia, 7, 35, 166, 214–225, *220*, 251
urban resilience, 12, 26, 33, 75, 102, 182, 198, 208, 218, 228, 237, 242
urban rewilding, 35, 38, 83, 192, 224, 258
urban-rural distinctions, 13, 15, 149, 208, 229, 230
urban-rural gradient, 142
urban sensorium, 9, 33, 159, 258
urban soils, 9, 74, 225, 313 (n. 33)
urban soundscapes, 8, 9, 188, 241
urban transect, 140–146, 188, 308 (n. 74). *See also* botanical transect
urban trees, 10, 16, 28, 50, 92, 100, 105, 117, 119, *121*, 122–123, 130, 138, 149, 153, 185, *186*, 191–192, 193, 216, 218, 225, 226, 298 (n. 72), 301 (n. 8)
urban wastelands, x, 1, 3, 7, 12, 22, 34, 85–116, 117–133, 156, 165, 166, 291 (n. 4), 296 (n. 54), 301 (n. 4), 304 (n. 31)
urban wetlands, 52, 161, 163, 170, 172, 180, 195, 196, *197*, 237, 264 (n. 52)
Utrecht, 185

Van Cortlandt Park, New York, 8
VanderMeer, Jeff, 16, 230, 238
Varela, Francisco, 23
Vasset, Philippe, 188
Venezuela, 127
Vera, Frans, 224
vernacular spaces, 8, 34, 85, 109, 137, 151, 247, 250
Vicenzotti, Vera, 14

Vidal, Fernando, 166
Vidler, Anthony, 87
Vienna school of social ecology, 23, 25, 266 (n. 71)
Vietnam, 52, 163, 183
virology, 163
vitalism, 26, 103–104, 155. *See also* neovitalism
Vizela, 188
vultures, 63

Warsaw, 6, 29
wasps, x, 66, 102, 296 (n. 54)
Wasteland Twinning Network, 110, *112*
Watts, Michael, 29–30
Weber, Max, 182
weeds, 3, 8, 13, 84, 91, 99, 106, 119, 139, 250, 301 (n. 6)
weeds ordinances, 106
"weird ecologies," 230–231
Weisman, Alan, 227
Weizman, Eyal, 153, 158–159, 160, 161, 167
West Nile virus, 226
Whittaker, Robert, 214, 302 (n. 16)
Wild (Krebitz), 54–55, *56*
wild boar, 83
Wildwuchs (wild growth), 137
Wilson, Edward O., 26, 247
Wolch, Jennifer, 34, 39, 41–44, 84, 252
Wolf, Benno, 126
Wolman, Abel, 91
wolves, 53, 54, 55, *56*, 57, 63, 65, 83, 227, 280 (n. 76)
Woods, Derek, 238
Woolfson, Esther, 9
Wordsworth, William, 148
World Health Organization, 78
world without us, The (Weisman), 227
Wrigley, Anthony E., 202
Wuhan, 52

xenophobia, 3, 52, 126, 147, 148, 149, 233
xerophytic planting schemes, 251